BAR BRITISH SERIES 681, 2023
ARCHAEOLOGY OF ROMAN BRITAIN
VOLUME 10

MAKING FLOUR THE GERMAN WAY IN ROMAN BRITAIN

Distribution, use and deposition of imported lava quernstones and millstones

Lindsay Banfield

BAR
PUBLISHING

Published in 2023 by
BAR Publishing, Oxford, UK

BAR British Series 681

Archaeology of Roman Britain, 10
Making Flour the German Way in Roman Britain

ISBN 978 1 4073 6081 2 paperback
ISBN 978 1 4073 6082 9 e-format

DOI https://doi.org/10.30861/ 9781407360812

A catalogue record for this book is available from the British Library

COVER IMAGE *Upper stone of a lava quern recovered from Vindolanda. Image ©Vindolanda Trust.*

BAR
PUBLISHING

BAR titles are available from:

BAR Publishing
122 Banbury Rd, Oxford, OX2 7BP, UK
info@barpublishing.com
www.barpublishing.com

ARCHAEOLOGY OF ROMAN BRITAIN

A specialist sub-series of the BAR British Series

Roman Britain presents a dynamic and exciting field of study, with an abundance of data amenable to multi-disciplinary approaches, 'big data' studies, the application of theoretical approaches, and a variety of visually stimulating artefacts and reconstructions that speak to our own age in a remarkably direct way. This series promotes research relating to the Roman province of Britannia, spanning a broad period from the late Iron Age to post-Roman Britain (roughly from the 1st century BC to the 5th century AD), as well as encompassing studies that examine the interaction between the British Isles, the nearby Continent, and other parts of the connected Roman empire.

If you would like to submit a proposal for the *Archaeology of Roman Britain* series, please contact editor@barpublishing.com

TITLES IN THE ARCHAEOLOGY OF ROMAN BRITAIN SUBSERIES

The Clayton Collection
An archaeological appraisal of a 19th century collection of Roman artefacts from Hadrian's Wall
Frances McIntosh
BAR British Series **646** | 2019 — Volume 1

Worcester Magistrates Court
Excavation of Romano-British homes and industry at Castle Street
Andy Boucher
BAR British Series **658** | 2020 — Volume 2

London's Roman Tools
Craft, agriculture and experience in an ancient city
Owen Humphreys
BAR British Series **663** | 2021 — Volume 3

Belonging and Belongings
Portable artefacts and identity in the civitas of the Iceni
Natasha Harlow
BAR British Series **664** | 2021 — Volume 4

On the Edge of Empire
Society in the south-west of England during the first century BC to fifth century AD
Siân Alyce Thomas
BAR British Series **667** | 2021 — Volume 5

Roman Rural Settlement in Wales and the Marches
Approaches to settlement and material culture through big data
Leah Reynolds
BAR British Series **670** | 2022 — Volume 6

Dying Young
A Bioarchaeological Analysis of Child Health in Roman Britain
Anna Rohnbogner
BAR British Series **673** | 2022 — Volume 7

The Rise and Decline of Druce Farm Roman Villa (60-650 CE)
Excavations 2012-2018
Lilian Ladle
BAR British Series **676** | 2022 — Volume 8

Facing the Enemy?
A GIS Study of 1st Century Roman Fortifications in the Scottish Landscape
Andrew Tibbs
BAR British Series **677** | 2022 — Volume 9

Making Flour the German Way in Roman Britain
Distribution, use and deposition of imported lava quernstones and millstones
Lindsay Banfield
BAR British Series **681** | 2023 — Volume 10

Of Related Interest

Grinding and Milling
A study of Romano-British rotary querns and millstones made from Old Red Sandstone
Ruth Shaffrey
BAR British Series **409** | 2006

Bread for the People: The Archaeology of Mills and Milling
Proceedings of a colloquium held in the British School at Rome 4th - 7th November 2009
Edited by David Williams and David Peacock
BAR International Series **2274** | 2011

Whetstones from Roman Silchester (Calleva Atrebatum), North Hampshire
Character, manufacture, provenance and use. 'Putting an edge on it'.
J. R. L. Allen
BAR British Series **597** | 2014

Roman Quarrying and Stone Supply on the Periphery - Southern England
A geological study of first-century funerary monuments and monumental architecture
Kevin M. J. Hayward
BAR British Series **500** | 2009

North Sea and Channel Connectivity during the Late Iron Age and Roman Period (175/150 BC-AD 409)
Francis M. Morris
BAR International Series **2157** | 2010

Agricultural Implements in Prehistoric and Roman Britain, Parts i and ii
Sian E. Rees
BAR British Series **69** | 1979

Agricultural Buildings in Roman Britain
Pat Morris
BAR British Series **70** | 1979

Roman Military Stone-built Granaries in Britain
Anne P. Gentry
BAR British Series **32** | 1976

The Romano-British countryside, Parts i and ii
Studies in Rural Settlement and Economy
Edited by David Miles
BAR British Series **103** | 1982

Iron Age Societies in the Severn-Cotswolds
Developing narratives of social and landscape change
Tom Moore
BAR British Series **421** | 2006

The Vindolanda Tablets and the Ancient Economy
Kasper Grønlund Evers
BAR British Series **544** | 2011

The Roman Pottery Production Site at Wickham Barn, Chiltington, East Sussex
Chris Butler and Malcolm Lyne
BAR British Series **421** | 2006

For more information, or to purchase these titles, please visit **www.barpublishing.com**

Acknowledgements

Many people have been involved in this project and I am very grateful for the help and assistance throughout. The complexity of conducting research during a global pandemic created a range of difficulties in accessing materials and information. Without the support of the people named here, this research may not have been possible. First and foremost, I must thank the Arts and Humanities Research Council South-West and Wales Doctoral Training Partnership. I would like to thank my supervisors; Hella Eckardt and Ruth Shaffrey, who were responsible for the initial research idea for this project that I then helped to develop, and then helped me the whole way through the PhD. They have both been incredibly supportive, full of good advice, guidance, and information. I cannot emphasise enough how much I have enjoyed working with them and how much I have learnt from being their student. I would also like to thank Martin Pitts who supervised and contributed to the development of the project whilst providing invaluable feedback that made me consider aspects of my research using new perspectives, and John Creighton who helped me navigate some of the more complex aspects of the Roman economy and offered guidance as a mentor throughout the PhD.

Extra thanks to John Cruse and the Yorkshire Quern Survey who helped to sponsor this project and have provided support, information, and assistance throughout. I would like to extend my thanks to the University of Reading and all staff and students who have helped me over the years: offered work, lent books, or been a sounding board for quern-related theories and thoughts. This includes the administration teams in SAGES and the Graduate School, especially Heather Browning, Elizabeth Wyeth, Susie Fullarton, Marie Taylor, and Jen Baker. Thanks also to Sara Machin and Kev Williams who provided lab training. Although geochemical analysis was not completed due to Covid restrictions, their time has been much appreciated. Thanks to BAR publishing, with special thinks to Tansy Branscombe who has provided much needed advice for bringing this work to publication.

I must also thank the museums, archives and individuals who helped and provided access to material for my data collection. Much of this took place between Covid lockdowns and I am indebted to those who enabled on-site visits under difficult circumstances, and then followed up with context and site information. Thanks, therefore, to Frances McIntosh and English Heritage/Historic England, David Petts and University of Durham, David Thurold and Verulamium Museum, Barbara and Andrew Birley and the Vindolanda Trust, Rob Symmons and Fishbourne Palace Museum, and Portia Tremlett and the Novium Museum. I would also like to thank those that provided data or access to information in other ways, specifically Michael Marshall at Museum of London Archaeology, Stefan Wenzel at Römisch-Germanische ZentralMuseum (now Leibniz-Zentrum für Archäologie), John Cruse and the Yorkshire Quern Survey, and David Buckley.

On a personal level, I would like to thank John Kane for his endless support and patience, Suzanne Banfield, Lucy Staniforth and Lesley Goodair for providing much valued advice and encouragement, and all the friends who have had my back, offered kind words and completely understood periods of absence or distance during a challenging 3 years. Thanks also to my long-time mentors Rob Wallace, David Millum and Mark Roberts, from whom I have learnt so much. I would like to dedicate this work to Joseph Anthony Goodair- he would have been immensely proud.

Please note that all best efforts have been made to contact the copyright holders of any third-party images, but in some cases the rightsholder could not be contacted. If any issues arise with the inclusion of images in this book, please contact the publishers who will be happy to assist.

Contents

List of Figures ..... **xiii**
List of Tables ..... **xxii**
Abstract ..... **xxiv**

1. Introduction **1**
1.1. Context of Study 1
1.2. Querns as Artefacts 1
1.3. Mills and Millstones 3
1.4. Differentiating Between Mills and Querns 4

2. From Quarry to Quern: Previous Research on Lava Milling Tools **5**
2.1. Lava Querns: Terminology 5
2.1.1. Surface Dressing 5
2.1.2. Quern Mechanisms and Components 6
2.2. Lava Quern Typology 10
2.3. Lava Millstones: Disc Types 13
2.4. Pompeian Style Mills 16
2.5. Quarrying and Organisation of Labour 18
2.6. Stone extraction: Quarrying process 19
2.7. 'Im Winkel': A Possible Quern Finishing Location 22
2.8. Distribution Analysis and Mechanisms of Exchange 23
2.9. Transporting Goods 27
2.10. Trade Routes for Imported Goods: North Sea Trade 30
2.11. The Economy of Stone Exchange in the Roman Empire 31
2.12. The Lava Quern Trade in Continental Europe and Britain 34
2.13. Distribution of Provenanced Lava Milling Tools in Mainland Europe and Britain 34
2.14. Applying Economic Considerations: Roman Britain and the Lava Quern Trade 35

3. Volcanic Beginnings: Lava Sources **37**
3.1. Mayen Lava as a Material 37
3.2. Geological Background 37
3.3. Geological Description and Properties 37
3.4. Petrographic and Geochemical Analysis 38
3.5. Future Research 39

4. Methodology **41**
4.1. Practical Aspects of Data Collection, Recording and Analysis 41
4.1.1. Quantification 41
4.1.2. In-Person Quern Recording 42
4.1.3. Digitising the Data and Completing the Analysis 42
4.2. Object Biographies 42
4.3. Imported Lava Querns and Millstones: Applying an Object Biography Approach 43
4.3.1. Manufacture 44
4.3.2. Distribution 44
4.3.3. Primary Use 45
4.3.4. Re-use/Modification 46
4.3.5. Deposition/Disposal 47

5. Making Flour the German Way: Analysis of Lava Quern and Millstone Use Across the Province **49**
5.1. The Dataset 49
5.2. Manufacture 51
5.2.1. Diameter of Stone 51
5.2.2. Width of Kerb 52
5.2.3. Handle Fittings 53

5.2.4. Rynd Chase Types 55
5.3. Distribution 55
5.3.1. Overall Spatial Distribution 55
5.3.2. Distribution by Site Type 57
5.3.3. Chronological Change in Distribution 57
5.4. Primary Use 60
5.4.1. Stone Thickness for the Whole Dataset 60
5.4.2. Stone Thickness by Region 61
5.4.3. Stone Thickness by Site Type 62
5.5. Reuse 65
5.6. Deposition 66
5.7. Conclusion 70

6. Foreign Stone in the Hands of Farmers: Lava Querns and Millstones in Rural Roman Britain 73
6.1. Introduction 73
6.2. The Dataset 73
6.2.1. Millstones vs Quern stones 74
6.2.2. Rural Quern Stone Thickness 76
6.3. Distribution Analysis 78
6.3.1. General Distribution: Examining Presence/Absence of Lava Milling Tools 79
6.3.2. Close-up Analysis of Distributions in East Anglia and the South-East 83
6.3.3. Distribution by Site Type 85
6.3.4. Chronological Distribution 89
6.4. Conclusion 91

7. Day to Day Grind of Urban Life: Lava Quern and Millstone Use in Romano-British Towns 93
7.1. Verulamium: Introduction and Background 93
7.2. Verulamium: The Dataset 95
7.3. Verulamium: Manufacture 95
7.3.1. The Millstones 96
7.3.2. The Miniature Milling Tool 98
7.3.3. The Querns 100
7.4. Verulamium: Distribution 101
7.5. Verulamium: Primary Use 109
7.6. Verulamium: Reuse/Modification 110
7.7. Verulamium: Deposition 113
7.8. Verulamium: Conclusions 114
7.9. Silchester: Introduction and Background 114
7.10. Silchester: The Dataset 116
7.11. Silchester: Manufacture 116
7.11.1. Millstones vs Querns 116
7.11.2. The 'Pompeian' Style Millstones 118
7.11.3. Identifiable Typological Features 119
7.12. Silchester: Distribution 120
7.12.1. Distribution by Stone Type 120
7.12.2. Chronological Distribution- Lava Volume and Comparison with Other Stone Types 121
7.13. Sichester: Primary Use 124
7.13.1. Stone Thickness 124
7.13.2. Dressing Style 124
7.14. Silchester: Reuse/Modification 125
7.15. Silchester: Deposition 126
7.16. Silchester: Conclusions 128
7.17. Urban Sites: Conclusions 129

8. Milling at the Edge of an Empire: Lava Quern and Millstone Use at Northern Military Sites 133
8.1. Corbridge: Introduction and Background 133
8.2. Corbridge: The Dataset 135
8.3. Corbridge: Manufacture 135
8.3.1. Stone Diameter 135

8.3.2. Kerbs 136
8.3.3. Handle Fittings 136
8.3.4. Rynd Fittings 137
8.3.5. Surface Dressing 137
8.4. Corbridge: Distribution 137
8.5. Corbridge: Primary Use 139
8.5.1. Upper Quern Stone Thickness 139
8.5.2. Lower Quern Stone Thickness 140
8.6. Corbridge: Reuse/Modification 140
8.7. Corbridge: Deposition 142
8.8. Chesters: Introduction and Background 142
8.9. Chesters: The Dataset 143
8.10. Chesters: Manufacture 144
8.10.1. Lower Stones 144
8.10.2. Upper Stones 145
8.10.3. Unidentified Stone 145
8.10.4. Stone Diameters 145
8.11. Chesters: Primary Use 146
8.12. Chesters: Reuse/Modification 146
8.13. Housesteads: Introduction and Background 146
8.14. Housesteads: The Dataset 148
8.15. Housesteads: Manufacture 149
8.15.1. Diameter of Stone 149
8.15.2. Kerbs and Handle Fittings 150
8.15.3. Dressing 151
8.16. Housesteads: Distribution 151
8.17. Housesteads: Primary Use 154
8.17.1. Upper Stone Thickness 154
8.17.2. Lower Stone Thickness 154
8.18. Housesteads: Reuse/Modification 154
8.19. Housesteads: Deposition 157
8.20. Vindolanda: Introduction and Background 157
8.21. Vindolanda: The Dataset 158
8.22. Vindolanda: Manufacture 160
8.22.1. Kerbs 160
8.22.2. Diameter of Stone: Quern vs Millstone 160
8.22.3. Quern Diameter 161
8.22.4. Rynd 162
8.22.5. Handle Fittings 162
8.23. Vindolanda: Distribution 163
8.23.1. Spatial Distribution 163
8.23.2. Chronological Distribution 164
8.24. Vindolanda: Primary use 165
8.24.1. Wear Traces 165
8.24.2. Quern Thickness: Upper Stones 166
8.24.3. Quern Thickness: Lower Stones 167
8.24.4. Inscriptions 168
8.25. Vindolanda: Re-use/Modification 169
8.26. Vindolanda: Deposition 172
8.27. Vindolanda: Conclusion 173

9. Conclusion 177

10. Bibliography 179

List of Figures

Figure 1.1. Bronze Age milling stone and saddle quern. 2

Figure 1.2. Rotary quern in use in Tibet. 2

Figure 1.3. Diagram of general components of a quern of indigenous British form. 2

Figure 1.4. Disc-type Roman millstone and reconstruction of a Pompeian style mill from millstone fragments recovered from Princess Street, London. 3

Figure 1.5. Location of identified and possible mill structures from structural evidence in Roman Britain, also showing the main urban centres. 3

Figure 2.1. Diagram of typical lava quern stones showing the terminology used for different parts of the tool. 5

Figure 2.2. Stone remains of a pair of complete Roman period rhyolite quern stones with perforated meta. 6

Figure 2.3. Different styles of dressing for lava quern stones: pecked, radial, harped or segmented radial 6

Figure 2.4. Catillus from Newstead Fort with complete metal components showing how an elbow handle was assembled 7

Figure 2.5. Top and bottom stones of a rotary quern retrieved from the Roman shipwreck De Meern 1. 7

Figure 2.6. Two women grinding corn on a quern mill on Pine Mountain. The photograph shows relatively modern use of the swivel pole method for grinding and how two people could work the quern simultaneously. 8

Figure 2.7. An example of how a swivel pole quern mechanism would have operated. 8

Figure 2.8. Spindle and wooden spindle plug from quern recovered from De Meern 1. 8

Figure 2.9. Meta from quern recovered from De Meern 1 showing how the spindle components fit together. 9

Figure 2.10. Two of four spindles recovered intact from lava querns at Newstead. The shape of the objects indicate how they may have contributed to the mechanics of rotary quern movement 9

Figure 2.11. Diagram to show the configuration of a Scottish rotary quern that is assumed to be the same or similar to the Roman lava quern tentering mechanism 10

Figure 2.12. Neolithic to Middle Hallstatt; Late Hallstatt (600–400 BC); Napoleon hat (Early and Middle La Tène); oscillating rotary quern (from La Tène onwards); completely rotating quern (Roman period); Early Medieval; about AD 1000 11

Figure 2.13. The main features of a Röder type 4 lava quern as the most prevalent type in Romano-British contexts. 12

Figure 2.14. The three types of rynd fitting for millstones. Rynd cut into the upper surface (overdrift); middle-Rynd cut into the grinding surface (underdrift); rynd cut through the stone (probably overdrift). 13

Figure 2.15. Mechanical parts of a powered mill from the Roman period reconstructed using literary and archaeological evidence. 14

Figure 2.16. Mill set-up with mechanical workings as shown in figure 2.15 but powered through different means: an animal powered mill that could also be turned by people; a water powered mill. 15

Figure 2.17. Possible footstep bearing used to hold the weight of the upper millstone via the spindle, whilst still providing the means for the spindle to pivot. This example is made of basalt and recovered with the remains of a probable mill at Haltwhistle Burn, Hadrian's Wall. 15

Figure 2.18. Cross-section of a typical Pompeian style mill. 16

Figure 2.19. Bakery with *in-situ* Pompeian-style mills, at Pompeii. These show how the stones fitted together and how they were fixed in position. Trough-like flour catchers (gremia) would have been placed on top of the structure that held the meta in place 17

Figure 2.20. Part of a Roman period relief depicting bread-making, found over the oven in the bar of the restaurant 'Romolo' in Rome. The detail shown here shows the Pompeian style mill in use, with a donkey or mule being used to rotate the device. 17

Figure 2.21. Drawing of part of the frieze from the tomb of Eurysaces the baker, at the Porta Maggiore. This section of the frieze shows the grinding of grain to flour using Pompeian-style mills powered by donkeys or mules. 18

Figure 2.22. Location of Mayen within Germany 19

Figure 2.23. Reconstruction of the quarrying process and the division of the quarry face into 'plots' for extraction 20

Figure 2.24. Photograph of quarry face at Etringer Lay showing the naturally formed lava columns 20

Figure 2.25. Lava quernstone quarrying: Preparatory work - visual inspection, sounding the stone, checking for fissures. Marking out - measuring the top, marking the circumference, making a notch. Stone breaking - creation of wedge pockets, adding iron wedges, hitting iron wedges. 21

Figure 2.26. Remnants of stone breaking technique in Mayen quarries to show how the iron wedges were used in the process. 22

Figure 2.27. Lava quernstone quarrying: Stone lifting, measuring the diameter, roughly shaping the stone. Shaping the surfaces, shaping the sides, smoothing the surfaces. Hollow out. 22

Figure 2.28. Example of a typical roughout recovered from Mayen quarries. All three examples show that the quern stone shape is beginning to emerge, but that the surfaces are still roughly shaped. 23

Figure 2.29. Smoothed roughout of a meta retrieved from Mayen quarry site. Tool marks show the direction of working and the smaller size of the tool blade used. 23

Figure 2.30. Organisation of production and quern finishing workshops. 24

Figure 2.31. Slot dug through cellar at 'Im Winkel' showing querns in section in the basement fill. 25

Figure 2.32. Some of the partially finished querns recovered from the site an 'Im Winkel' 26

Figure 2.33. Distribution of BB2 pottery in Roman Britain, including production locations. 28

Figure 2.34. Maritime systems in the North Sea and English Channel. 32

Figure 2.35. Provenanced lava quern distribution. 35

Figure 3.1. High resolution image of Mayen lava to show the vesicles (holes) that are an identifying feature of the material. It is also possible to see some of the sanidine phenocrysts, visible as white crystalline material 38

Figure 3.2. Photomicrograph of Mayen lava in thin section with cross polarised light. 38

Figure 3.3. Photomicrograph of basalt lava sample from Banne d'ordanche, Monts d'Ore in the Massif Central region of France. 39

Figure 4.1. Upper stones of two lava querns recovered from Vindolanda. 44

Figure 4.2. Typical indigenous pre-Roman rotary quern top stone with uneven wear, possibly from oscillating motion. 45

Figure 5.1. Chart showing the breakdown of querns, millstones, and unidentified lava in the assemblage according to whether they are upper or lower stones (T= 1271). 49

Figure 5.2. Chart showing the number of sites recorded within each region in Britain (T=1165). 50

Figure 5.3. Chart showing the breakdown of the site types included within the dataset (T= 1167) 51

Figure 5.4. Chart showing the diameters for all lava milling tools for which this measurement has been taken, with a line marking where the difference between a quern and a millstone is most clear. (T= 607) 52

Figure 5.5. Chart showing the diameters of all lava querns and unidentified milling tools from Britain. The 25th and 75th percentiles have been indicated with lines to demonstrate the extent of standardisation in the dataset (T= 553). 52

Figure 5.6. Chart showing the variation of kerb widths for all lava milling tools from Roman Britain (T= 137). 53

Figure 5.7. Chart showing the relationship between lava milling tool diameter and kerb width for the British assemblage (T= 103). 53

Figure 5.8. Kerb width from lava milling tools of early Roman date (T= 36). Kerb width from lava milling tools of mid-Roman date (T= 23). Kerb widths for lava milling tools of late Roman date (T= 11). 54

Figure 5.9. Presentation of different handle types in the whole British lava assemblage (T= 100). 54

Figure 5.10. Fragments of CAT 0032 showing the pieces of iron band adhered to the outer circumference. 54

Figure 5.11. Upper quern stone from Vindolanda showing a worn rynd chase that cannot be identified as having either an under or overdrift fitting. 55

Figure 5.12. Distributions of lava and sites with milling tools but none of lava based on the whole dataset. 56

Figure 5.13. Map showing the social distribution at sites with lava and only non-lava milling tools, alongside the number of objects. 58

Figure 5.14. Chronological change in lava milling tool distributions from the first to fifth centuries in Roman Britain. 59

Figure 5.15. Upper stone thickness for whole dataset with 25th and 75th percentiles shown as lines (T= 248). 60

Figure 5.16. Lower stone thickness for whole dataset with 25th and 75th percentiles shown as lines (T= 138). 61

Figure 5.17. Charts showing the relative thickness of lower stones by region (excluding the south-west, east and west midlands regions): East T=10, north-east T= 44, north-west T= 20, Scotland T= 5, south-east T= 45, Wales T= 11 62

Figure 5.18. Charts showing regional difference in upper stone thickness. North-east T= 60, north-west T= 22, south-east T= 111, south-west T= 9, Wales T= 23, Scotland T= 8, East Midlands T= 6, East T= 33. 63

Figure 5.19. Chart showing the 25th and 75th percentiles by region for upper stone thickness. 64

Figure 5.20. Analysis of the thickness of lower stones by site type. Military T= 62, rural T= 35, urban T= 35. 64

Figure 5.21. Analysis of the thickness of upper stones by site type. Military T= 98, rural T= 83, urban T= 52. 65

Figure 5.22. Chart showing the most common archaeological features where lava milling tools were deposited and the frequency that this occurred for each (T= 1241)........ 66

Figure 5.23. Chart showing the frequency that lava querns were recovered from road, floor, surface, or levelling contexts, and those from an unspecified surface (unknown) (T=77). The data excludes the querns recovered from 1 Poultry........ 67

Figure 5.24. Simple correspondence analysis completed on the frequency of occurrence for deposition of lava in specific contexts at different site types. 67

Figure 5.25. Chart showing the frequency at which lava milling tools are deposited in different archaeological features by region (T= 794)........ 68

Figure 5.26. Simple correspondence analysis completed on the frequency of occurrence for deposition of lava in specific contexts at different site types. 69

Figure 5.27. Chart showing the frequency of deposition of lava in specific archaeological features by site type........ 69

Figure 5.28. Results of correspondence analysis showing the relationship between deposition context type and site type. 70

Figure 6.1. Percentage of sites in the dataset with lava milling tools present compared to those with milling tools, but none of lava (T=920)........ 73

Figure 6.2. Percentage proportions of millstones to quern stones for sites with lava and non-lava milling tools........ 74

Figure 6.3. Conglomerate Puddingstone quern from Chichester district that shows excellent preservation. 74

Figure 6.4. Frequency of different diameter groups for lava quern stones/millstones (T=177). 75

Figure 6.5. Frequency of different diameter groups for lava quern/millstones up to 580mm (T=165)........ 75

Figure 6.6. All measured and estimated diameters of lava querns and millstones (T=176)........ 76

Figure 6.7. Max Thickness at Edge for Upper stones (T= 114)........ 76

Figure 6.8. Individual thickness for each upper quern stone in the dataset, identified according to the site type from which it was recovered (T=114)........ 77

Figure 6.9. Max thickness of lower quern stones at edge (T=62)........ 78

Figure 6.10. Individual thickness for each upper quern stone in the dataset, identified according to the site type from which it was recovered (T= 62)........ 78

Figure 6.11. Rural distribution map to show the presence of lava milling tools in Roman Britain and their relative volumes, alongside locations of sites with milling tool, but none of lava........ 80

Figure 6.12. Heatmap showing quantities of lava querns at rural sites. 81

Figure 6.13. Heatmap distribution of non-lava querns and millstone volumes at sites with milling tools present, but none of lava........ 82

Figure 6.14. Heatmap distribution and volumes of non-lava querns at sites with no lava present and the location of known or assumed quern quarry sites of indigenous stone types. 83

Figure 6.15. Distribution map to show the rural presence of lava milling tools in the eastern and south-eastern region of Britain, and rural sites with milling tools, but none made of lava. 84

Figure 6.16. Distribution of lava milling tools by rural site type. 85

Figure 6.17. Distribution map showing rural sites with milling tools present, but none of lava by rural site type........... 86

Figure 6.18. Distribution by rural site type of non-lava milling tools and lava.. 87

Figure 6.19. Location of rural religious/ritual sites with lava and non-lava milling tools in the east and south-east of Britain.. 88

Figure 6.20. Chronological distribution of lava at rural sites. 90

Figure 6.21. Sites with occupation dating after AD 250 showing percentage with lava (T=7) or only non-lava (T=16) milling tools present. 91

Figure 7.1. Location of Verulamium and other nearby important sites. 94

Figure 7.2. Percentage of assemblage identified as quern or millstone T= 126. Percentage of millstones identified as upper or lower stones T=6.. 95

Figure 7.3. Histogram to show the diameters of the lava millstones recovered from Verulamium (T= 6) compared with those of Roman Britain (T= 45). 96

Figure 7.4. Fragment of decorated lava millstone; reconstructed view of the decoration on the millstone if complete......... 97

Figure 7.5. Millstone of lava from insula XXVIII showing a hole in the upper surface that may have allowed for a lifting mechanism to be fitted. 98

Figure 7.6. Miniature lava millstone upper stone (CAT 2679) viewed from the top and bottom surfaces........................ 98

Figure 7.7. Miniature upper lava millstone recovered as part of the A14 road development scheme in Cambridgeshire (CAT 1483). Provided here as a parallel to CAT 2679. 99

Figure 7.8. Miniature upper millstone recovered from the local area of Huntingdonshire and stored at the Norris Museum... 99

Figure 7.9. Verulamium quern assemblage showing percentage of upper and lower stones present (T= 32)................ 100

Figure 7.10. Graph showing the different diameters for lava querns from Verulamium compared with Britain. 101

Figure 7.11. Fragment of lava quern (CAT 2594) showing an irregularly shaped rynd chase that may have been the result of mending or reuse... 101

Figure 7.12. Map of Verulamium showing the different insulae of the town and areas of excavation covered in the data analysis. Locations where lava was recovered is shown in shaded. 102

Figure 7.13. Site plan of Folly Lane excavations indicating the location of different areas/zones. 103

Figure 7.14. Location of lava and non-lava milling tools at King Harry Lane site. 105

Figure 7.15. Processing of rice at the side of the road in the central highlands' region of Dalat in Vietnam, specifically drying, de-husking and bagging of rice. 106

Figure 7.16. Plan of Verulamium showing the locations where lava millstones were recovered in red. 108

Figure 7.17. Graph showing the edge thickness of upper stones from Verulamium (T= 13) compared with those from all Roman Britain (T= 248) excluding those from 1 Poultry. 109

Figure 7.18. Graph showing the edge thickness of lower stones from Verulamium (T= 9) compared with those from all Roman Britain (T= 138) excluding those from 1 Poultry. 109

Figure 7.19. Lower quern stones that have been chipped at the bottom edge. 110

Figure 7.20. Examples of quern stones from Verulamium that show possible signs of reuse. 112

Figure 7.21. Body fragment of lava quern that may have been part of a quern deliberately broken as it does not correlate with normal breakage due to wear or other vulnerabilities. 112

Figure 7.22. Underside of CAT 2559 showing the tooling marks where the underside had started to be hollowed out. 112

Figure 7.23. Location of Silchester Roman town. 115

Figure 7.24. The number of querns or millstones of lava in the assemblage compared with those of other stone types (T= 610). 116

Figure 7.25. Number of lava fragments identified as derived from querns or millstones, and the number of fragments identified as upper or lower stones (T=62). 117

Figure 7.26. Diameters of both upper and lower quern stones and millstones from Silchester (T=24). 117

Figure 7.27. Lava millstone from Silchester showing holes that would have been used to accommodate iron fittings, possibly for lifting the stone as part of its function (CAT 0919). 117

Figure 7.28. Frequency of mill and quern stones within different diameter ranges. 117

Figure 7.29. Distribution of milling tools of different stone types in Silchester. 119

Figure 7.30. Change in number of querns recovered from insula IX by period. 121

Figure 7.31. General trend showing decrease in milling tools numbers at other parts of Silchester with smaller datasets. 122

Figure 7.32. Change in the number of lava milling tools from insula IX over time. 122

Figure 7.33. Graph showing the edge thickness of upper stones from Calleva compared with those from all Roman Britain. 124

Figure 7.34. Graph showing the edge thickness of lower stones from Calleva compared with those from all Roman Britain. 125

Figure 7.35. The types of feature where lava milling tools were deposited classified by chronological period. 126

Figure 7.36. Plans of excavated area in insula IX showing the features where lava was recovered and corresponding excavated structures and features. 127

Figure 7.37. Map of central southern Britain showing the breakdown of stone types used for querns in the Roman period at various urban sites. 130

Figure 8.1. Location of Corbridge on Hadrian's Wall and with respect to the site at Red House and the River Tyne and bridge crossing. 134

Figure 8.2. Chart showing the number of upper and lower stones in the dataset, alongside those unidentified (T= 36). 135

Figure 8.3. Corbridge quern diameters compared with the diameters of milling tools for the whole of Roman Britain (T=8). 136

Figure 8.4. Complete upper quern stone showing a faint kerb and no sign of the standard elbow handle fitting (CAT 2192). 136

Figure 8.5. Upper quern stone showing recessed rynd chase in the upper surface. (CAT 2179). 137

Figure 8.6. Plan of the site at Corbridge showing the distribution and volumes of lava querns by findspot. 138

Figure 8.7. Chart showing the upper stone thickness for lava querns from Corbridge (T= 6) in comparison with those from the rest of Roman Britain (T= 248) excluding the querns from 1 Poultry.. 139

Figure 8.8. Chart showing the thickness of lower stones (T= 7) compared with the lower stone thickness of data from the whole of Britain (T= 138) excluding the querns from 1 Poultry.. 140

Figure 8.9. Upper quern stone with broken thick edges that may have been fragmented deliberately for the purposes of reuse (CAT 2180). 141

Figure 8.10. Modified lower stone that has had its sides removed near the lower edge (CAT 2190) that is similar to reworked lower stones from Verulamium and Silchester. 141

Figure 8.11. Lower quern stone of indigenous stone that has been similarly worked to round the lower surface, possibly to fix it into a floor (Acc No CH266)... 141

Figure 8.12. Possible modified lower stone that has been more heavily worn and has a decreased diameter (CAT 2254)... 141

Figure 8.13. Chart showing the feature type where lava querns were deposited at Corbridge (T= 12). 142

Figure 8.14. Plan of the location of Chesters Roman fort. 143

Figure 8.15. CAT 2172, lower quern stone showing all characteristics of a Mayen lava product. 144

Figure 8.16. Part of the complete lower millstone showing top and bottom view (CAT 2176).................................... 144

Figure 8.17. Complete upper stone that has been weathered and heavily worn around the central hole. A small part of the rynd chase remains on the right side of the central hole. No handle fitting exists on the stone (CAT 2177)...145

Figure 8.18. Small fragment of unidentified lava stone that has been pecked on the grinding surface (CAT 2178).........145

Figure 8.19. Upper quern stone from Chesters (CAT 2177) that shows evidence for modification in the form of holes that would have accommodated iron fittings not associated with manufacture... 146

Figure 8.20. Upper quern stone from Housesteads (CAT 2284) that retains the iron fittings and provides a parallel for the incomplete stone from Chesters above... 146

Figure 8.21. Location map of Housesteads Roman fort. 147

Figure 8.22. Oven at Housesteads in the bakehouse in the eastern rampart area of the primary fort. The layout and structure changes considerably over subsequent phases.. 148

Figure 8.23. Chart showing the diameters for lava querns from Housesteads (T=10) compared with those from the whole of Roman Britain (T=553). 149

Figure 8.24. CAT 2284 showing the faint shallow and wide kerb on the upper surface with upper surface dressing, and iron corrosion product where the iron band would have been fitted around the circumference of the stone. 151

Figure 8.25. Phased plan of the north-east area of Housesteads Roman fort showing the placement of bakehouses, ovens, and granaries. This shows the general locations where lava milling tools and non-lava milling tools were recovered... 152

Figure 8.26. Chart showing the chronological change in lava and non-lava milling tool deposition in the north-east area of Housesteads fort (T=26). 153

Figure 8.27. Chart showing the thickness for upper lava querns from Housesteads compared with those from the whole of the British dataset excluding those from the site at 1 Poultry. 154

Figure 8.28. Chart showing the thickness for lower lava quern stones from Housesteads compared with those from the whole British dataset excluding those recovered from the site at 1 Poultry. 155

Figure 8.29. Site photo of mortar, with a glimpse of the upper lava quern CAT 2269 that has been interpreted as reused as a possible lid. 155

Figure 8.30. Upper stone CAT 2284 that has an iron bar fitted into the upper surface as a possible modification of the original manufacture design. The upper image shows the widened central hole that may have removed the original rynd chase. The end of the iron bar that is fully bedded into the stone has interfered with the kerb. The opposite end is held in place with an iron loop that allows for the bar to be adjusted and may have altered the angle of the upper grinding surface. The iron loop is fixed into the stone via two holes in the surface which have then been filled with lead 156

Figure 8.31. Map showing geographical location of Vindolanda and its relative position to other Roman period frontier fortifications. 157

Figure 8.32. Plan of the excavated area of Vindolanda showing the locations of the different period forts. 159

Figure 8.33. Upper lava quern stone showing annular raised kerb on the top surface (CAT 0014). 160

Figure 8.34. Distribution of all upper and lower stone diameters for Vindolanda lava milling tools, T=20. 161

Figure 8.35. Chart showing the diameters of querns from Vindolanda (T=44) compared with those from the wider British dataset. 161

Figure 8.36. Three lava quern upper stones from Vindolanda showing differing levels of preservation and wear. Intact iron rynd *in-situ* within rynd chase (CAT 0002), intact rynd chase with iron rynd removed (CAT 0045) and fully worn rynd chase with absent rynd (CAT 0001). 162

Figure 8.37. Fragmented upper stone with iron band handle fitting (CAT 0032). 163

Figure 8.38. Upper quern with signs of wear but no evidence for a handle fitting (CAT 0002). 163

Figure 8.39. Chronological distribution of lava querns, T=30, by site period at Vindolanda. 164

Figure 8.40. Chronological distribution of non-lava querns, T=55, by site period. 164

Figure 8.41. Possible pair of lava quern stones (CAT 0047 & 0048). 165

Figure 8.42. Upper quern stone (CAT 0048) showing heaviest wear after the original handle, showing it was probably rotated in a clockwise direction. 165

Figure 8.43. Quern upper stone with wear traces showing the direction of movement when in use (CAT 0045). 166

Figure 8.44. Chart showing the thickness of upper quern stones from Vindolanda compared with those from all Roman Britain, excluding the querns from the site at 1 Poultry (T=13). 167

Figure 8.45. Thickness for lower stones from Vindolanda (T=14) compared with those from the wider British dataset, excluding those from the site at 1 Poultry. 167

Figure 8.46. Inscription on side of lava quern from Vindolanda (CAT1191). 168

Figure 8.47. Fragment of lava upper stone with inscription on outer circumference. A possible 'VI' is visible, though any other figures are indistinguishable (CAT 0021). 169

Figure 8.48. Inscribed upper lava quern stone from Vindolanda (CAT 0048). 169

Figure 8.49. Upper quern showing wear at the grinding surface, which is worn down to the elbow handle fitting, making it unusable (CAT 0014). 170

Figure 8.50. Upper quern stone showing two elbow handle fittings, almost opposite each other.................................. 170

Figure 8.51. Upper stone CAT 0048 that has 2 handle fittings and one abandoned attempt at adding a handle............. 171

Figure 8.52. Side view of handle fittings,1 and 3, and attempted fitting 2 on CAT 0048.. 171

Figure 8.53. CAT 0025 that has a recessed rectangular slot that may have accommodated an iron bar fitting like that at Housesteads (2284). 172

Figure 8.54. CAT 0030 showing shallow holes in the upper surface of the stone that may have housed the ends of an iron loop as part of an iron bar fitting.. 172

Figure 8.55. CAT 0032 showing possible iron bar adhered to the upper surface of the stone. A lead plug is visible under the iron on the stone as a white/grey shape to the left of the iron corrosion product. A further two lead plugs may have fitted the iron loop ends.. 173

Figure 8.56. CAT 0041 showing end of possible iron loop for an iron bar fitting... 173

Figure 8.57. Number of querns recovered from different types of feature and split into lava or non-lava types, T=59. 174

Figure 8.58. Percentage of upper to lower stones found in structural or flooring contexts, (T=13), compared to that of the whole assemblage (T=55)... 174

List of Tables

Table 3.1. Physical properties of Mayen lava. 38

Table 4.1. Summary of wear traces and possible ranges of movements that they might reveal. 46

Table 5.1. Data for figure 5.1, showing the breakdown of querns, millstones and unidentified lava in the assemblage according to whether they are upper or lower stones. 50

Table 5.2. Data for figure 5.3, the breakdown of the site types included within the dataset. 51

Table 5.3. The 25th and 75th percentiles of lower stone thickness by region, with the range between the two values. ... 62

Table 5.4. The 25th and 75th percentiles for upper stone thickness by region, alongside the range for these values. 64

Table 5.5. Data for the thickness of lower stones by site type. 65

Table 5.6. Data for the thickness of upper stones by site type. 65

Table 5.7. Sites, locations, and site types where reuse of lava querns has been observed and recorded. 65

Table 6.1. Comparison of quern thickness by site type. 77

Table 7.1. Summary of other miniature lava milling tools recovered in Britain. 99

Table 7.2. Number of lava groups recovered from sites in and around Verulamium, quantified by site zone where information is available 104

Table 7.3. All fourteen querns recovered from insula XIV with their allocated date period. 107

Table 7.4. Summary of lava querns showing signs of reuse in Verulamium. 111

Table 7.5. Comparison for volumes of pecked and harp-dressed querns in the Silchester and British datasets. 125

Table 7.6. Summary of the number of millstones recovered from urban case study sites. 130

Table 8.1. Number of upper and lower stones in the dataset. 135

Table 8.2. All sites with upper stones thicker than 100mm. 140

Table 8.3. Diameters of the lava milling tools in the assemblage with their associate catalogue numbers. And values of the upper and lower quartiles for lava quern diameter at Chesters and for the complete British dataset 145

Table 8.4. Upper and lower stone thickness for the querns from Chesters, and the upper and lower quartiles for the lava querns from the whole British dataset. 146

Table 8.5. Breakdown of the data in terms of stone part and milling tool type. 149

Table 8.6. Upper and lower quartiles for quern diameter at Housesteads compared with the full British dataset. 150

Table 8.7. Chronological distribution of milling tools of lava and non-lava types 153

Table 8.8. Upper and lower quartiles for quern diameter for stones from Vindolanda compared with the wider British dataset. 162

Table 8.9. Number of querns recovered from the fort compared to the vicus, with a further breakdown of the lava and non-lava examples. 164

Table 8.10. Table showing how the upper and lower quartiles for Vindolanda and Britain compare. 167

Table 8.11. Comparison between upper and lower quartiles for Britain and Vindolanda. 168

Table 8.12. Comparison of upper and lower quartiles for quern diameter across the four sites compared with that of the wider British dataset. 176

Abstract

The importance of food processing in the past is evident from the frequent occurrence of milling tools on archaeological sites. During the Roman period in Britain, imported lava milling tools complement those of indigenous stone types, but distribution and use of lava in the province has never been systematically examined. This research presents the first study of this material in Roman Britain, cataloguing and analysing 2,707 lava milling tools from 564 sites. A further 601 sites where lava was absent but other lithologies occurred have also been recorded. Analysis was completed using an object biography approach to investigate the key stages of manufacture, distribution, primary use, reuse/modification, and deposition at various case study sites to reflect lava milling tool use in rural, urban, and military contexts.

Quarrying and manufacture of lava milling tools at Mayen, Germany, was examined in detail, and distribution analysis was undertaken to explore spatial, chronological, and social distribution. This has demonstrated that lava consumption was limited geographically and socially, with regions in the south-east and east of England, alongside military and urban sites being most likely to use lava milling tools. Lava millstones were most common in southern urban and northern military contexts, showing that access and use of centralised food processing technology was unequal. Chronological analysis of lava distribution has shown that the earliest lava imports occurred in the first century, with a peak in deposition during the second to third centuries.

Evidence for the modification of lava querns has provided insight into innovation and specific modes of lava quern use, while varying use wear prior to deposition points to privileged access to imported lava at urban and military sites compared to rural ones. Analysis of deposition suggests that lava querns were ritually deposited in pits in urban contexts, and that reuse in construction and road surfaces was common at military sites.

Overall, this study has generated a large corpus of new data, the analysis of which has delivered previously unknown detail of the role of lava milling tools in Roman Britain. The dataset also creates the potential for future analysis to help build a more complete view of the complex biographies of these significant objects and their relationship with past peoples.

1

Introduction

Processing grain is an activity fundamental to agrarian societies and often imbued not just with economic, but also social and ritual significance. The need to feed a growing population in Britain from the Iron Age onwards resulted in an increase in the production and consumption of cereals and the role of the quern for grain processing grew significantly as a result. Querns saw widespread use and are frequently present on most archaeological sites dating from the Iron Age onwards. Although not entirely ubiquitous after the introduction of mill technology, for a long period of time querns were essential tools in food production. Their use within domestic contexts for small-scale immediate food consumption is evident at archaeological sites, but hand operated milling tools may also have been utilised on a larger scale to feed entire communities or to provide surplus flour for wider regions. Querns were probably used to feed a mobile army during the invasion of Britain and stationary troops in garrisons and forts. The significance of grain to the military diet is well-attested and quern use would have played an important role in the everyday life of the Roman soldier, who literary evidence suggests, was responsible for preparing his own daily food ration (Davies, 1971).

Imported lava was one of the most popular choices for querns and millstones in Roman Britain, but distribution and use of this material in Britain has never been systematically studied. This work will, therefore, focus on a neglected body of material culture to enhance our understanding of Romano-British food production, of the rural economy and of the social, cultural, and religious identities of the people who used them.

1.1. Context of Study

The aim of this research was to investigate what lava quernstones and millstones reveal about life in Roman Britain; their use in food production, involvement in trade and exchange networks, ritual activities, and the role they played in the formation and maintenance of socio-cultural identities. This was completed using a theoretically informed approach, applying the concept of object biography to analyse the various stages of the lava quern or millstone lifecycle (Kopytoff, 1986). For each stage, the interactions between human and object were identified, with consideration given to the possible social and economic implications of these interactions. Case studies were selected to specifically address variation across different site types: military, urban and rural. Overall, 2,707 lava milling tools from 564 sites in Roman Britain were captured in the dataset, alongside 601 sites where milling tools were present, but none of lava: making 1,165 sites in total.

While major advances to our understanding of Roman rural life have recently been made, artefactual evidence has often been ignored. Stone artefacts, such as querns, have largely been neglected. Earlier studies of milling tools have focused mainly on petrology, with little consideration given to other aspects of the artefact life cycle and socio-cultural significance (e.g., Peacock, 1980). However, routine practices, especially those connected with food production, can offer profound insights into ancient societies. For this reason, there has recently been an increased emphasis on contextualising milling tools, for example by distinguishing between querns and millstones and by linking petrology explicitly to trade and consumption (e.g., Shaffrey, 2006; Williams & Peacock, 2011). There has also been an increased recognition of deposition practices, which can have ritual aspects (e.g., Watts, 2014; Buckley & Major 2016, p. 135).

While easily recognised by non-specialists as volcanic, the substantial body of lava milling tools from Roman Britain has never been studied as a group. Until now, even the most fundamental information on typology, chronology, and distribution has been missing from current research. This study presents data that allows for this information to be explored, and from this provides new insights into trade relationships (cf. Morris, 2010) and consumption preferences with consideration of socio-ritual practices. The new data on lava querns has been compared to published data on other types of Romano-British querns (Shaffrey, 2006; Shaffrey & Roe, 2011) to create a holistic picture of the role of lava milling tools in Roman Britain.

1.2. Querns as Artefacts

Archaeologically, the introduction of the quern as a means of food processing is strongly associated with the introduction of agriculture during the Neolithic period (Peacock, 2013, p. 17–21: 26), though the tool itself has precursors dating back to the Palaeolithic (de Beaune, 2004). As European populations became increasingly reliant on grain as a foodstuff, a dependence on the tools used to transform grain into edible food products also developed. Querns may also have been used to crush other materials, such as metal ores and pottery tempers (Watts, 2014, p. 22–23), though interpretations relating to these methods of use are highly contextual and most querns are ordinarily associated with food processing activities. The earliest quern forms in Britain are of the 'saddle' type, consisting of a large slightly dished lower stone that can hold the grain, and a smaller hand-held stone that was moved back and forth, crushing the grain between the two stones (Peacock, 2013, p. 18–19) (figure 1.1). This form continued to be used in Britain well into the Iron Age and examples have been recovered in small numbers from Roman contexts, showing a long period of use (for example, see Williams & Zeepvat (1994) and Stead (1976)).

Figure 1.1. Bronze Age milling stone and saddle quern (Todd, G., 2016: CC0 1.0).

The introduction of the rotary quern brought huge change to an everyday activity, reducing the amount of work needed to process the same volume of grain (Peacock, 2013, p. 120–130). Despite the name, there is a high possibility that rotary querns were oscillated to imitate the movement of a saddle quern, and not rotated (Heslop, 2008, p. 55; Watts, 2014, p. 21). The precise chronology relating to the introduction of the rotary quern in Britain is not known and is likely to have been regionally different (Shaffrey, pers. comm.), but is thought to have occurred sometime between the middle to Late Iron Age (Watts, 2002, pp. 27–29; Peacock, 2013, pp. 54–58).

Rotary querns take the form of a pair of stones of relatively equal sizes (figure 1.2). Different quern forms or types exist, and these can be regionally or chronologically distinct (Watts, 2002, pp. 31–38; Peacock, 2013, pp. 63–76), but all tend to have circular grinding surfaces; one on the upper (*catillus*) and one on the lower stone (*meta*). The lower stone remains stationary, while the upper stone is either rotated or oscillated via a spindle fitted into the lower stone that is pivoted on a 'rynd', usually of iron, fixed into the upper stone (figure 1.3). Grain is fed into the quern via an opening in the upper stone, sometimes widened at the top to form a hopper. Various handle fittings exist across different rotary quern types, used to move the upper stone. Querns are typically made from sandstone, conglomerate, limestone, or lava, though any suitable geology can be utilised. In regions with access to suitable stone types, these were locally sourced, while good stone for querns was often moved many miles to supply regions without (Parkhouse, 1997; Moore, 2007, pp. 84–85; Morris, 2010, p. 78; 107). The stone needed to be hard enough to withstand the friction of movement associated with grinding without excessive breakdown of the material, while remaining suitably rough to maintain the friction needed to cut or crush the grain (Watts, 2002, p. 29; Peacock, 2013, p. 2).

Figure 1.2. Rotary quern in use in Tibet (Beger, B., 1938: CC BY-SA 3.0 DE via Wikimedia Commons).

Most modern interpretations of ancient quern use tend to focus on utilitarian aspects of function, form, and mode of operation as well as the relationship between querns and the economy of the stone trade and food production. However, the social significance of these objects has come increasingly to the forefront of investigations. The important and diverse role that querns held in past cultures is well recognised in prehistory (Hill, 1995, p. 108; Hamilton, 2002, p. 40; Brück, 2006; Moore, 2007, pp. 90–91; Watts, 2014, pp. 5–12;), and these ideas are supported by anthropological studies involving quern use in modern-day traditional societies (Watts, 2014, p. 31; 33–35). Less focus has been given to such relationships between people and querns in the Roman period, which might relate to the fact that there is such a large volume and variety of material culture available for study. This makes it much harder to establish patterns in deposition or use. However, the socio-cultural significance of these objects appears to span time and space and it is highly likely that querns in Roman Britain also held more value than that of their function alone.

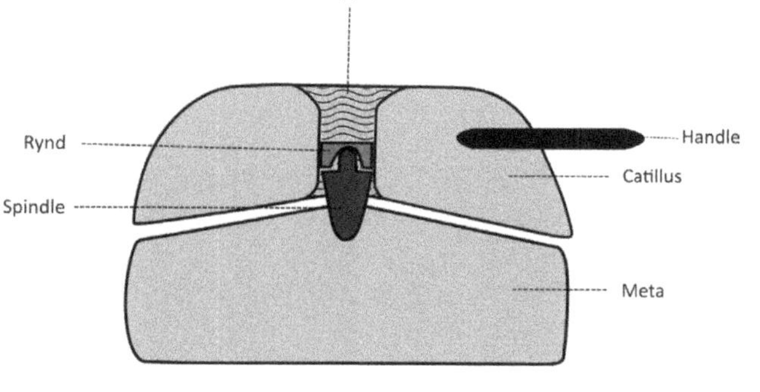

Figure 1.3. Diagram of general components of a quern of indigenous British form.

1.3. Mills and Millstones

Roman period mills in Britain existed in two main forms: disc or 'Pompeian' types (figure 1.4) (see Chapter 2). Both could be powered by humans or animals, while disc types could also be turned using water. Structural evidence for mills within the province is a rare occurrence (figure 1.5), though it is highly likely that a greater number of mills were in use than have currently been recorded. Shaffrey (2015) has named known Roman period mills at Fullerton, Ickham, Stanwick and Silchester, with possible mill structures at Darenth, Kenchester and Dickets Mead. The Rural settlement of Roman Britain Project has identified a further 13 possible mill structures (Allen, et al., 2018), while two probable watermills are known at military sites at Chesters and Haltwhistle Burn on the northern frontier (Simpson, 1976, p. 26–49). Current distribution of identified mill structures suggests that there was a greater number in the south of the province, which might relate to population, but may also be indicative of more modern-day development in these areas.

Utilisation of mills can also be identified by the presence of millstones, which have a much better survival rate than other evidence types, meaning they have a higher probability of recovery than timber mill structures.

Figure 1.4. Disc-type Roman millstone (right) (Bjoertvedt, CC BY-SA 4.0 via Wikimedia Commons) and reconstruction of a Pompeian style mill (left) from millstone fragments recovered from Princess Street, London (photo by author reproduced with kind permission of the Museum of London).

Despite there being some debate about whether millstone presence constitutes the existence of a mill (Dawson, 2019, p. 239–240), the occurrence of millstones, especially those with clear signs of wear, can and should be connected to a probable mill in the area; even if this is not necessarily within the immediate vicinity (Shaffrey, pers. comm.). Sometimes, these provide the only means for identifying

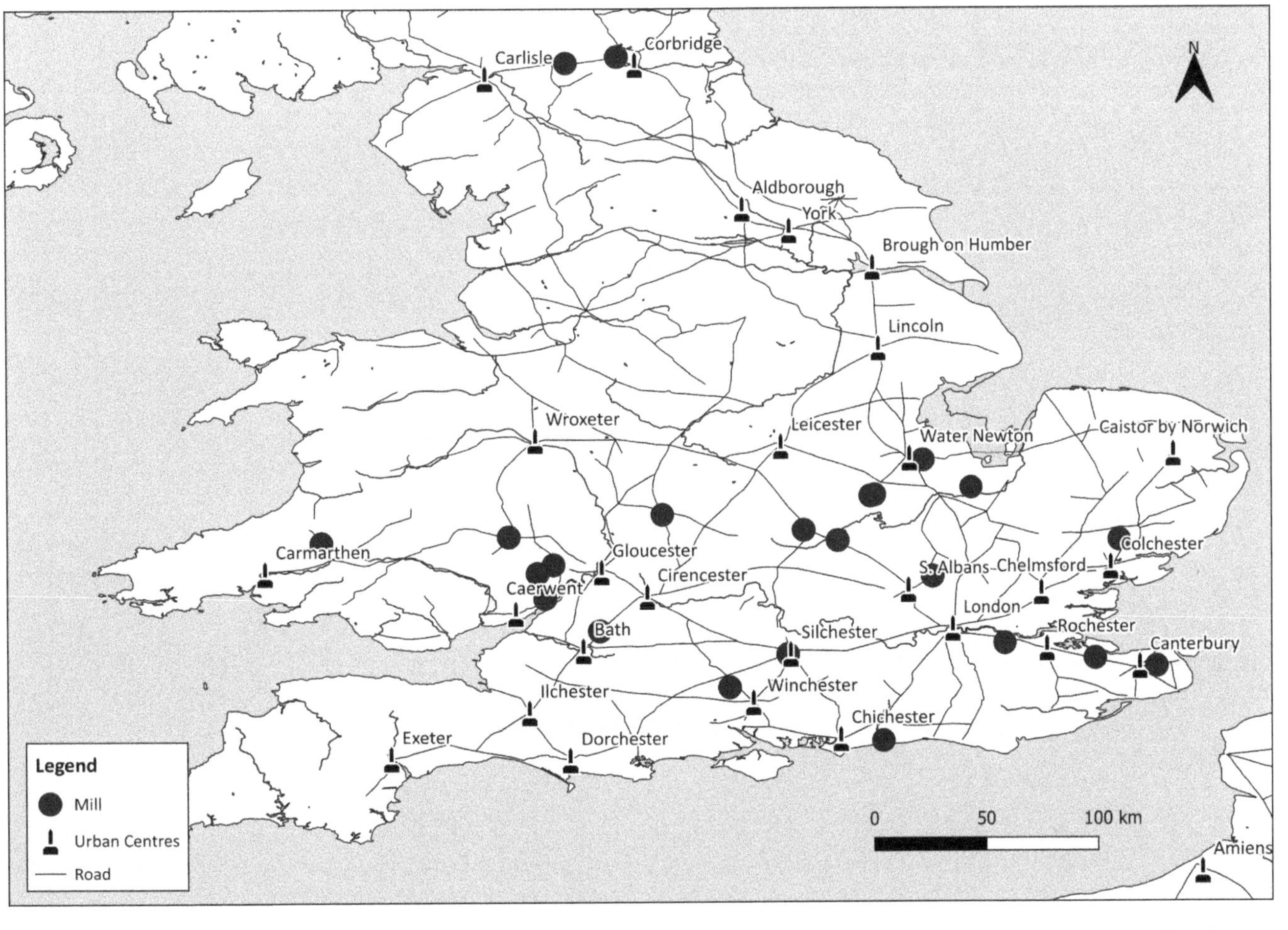

Figure 1.5. Location of identified and possible mill structures from structural evidence in Roman Britain, also showing the main urban centres (Background GIS data courtesy of the Ancient World Mapping Center).

the existence and extent of centralised grain processing within specific regions and chronologies; they act as an important indicator of population, status, organisation of labour, centralised food production, and access to technology and innovation.

1.4. Differentiating Between Mills and Querns

The ability to turn or oscillate a quern comfortably and use it efficiently as a hand-operated tool is highly dependant on ergonomics. The mechanics of grinding influence tool design heavily, whilst human requirements such as physical range of movement and the impact or stress on muscular and skeletal systems are also an important factor (Sefryn, 2013, p. 692). Similarly, the concept of 'affordance', as taken from design theory, has been applied to archaeological material to help interpret how form might be related to function using evidence of actual use (Swift, 2014, p. 203–204). Although this approach also considers the design of an object as being conducive to its intended use, it actively examines differential use and adaptation of purpose through alterations in the material and signs of use-wear. This is something that should be applied to quern and millstone identification too.

As millstones are not constrained by size, there can be a much wider variation in stone diameter than for querns, and it is generally assumed that millstones would be larger and more productive than querns to justify the extra investment in the mechanics of the device and any associated expense of running them. As querns or millstones may have been used differently to that of the intended design, there could be some 'blurring' between these definitions. Shaffrey (2015) has determined that anything larger than 570mm should be defined as a millstone, while Peacock (2013) suggests a slightly larger size of 600mm. However, these definitions encompass milling tools of all stone types and more specific divisions may exist for particular types or classes of millstone/quern. Where this separation exists for lava milling tools will be discussed further in Chapter 2.

2

From Quarry to Quern: Previous Research on Lava Milling Tools

2.1. Lava Querns: Terminology

The import of rotary lava querns into Roman Britain did not commence until the Roman era, which was when only rotary forms were being produced in the Mayen region (Watts, 2002, p. 33; Fitzpatrick, 2017). As previously described in the introductory chapter, rotary querns consist of two paired stones, the *catillus* (figure 2.1), and the *meta*. These tend to be the only surviving remains of a lava quern, are often fragmentary, and are rarely uncovered as pairs (figure 2.2). For lava querns, both the upper and lower stones were disc shaped and, prior to use, were around 100–120mm thick (Mangartz, 2008, p. 94). The grinding surface of the upper stone was manufactured to be concave in profile, while the lower stone was convex to the same degree. Both stones were fully perforated through the centre, with the central hole in the upper stone often being termed 'the eye'. For the upper stone, this was usually circular in plan with a cylindrical profile, while the central hole for the lower stone widened from the opening in the grinding surface to where it exited the lower surface of the stone creating a partial conical profile.

The eye was bridged with a strip of iron, called the rynd, which was bedded into the top surface of the upper stone within a cavity on either side of the eye, known as the rynd chase, and fixed into position at either end using lead. This strip provided the pivot point for the upper stone, which rotated on a spindle. The spindle was fed through the upper and lower stone and fixed into position in the lower stone, probably using a bung (see example from de Meern 1, this chapter). The upper stone was moved via a handle fitting, the most common type being an 'elbow' socket. This was an angled perforation of the upper stone that began on the upper surface and ended at the side, creating an elbow shaped profile. The handle fitting was threaded through this hole, allowing it to be fixed to the side of the upper stone. It was reinforced with a raised annular kerb, or rim, which was created around the outer circumference of the upper stone.

2.1.1. Surface Dressing

The grinding surfaces of Roman period lava querns tend to be dressed, which describes a way of etching, scoring, or pocking the surface of a milling stone to alter the texture of the surface. Dressing appears in several forms, with harped, radial, and pecking being the most common (figure 2.3). Harped dressing, sometimes referred to as segmented radial, is a linear style that produces a repeated

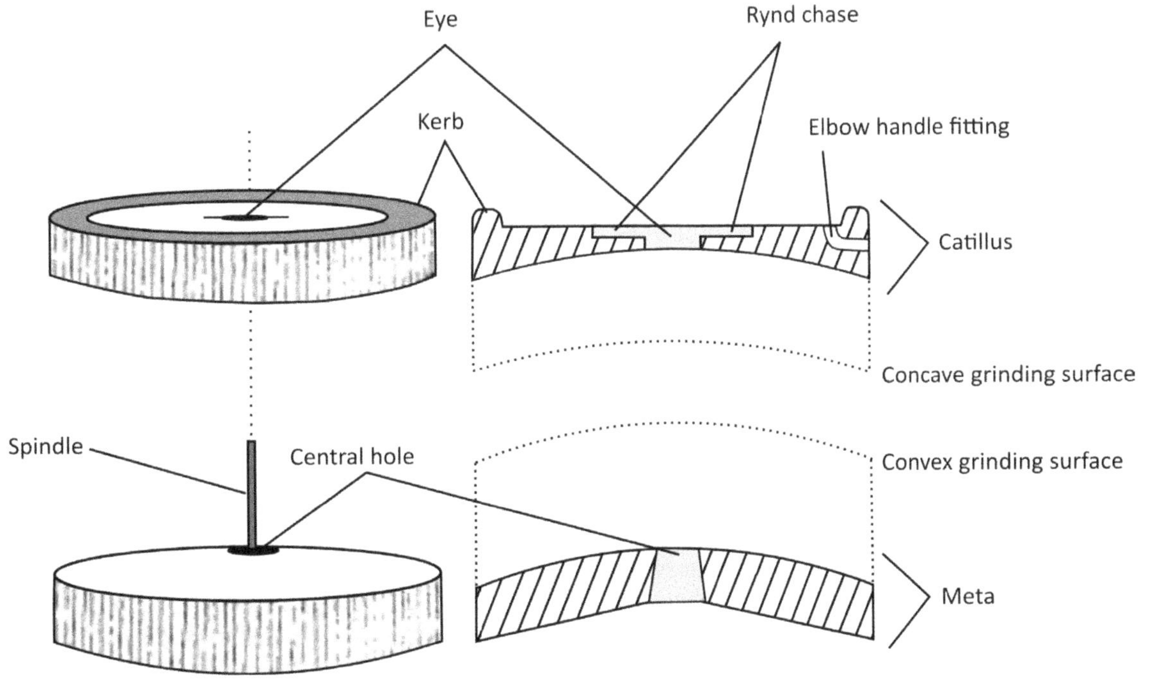

Figure 2.1. Diagram of typical lava quern stones showing the terminology used for different parts of the tool (illustration by author).

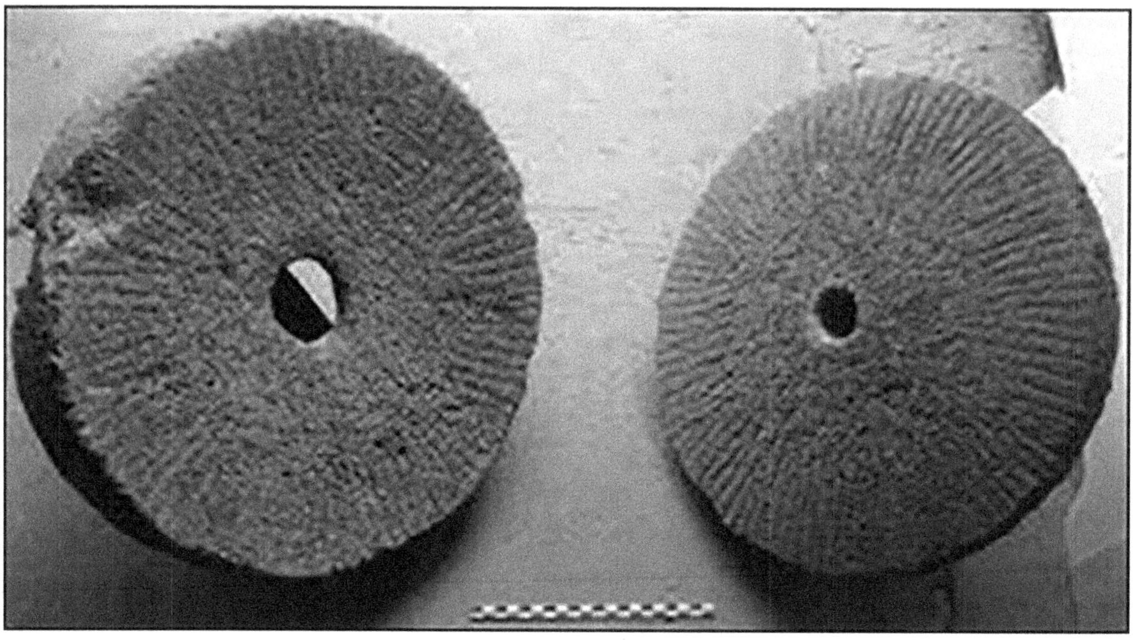

Figure 2.2. Stone remains of a pair of complete Roman period rhyolite quern stones with perforated meta (right) (Lepareux-Couturier, 2014, p. 151, fig. 6).

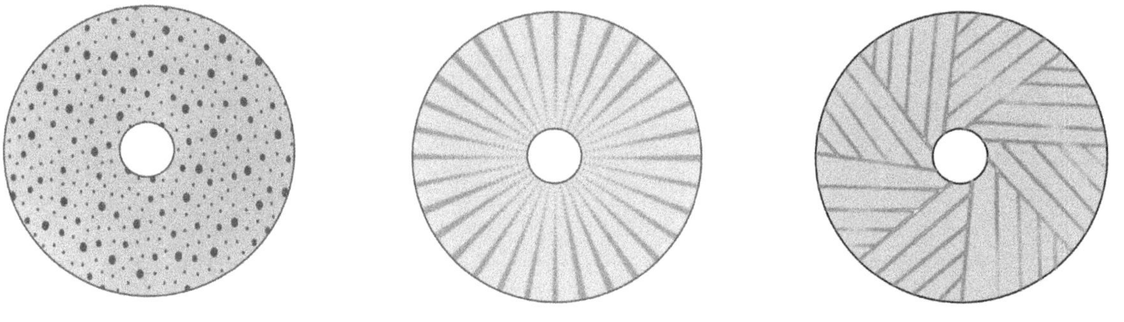

Figure 2.3. Different styles of dressing for lava quern stones. Left- pecked, centre- radial, right- harped or segmented radial (illustration by author).

grouped pattern of lines, one group of which resembles a harp. This is the most recognised form associated with lava querns, which were probably originally dressed to this specification The form was often reproduced on worn stones that were redressed with varying degrees of success. Radial dressing patterns are also linear, but they consist of straight lines that radiate out from the central point of the stone, and these also commonly occur on lava querns. Linear tooling on the grinding surfaces is often described as radial by default, even when the dressing is harped or unidentifiable on fragmented stones. The final form, pecking, is a method whereby the surface of the stone was periodically indented using a chisel or hammer. Though this type of dressing would seem redundant on vesicular lava, which is already pocked with small holes, it is common on indigenous quern types and is also known to occur on lava stones.

Mayen lava querns typically also carry 'decorative' linear dressing on the upper surface of the upper stone. This continues across the kerb and appears as vertical striae on the sides of the upper stone. These are also found on the sides of the lower stones. Although often interpreted as decoration (Jodry, 2011, p. 88), there is some suggestion that this represents a way of 'branding' the querns as products of the Bellerberg quarries and may have been a way of visually uniting querns from different quern manufacture sites as products of a single region and standard (Mangartz, 2012, p. 12; Wenzel, 2020, pp. 2–4). They may also have been added to provide interval measurements for redressing worn grinding surfaces in the distinctive, yet complex harp style.

2.1.2. Quern Mechanisms and Components.

There is some debate as to what the functional component parts of a quern consisted of and how these may have fitted together. This is because the mechanical elements of querns are not usually found in association with stone remains and are often misidentified or unrecognised. The sparse evidence we have of non-stone components associated with Iron Age and Roman period querns show that metal, predominantly iron and lead, was often used to form these components. Few examples of quern handles are known to exist and it is likely that these were formed of organic materials, such as wood (Reniere, 2018, p. 20), evidenced by a fragment of a wooden handle of ash recovered at Newstead Fort (Curle, 1911, p. 145). Heslop (2008, p. 52) has also discussed the possibility of iron handles from evidence of wear in Iron Age quern handle holes/slots.

However, it should not be assumed that all querns operated the same way or had the same constituent parts and more complete examples cannot be taken as representative of all quern mechanisms. A larger number of quern production sites would create a wider range of variability in how these objects were assembled. In the case of lava querns, the high level of standardisation present and the fact that they were centrally produced at Mayen translates into greater uniformity in the mechanical workings for lava querns across Romano-British assemblages. Very few Mayen lava quern unworked stone blanks were exported, and so this greatly reduces the variability present in lava quern assemblages (Wenzel, 2020, p. 3). If blanks were imported into Britain, deviation from the standardised form might be expected, but this is unlikely to have occurred at any scale. Therefore, the few examples that we have of partially assembled lava querns or those which show mechanical elements provide a good guide for the kinds of lava quern set-up in existence during the Roman period.

An example of a complete elbow handle fitting on a lava quern was recovered from Newstead Fort (Curle, 1911, p. 145–146), and is described by Mackie (2007) (figure 2.4). The wooden handle, which had not survived, sat in an upright position within an iron loop formed from a band. This loop was attached to the side of the stone via a hole, through which the ends of the iron band were fed. The ends emerged at the top surface of the stone, just clear of the kerb, terminating as spikes that were hammered down to lie flat against the surface. The iron was fixed into position by pouring molten lead into the opening, which plugged the spaces between the band and the stone. This form of lava quern is a common type in Britain and most elbow type handle fittings would have been assembled in a similar way.

An alternative example can be found in an almost complete 'working' Roman period quern of Mayen lava that was recovered from a shipwreck in De Meern, Netherlands (Mangartz, 2007) (figure 2.5). This quern was being used as ship equipment. Most of the component parts were well preserved and the quern was fully assembled. No elbow handle fitting is perceptible, which implies that a greater degree of typological variability could be present in lava quern assemblages than first thought. A long wooden pole found with the quern may have served as a 'swivel pole' handle that fit into one of the two holes in the upper surface of the quern when in use (Mangartz, 2007, p. 246). These handles are well-known with regards to historical quern operation and can be found in medieval images of quern use, alongside more modern depictions (figure 2.6). There is currently no clear evidence to suggest that the mechanism was used as early as the Roman period, but the concept behind 'swivel pole' querns is not complex and could have been implemented during this era also.

The set-up was designed to be used on a table or at a similar working height. A swivel pole consisted of a long handle/pole, probably of wood that would be fitted into the top stone via a hole on the surface, or from a loop at the side. This would be used to rotate the mill. The longer handle length reduces the angle of rotation for larger stones and allows for multiple people to turn the handle at the same time in a more comfortable working position. The far end of the pole could be poked through a hole or in something fixed to the ceiling or other elevated fixed object (figure 2.7). The tool is likely to have been used in a permanent fixed

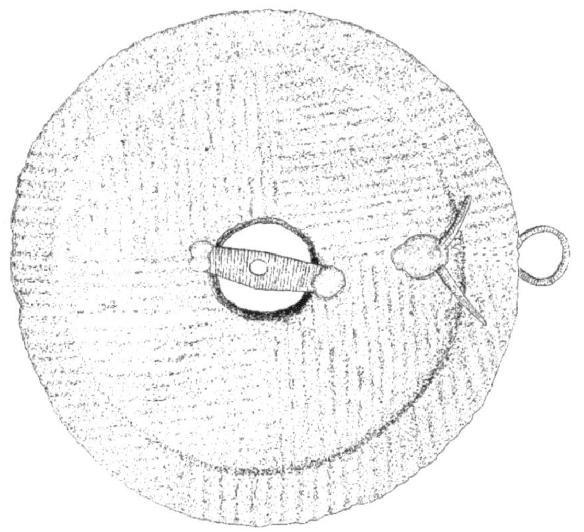

Figure 2.4. Catillus from Newstead Fort with complete metal components showing how an elbow handle was assembled (MacKie, 2007, p. 494). The upright wooden handle would have fit into the iron loop, seen here emerging from the right side of the quern.

Figure 2.5. Top and bottom stones of a rotary quern retrieved from the Roman shipwreck De Meern 1. This view shows the upper side of the catillus and two possible handle holes are visible. Also seen here is the rynd of iron which is embedded/recessed into the upper surface of the catillus and fixed in place using lead at both ends (Mangartz, 2007, p. 248, photo RGZM, Mainz).

Figure 2.6. Two women grinding corn on a quern mill on Pine Mountain, Kentucky (Gallery V Exhibits, 1998). The photograph shows relatively modern use of the swivel pole method for grinding and how two people could work the quern simultaneously. Though the top of the pole handle is not visible, it was probably threaded through a loop or hole in the ceiling to keep it fixed in place as seen in figure 2.7.

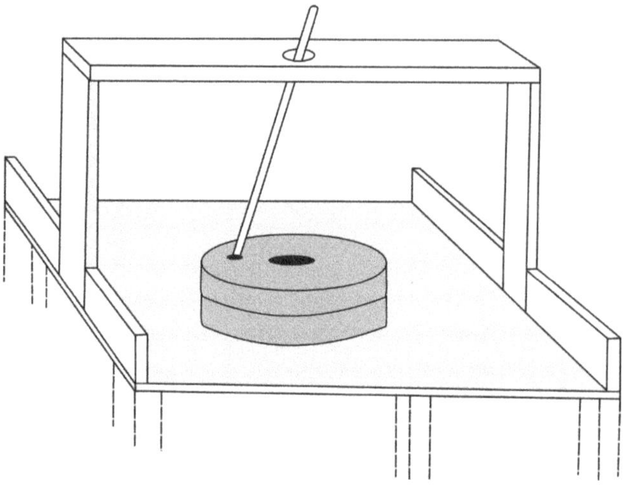

Figure 2.7. Illustration to show an example of how a swivel pole quern mechanism would have operated. The set-up would have worked best with equipment placed on a table or other raised surface and the length of pole handle is variable (illustration by author).

position, perhaps to accommodate for the listing of the moving ship. This method of rotation may have been replicated elsewhere. For the de Meern 1 example, this was clearly the intended method of use as expressed through the design of the quern. However, a long handle placed in an iron loop from an elbow handle slot would be similarly effective and indistinguishable from shorter handle forms.

Other aspects of the quern mechanism revealed from the De Meern 1 quern include the way that the spindle and rynd were assembled to provide the rotary motion. The rynd, as seen in figure 2.8 measures 150mm by 30mm and

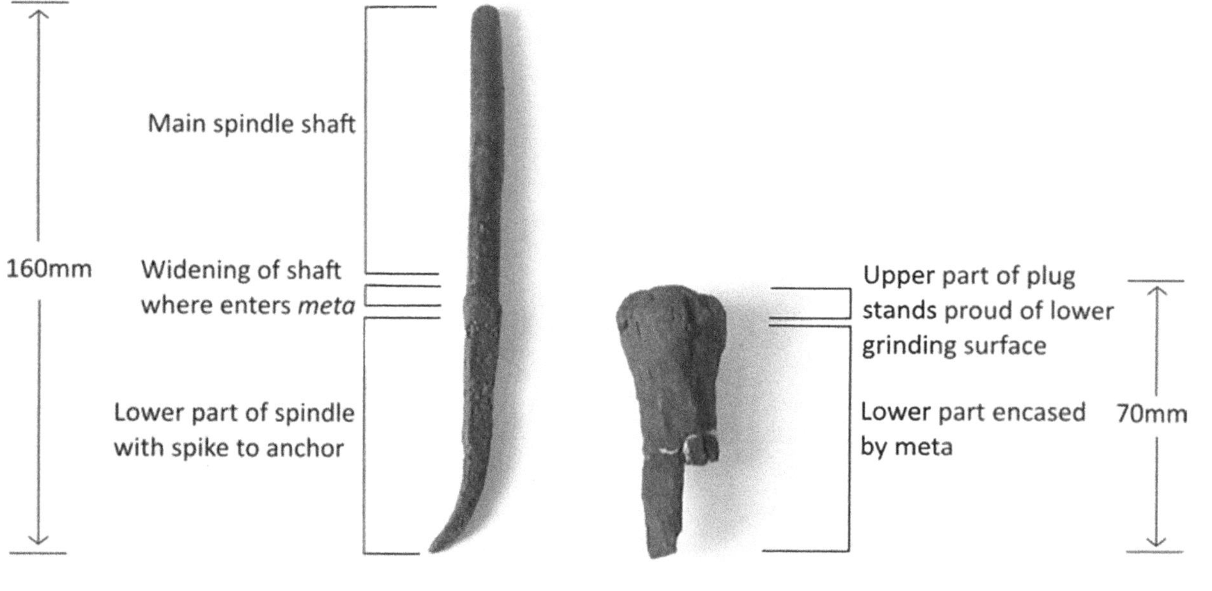

Figure 2.8. Spindle (left) and wooden spindle plug (right) from quern recovered from De Meern 1 (edited by author using images from Mangartz (2007, p. 246).

has been fixed into place at both ends. It spans the central hole and is perforated in the middle, with the perforation measuring approximately 10mm. The spindle would either feed through this hole or use it as a fulcrum, centring the top stone over the lower stone and providing the pivot on which the *catillus* could rotate. As the quern was fully assembled at the time of its discovery it has been possible to observe how the spindle was arranged, its dimensions and how it assisted in the operation of the device. This level of detail is rarely available for querns, and it is with some difficulty that accurate reconstructions of this part of the mechanism are produced. With the De Meern 1 example, the spindle comprises an iron rod with a square-shaped cross-section measuring 160mm in length which had been wedged into the lower stone's central hole using a 70mm long wooden plug (figure 2.9). The pivot end of the spindle is more rounded, whilst the other terminus ends in a slightly bent spike which may have helped stabilise the weight of the top stone.

Observations have shown that the spindle is not long enough to lower or raise the catillus using a tentering device (tentering will be discussed in more detail below). It might be assumed that the upper stone balanced on the end of the rounded spindle, although this arrangement must have been unstable considering the weight (15.5kg) of the *catillus* and the depth of the rynd (7mm). Whether or not the spindle fed completely through the rynd hole can only be speculated; current assumptions are that a separation needed to exist between the upper and lower grinding surfaces (also known as the 'light'), which would mean that the rynd took the full weight of the stone. However, experimental work using lava querns suggests that the upper stone can fully rotate without this separation, and that the stones can sit directly on top of one another with no restriction in movement (Chapman, unpublished). This would imply that the spindle could be safely threaded through the rynd, creating a more stable stone pairing.

Figure 2.9. Meta from quern recovered from De Meern 1 showing how the spindle components fit together (Mangartz, 2007, p. 246).

In Britain there have been at least four querns of both lava and non-lava, recovered with spindles *in-situ*, though these have since been dismantled with no record of how they were originally fitted together. The four known examples were recovered from Newstead Fort during excavations in the early 20th century. Curle (1911) describes one spindle as 7.5 inches in length (approx. 190mm), with one pointed end that was probably inserted into a wooden peg fixed into the lower stone. The other end of the spindle was rounded, which had been passed through the perforation in the rynd bridge. Unlike the De Meern 1 spindle, two from Newstead (figure 2.10) show an increasing diameter from the rounded top where it provided the pivot, to where the spiked end begins. Therefore, if the spindle did feed through the rynd perforation, the rynd would sit on the spindle where the spindle diameter became wider than the perforation; this may have prevented it from falling to the level of the *meta* grinding surface, creating a separation between the stones.

The main difference between an Iron Age and a Roman period rotary quern is that the *meta* is almost always fully perforated in the Roman period examples. This is frequently interpreted as a functional feature to enable adjustment of the 'light', known as 'tentering'. How this was achieved through use of the central hole in the lower stone is rarely explained in any detail, and it has become a statement often replicated with little supporting evidence. Medieval examples of querns could be tentered, but this was achieved through adjustment of the rynd. Moritz (1958) produced a diagrammatic version of a tentering mechanism using an example from a contemporary Scottish quern (figure 2.11) (see also Reniere (2018, p. 24, Fig. 4) and Amouric (1997)). Here, the spindle is attached to a moveable board that is lowered or raised by adjusting the length of the string at the side. The arrangement requires that the rynd and rynd chase are fixed within the grinding surface of the upper stone in what is referred to as an

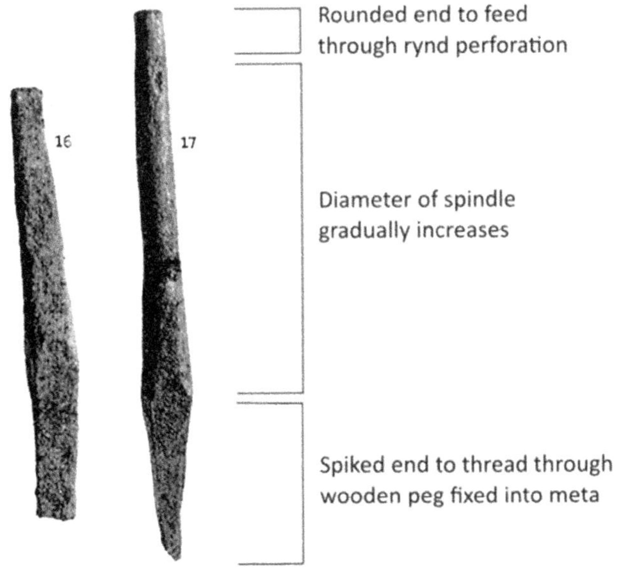

Figure 2.10. Two of four spindles recovered intact from lava querns at Newstead Fort (edited from Curle, 1911, plate LXVI). The shape of the objects indicate how they may have contributed to the mechanics of rotary quern movement.

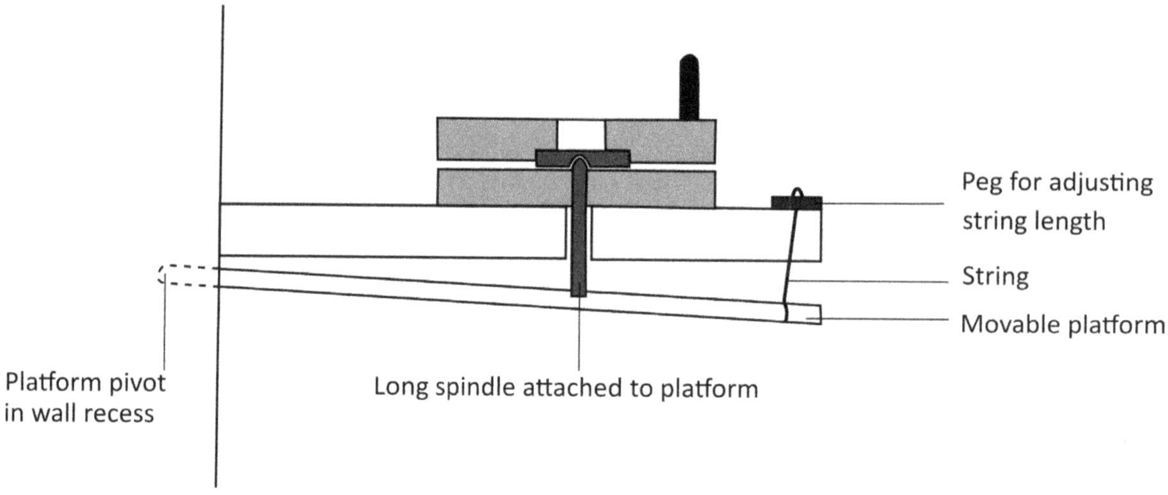

Figure 2.11. Diagram to show the configuration of a Scottish rotary quern that is assumed to be the same or similar to the Roman lava quern tentering mechanism (illustration by author after Moritz, 1958, p. 119, Fig. 11).

'underdrift' configuration. In Britain, lava querns tend to have an 'overdrift' rynd fitting, being placed in the upper surface of the upper stone (Watts, 2002, p. 153), as can be observed on the illustrated examples from De Meern and Newstead (figures 2.4 and 2.5). This is not to say that lava querns were not tentered, as this would provide an efficient means of dehusking grain (Watts, 2002, p. 63; Reniere, 2018, p. 22), but other explanations need to be sought for how this may have been achieved alongside identification of typological features on the lava querns that might relate to this form of operation.

2.2. Lava Quern Typology

Currently, there is only one working typology that has been created specifically for Mayen lava querns. The distinguishing features that mark the difference between quern types were first identified and recorded by Röder (1955) and developed into a typology using querns recovered from well-dated contexts at unspecified German settlement sites and waste dumps of roughouts at Mayen quarries (figure 2.12). The Mayen quarries saw a long period of exploitation to produce querns, stretching from the early Neolithic into the Medieval period (Röder, 1955, p. 69). Therefore, the typology covers a long chronology, leaving little room for nuance or detailed differentiation. According to this typology, the Roman period had two quern types, 5 and 6, differentiated by handle and rynd fittings. The early Roman quern (type 5) has a handle socket fitted vertically into the upper quern surface, while for the late Roman period, the handle is fitted laterally with a bowtie shaped rynd chase (type 6). Neither of these types have been identified in this study.

Most of the lava querns in Roman Britain are of type 4 (Chapter 4) and are classified by Röder as produced from La Tène period onwards. This is a somewhat ambiguous timeline for production with no specified end date. The fact that type 5 and 6 are classified as Roman suggests that type 4 was not in production during the Roman era despite its known presence at Romano-British sites. It is possible that types 5 and 6 were, at times, produced alongside that of type 4 in Germany, or that these were later Roman period innovations that did not arrive in Britain. It is a shame that a more precise chronology for these changes has not been determined, as the absence of type 5 and 6 in Britain could be used to identify when Roman Britain ceased to import lava querns. However, as the Röder sequence was derived from querns recovered from German settlements, some variation between the British and the German data should be expected, especially if only specific types were exported/imported. It is beyond the scope of this project to update the typology and to examine lava quern data from German settlement sites produced since the 1950s, and so the typology has been used in its current form.

Geochemical analysis of lavas from querns in north-west Europe by Gluhak & Hofmeister (2011, p. 1618) proved that different Mayen quarries were exploited to supply different areas, some of which had a wide reach. Whether the different quarries produced different types or styles of quern has not yet been discussed. The continental material would benefit from an in-depth study of typological variation alongside that of geochemical characterisation to clarify this further. It does also provide justification for developing a typology and chronological sequence for the Romano-British material, instead of simply applying Röder's scheme unquestioningly. The absence of a British lava quern typology is, however, understandable; the extent to which these objects are standardised has not incentivised such endeavours. Other material culture types, such as brooches or coins, show clear chronological and regional variation that can be used to generate interpretations relating to Roman period society, culture, and economy. This is less obvious when examining lava milling tools. There is a general agreement in excavation reports that catalogue and discuss lava querns that there is a 'regular' type that describes most examples, eliminating any perceived need for detailed description or further examination.

Furthermore, the fragmentary nature of many lava querns when recovered means that diagnostic features are often

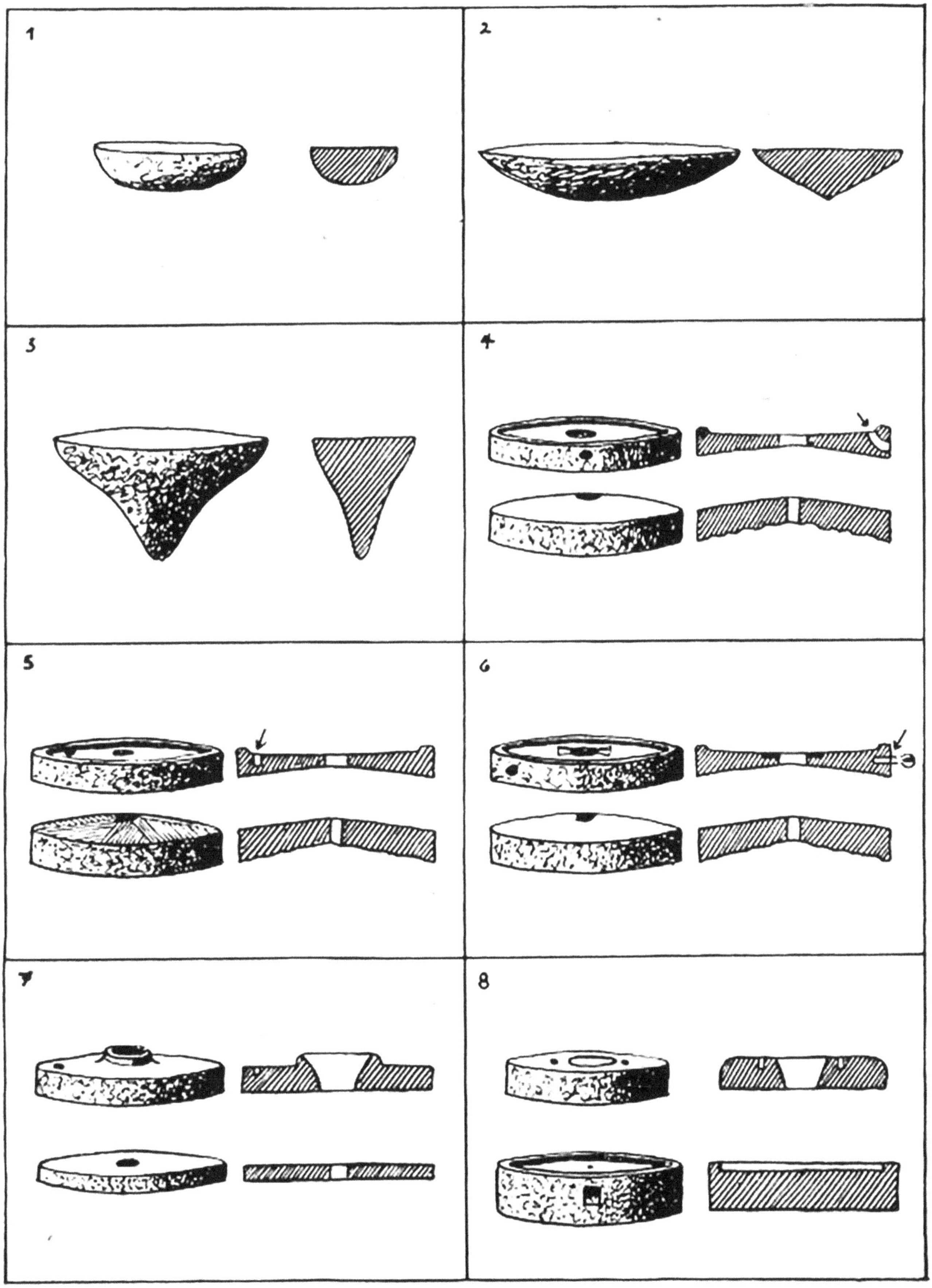

Figure 2.12. Neolithic to Middle Hallstatt, 2. Late Hallstatt (600–400 BC), 3. Napoleon hat (Early and Middle La Tène), 4. Oscillating rotary quern (from La Tène onwards), 5 and 6. Completely rotating quern (Roman period), 7. Early Medieval, 8. About A.D. 1000 (Röder, 1955, fig 1).

absent, and comparisons cannot be made. For example, no diagnostic features were present on lava recovered from 23 different contexts during excavation work on the Ewell Bypass, Surrey (Shaffrey, pers. comm.). To be able to develop a typology, single site analysis is unlikely to supply enough diagnostic pieces to yield much insight into the potential for variety. Therefore, if there are any typological differences, these would be better explored from an inter-site perspective, including a wide range of different material from different regions and different site types. The approach being applied as part of this research provides the best opportunity for exploring such similarities and differences, and for the generation of a more nuanced Romano-British lava quern typology.

There are some aspects of quern stones that are less suitable for the development of a typology, and these are the parts that are more likely to have been subject to repair. For example, the distinctive method of dressing the grinding surface of a lava quern that is very typical of the Roman era, known as 'harping'. As a quern would be redressed several times during its use-life, this is not a feature that can be consistently used to inform a typology. Furthermore, lava does not require dressing to function adequately in grinding activities, and so querns may not have necessarily been redressed once the original finish had worn away. Mending also creates some issues when identifying typological features and care must be taken to recognise evidence of alteration from the original type. This is especially challenging if the quern is incomplete, as original features may not always be presented in the remaining fragments, but mended features might. Original features may also become obscured by later reworking of the material, and this appears to be commonplace with regards to the rynd chase and eye which were subject to heavy wear. Nonetheless, so long as a suitable number of querns can be examined in detail, there should be sufficient data available for categorisation to be possible.

Upper stones of lava querns are the most diagnostic in terms of distinguishing features. Upper stones have more of the operational features necessary for a quern to function, and it is these aspects that are found to have changed over time when referring to the Röder typology. These comprise the handle and the rynd fitting, though the size and shape of the stone is also of significance. Stone diameter seems to be relatively variable, and further data collection will allow for this to be related to other characteristic features. Currently, features of the prevalent Röder type 4, illustrated in the typology, and observed on British examples, can be described as follows (figure 2.13):

1. Rynd fitted to the top of the upper stone, bridging the eye of the quern.
    a. Rectangular chase to accommodate the rynd that is set into the top of the upper stone and bridges the eye.

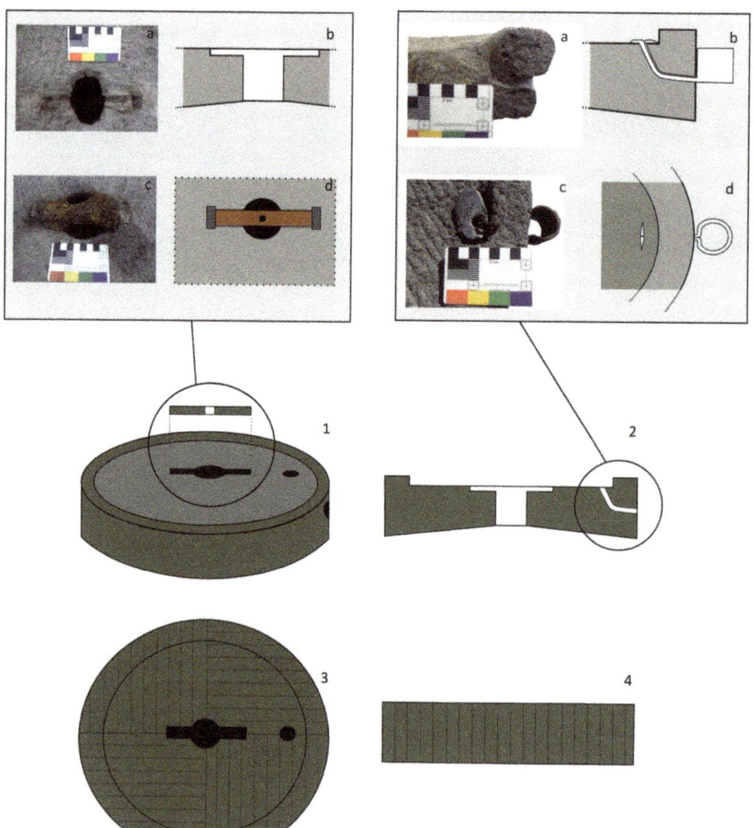

Figure 2.13. The main features of a Röder type 4 lava quern as the most prevalent type in Romano-British contexts (illustration by author).

   b. Rounded eye that is circular or sub-circular.
   c. Iron rynd fixed into the rynd chase by lead at the furthest ends.
   d. Perforated rynd to accommodate spindle.
2. The upper stone has a flat top, a concave grinding surface and a raised annular kerb that runs the circumference of the upper surface.
   a. The handle slot perforates the top of the upper stone, runs below the kerb, and emerges at the side of the quern.
   b. An iron loop used to accommodate the handle is fixed in place by two split pins that emerge from the bottom of the loop at either side and are fed through the handle slot.
   c. The split pins are fixed in position with lead on the top surface of the quern.
   d. The rounded iron loop probably held a handle made of wood or other organic material.
3. The top surface of the upper stone was 'decorated' with a linear design in quarter sections. Incised lines are at regular intervals and continue onto the kerb.
4. The incised lines also exist on the sides of the quern at regular intervals. Many of these appear to be a continuation of the top surface incisions.

There are few diagnostic features for lower stones. In the Röder typology there is nothing to differentiate between lower stones of types 4, 5 and 6. As a relatively featureless component, it would be difficult to group these typologically. A convex grinding surface is consistently present, as are incised vertical lines on the external edge. The lower surface tends to be roughly tooled and slightly concave. It may be possible to determine the presence of variation between examples using dimensions, such as diameter and central hole width, but this will need to be determined during the data analysis as any differences are not immediately apparent.

2.3. Lava Millstones: Disc Types

Millstones can be defined by their size as they tend to be larger than querns, though the precise size difference between querns and millstones is a matter of debate and should also involve evaluation of functional features. Lava millstones of disc types from Romano-British contexts are relatively rare, which is probably the result of poor recording, identification, and preservation. A high degree of fragmentation is also common with this material type. Disc types occur more frequently in Britain than 'Pompeian' types, which suggests that, in comparison, they were more prevalent (Chapter 4). Disc millstones may have been rotated using human, animal, or waterpower and the method of operation is not always apparent on the stones themselves. It seems probable that larger sized millstones were turned using water. However, there is no way of determining if this was the case or where any separation in size between water and animal/human powered millstones might lie.

The main feature that could be used to determine the mechanism type is the rynd, as this can be positioned on the underside (underdrift) or the upper side (overdrift) of the upper stone. Alternatively, there can be slots cut through the upper stone (probably also overdrift) (figure 2.14) (Wilson, 2010; Shaffrey, 2015, pp. 67–69). There is some debate as to whether this characteristic is distinctive of the mode of operation, as it is possible that water, animal, or human power could have been used regardless of whether the rynd were cut into the upper or lower surface. Current opinion suggests that overdrift stones were turned by people or animals, while underdrift mills were turned using waterpower (see Shaffrey, 2015 for details of this debate). In general, millstones in Roman Britain were predominately overdrift types, probably turned by people or animals. Underdrift types, probably turned by water, are only known from contexts of the early second century or later, implying a later date of introduction (*ibid*).

Disc type lava millstone blanks have been recovered from the quarries at Mayen, showing that they were sourced from Bellerberg quarries (Mangartz, 2012, p. 86), but blanks are not given functional features at this stage of the manufacture process. Meanwhile, finished or partially finished millstones at Andernach, where the final stage of manufacture was completed, were not retained, or

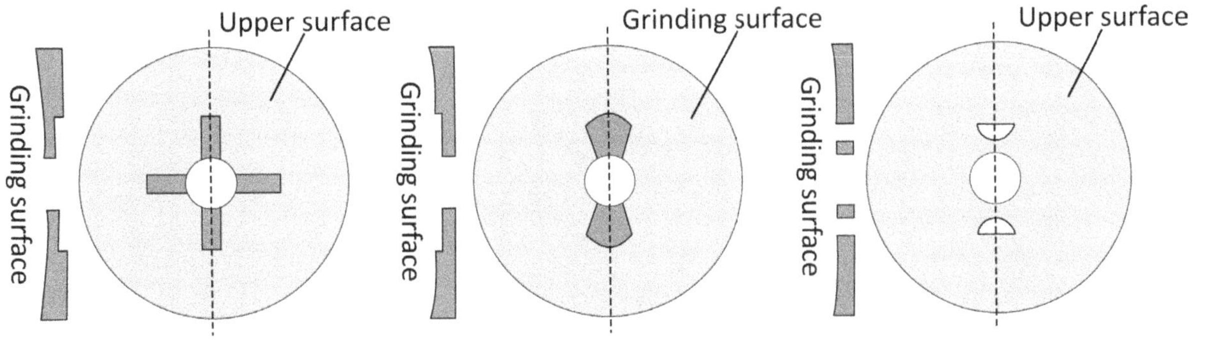

Figure 2.14. The three types of rynd fitting for millstones. Left-Rynd cut into the upper surface (overdrift). Middle- Rynd cut into the grinding surface (underdrift). Right- rynd cut through the stone (probably overdrift) (illustration by author after Shaffrey, 2015, p. 68).

fully recorded at the time of excavation (Röder, 1961; Schäfer, 2000). Consequently, there is no current detailed typology or chronology for Mayen lava disc millstones. The rynd fitting described above provides one possible way of differentiating between lava millstone types, but as the precise mechanics of a powered disc mill is not fully understood, nor the manufacture standard/varieties known, original features might be difficult to distinguish from modification or reuse. Complete millstones of other stone types may be used as a guide for the forms that would have been in use during the Roman era as powered mills were probably similarly operated regardless of the type of stone used. Reconstructions of complete mill assemblages have also been theorised using recovered millstones and mill components, literary descriptions, and knowledge of historical mill mechanisms.

A complete mill assemblage presented by Wenzel (2019), provides a standard underdrift mill arrangement that appears in a similar form across different published sources (figure 2.15). The form is based on a mill design described by Vitruvius (*de Architectura X*), though the translation and content of this description has presented some difficulties, especially with regards to the interpretation of terminology and mill components (Moritz, 1956). The literary account is partially supported by the discovery of a complete but disassembled lava mill recovered from a well in Saalsburg in 1912 (Jacobi, 1914; Baatz, 1995), which together provide an idea of how a Roman period geared mill operated. The set-up could have been powered by water or by animals/humans (figure 2.16).

An overview of the basic mill is provided here using modern-day English terminology. As these are different to those applied in the Roman era, and because Latin terms for mill components were not widely used in Roman literature, it is likely that some of the associated functional

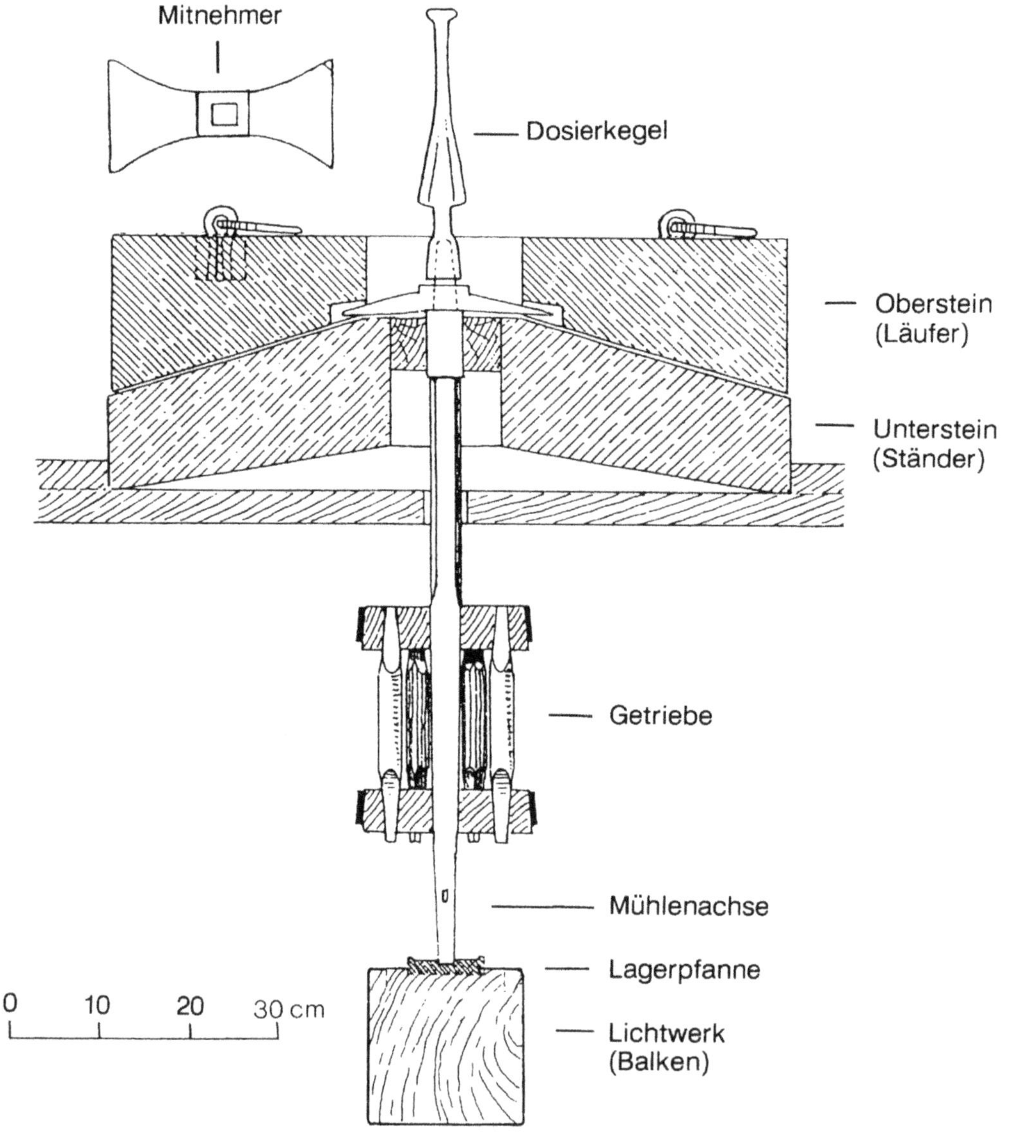

Figure 2.15. Mechanical parts of a powered mill from the Roman period reconstructed using literary and archaeological evidence (Baatz, 1995). Mitnehmer= Driver/rynd, Dosierkegel= Damsel/dosing cone, Oberstein= Upper stone, Unterstein= Lower stone, Getriebe= Lantern pinion, Mühlenachse= Spindle, Lagerpfanne= Bridging box (contains footstep bearings), Lichtwerk= Bridge tree.

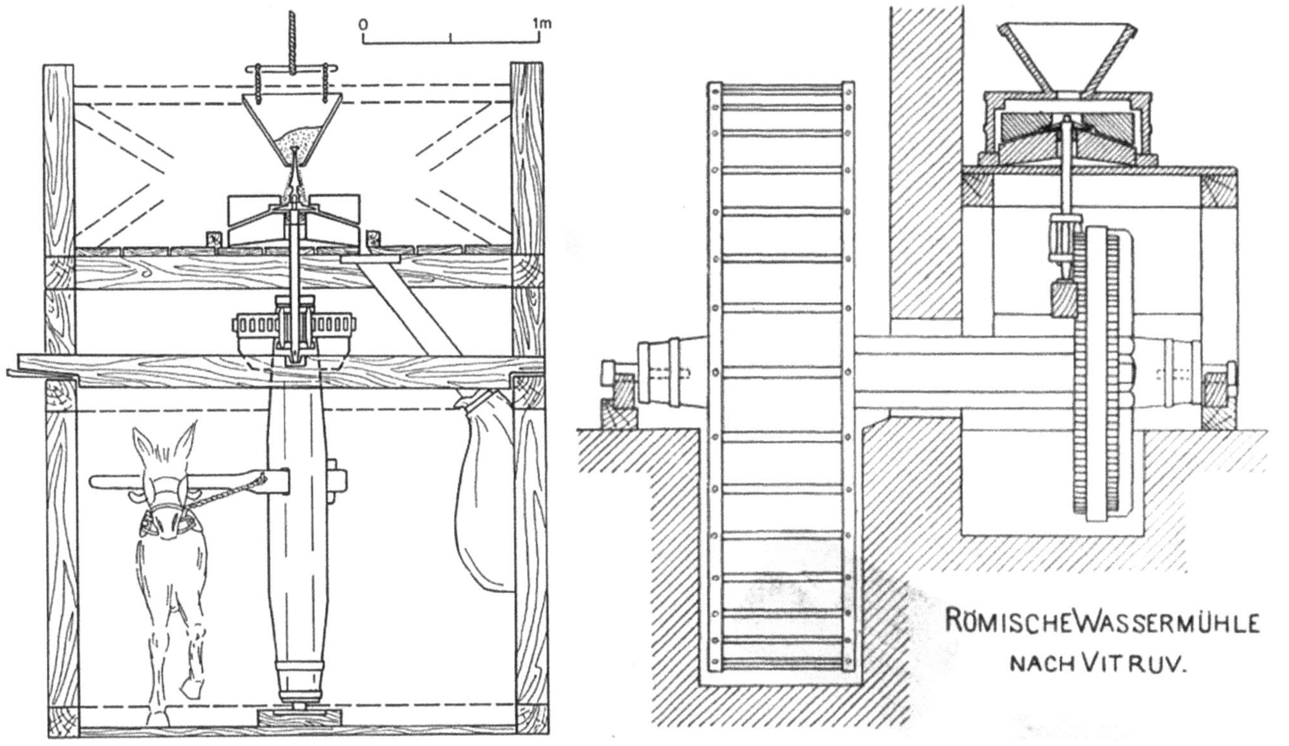

Figure 2.16. Mill set-up with mechanical workings as shown in figure 2.15 but powered through different means. Left, an animal powered mill that could also be turned by people (Baatz, 1995), right, a water powered mill (Jacobi, 1914, pp. 91, fig 45).

meaning for these terms has been lost in translation. It should be noted that constituent parts of a mill assemblage would ordinarily need to be recovered in association with millstones or structural mill remains for them to be used to identify the presence or use of mill technology. If found alone, the parts are rarely identified as associated with mills, if identified at all. These include the damsel, or dosing cone, which is a cone-shaped rod usually formed of iron that is positioned above the spindle and projects upwards to a funnel-like container for the grain, called the shoe (not illustrated).

The damsel vibrates or moves as the millstone is rotated, striking the shoe to release a steady flow of grain into the eye of the upper millstone for grinding. Roman period damsels have been recovered from archaeological sites in Belgium, but misidentification is common (Reniere, 2018, p. 32–33). Attached to the spindle below the millstone is the lantern pinion, which is a geared cylinder that provides the rotation for the runner stone via the waterwheel/animal wheel/crank. Below this, the weight of the upper millstone is supported at the bottom of the spindle by a footstep bearing housed within the bridge box. A piece of basalt recovered from the remains of a probable mill at Haltwhistle Burn has been identified as a bearing of this type (Simpson, 1976, p. 38–39) (figure 2.17). The bridge box itself is secured to the bridge tree, which is a horizontal beam fitted into the wall of the millhouse structure. This could be moved up or down to raise or lower the runner stone, thus providing a means for tentering the millstone configuration.

The most important typological features that can be identified on archaeological millstone remains include: the rynd chase, the stone diameter (which can be applied to both upper and lower stones), and the presence and placement of iron ring fittings to lift the upper stone. The rynd chase is an element of an upper millstone that is

Figure 2.17. Possible footstep bearing used to hold the weight of the upper millstone via the spindle, whilst still providing the means for the spindle to pivot. This example is made of basalt and recovered with the remains of a probable mill at Haltwhistle Burn, Hadrian's Wall (Simpson, 1976, pp. 38, fig 3a).

shared with upper quern stones, though the shape of the rynd can often be 'dovetailed'; different to the rectangular shaped rynds of lava querns in Britain. The dovetail shape mirrors the rynd chase on the lava millstone from the well in Saalsburg, along with others from the same site (Moritz, 1956), and this shape has also been noted for rynd chases on upper millstones of indigenous stone types recovered from Romano-British contexts. The extent to which it is present on lava millstones in Britain is yet to be determined, as are alternate forms. This feature might present the clearest and most identifiable typological feature for a lava millstone but can only be applied to upper stones that have this feature remaining.

In terms of chronology, Shaffrey (2015) suggests that lava disc type millstones are likely to have been imported into Britain around the same time as the Pompeian style donkey mills. They have been found in first-century contexts at Bath and Mucking and second-century contexts at Verulamium and Northfleet (Shaffrey, 2015, p. 60). Considering the extent of the movement of lava querns from Mayen, this seems highly probable. However, as lava millstones are likely to have been specially ordered, they did not necessarily travel in bulk to Britain in the same way that lava querns did. i.e., as possible saleable ballast on the return journey of grain shipments to the Rhineland (see section on trade/ economy below). Furthermore, production areas for lava millstones were separate to that of the querns at Mayen, occurring near to the port at Andernach (Mangartz, 2008, p. 77–78), and may have been part of very different production, distribution, and trade mechanisms. This has implications for the duration of their use in the province, as special orders for lava millstones may have extended beyond the period when lava quern stones were imported or ended much sooner.

2.4. Pompeian Style Mills

Roman period millstones of the easily identifiable hourglass shape, known generally as Pompeian millstones from their iconic presence at the site of Pompeii (Williams-Thorpe, 1988b, p. 275; Buffone, et al., 2003, p. 207), are some of the most extensively researched grinding tools in Europe. These were manufactured using lavas from a variety of sources depending on exchange networks and local supply. The unusual shape of these mills has helped archaeologists separate them easily from other lava grinding implements. Their impressive appearance as an iconic tool associated with Mediterranean culture has promoted the interest of many researchers.

The 'typical' Pompeian-style mill (figure 2.18) comprises two stones: the *catillus* (upper stone) and the *meta* (lower stone). The *catillus* is hourglass/hollow biconical-shaped and thus reversible, providing a new grinding surface once the first had been worn down. On either side of the narrowest part of the *catillus* sit two square projections with sockets/mortises for square timbers from which a 'yoke-like' frame could be fitted (Thurmond, 2006, p. 45). The *catillus* fit onto the *meta* which had a rounded conical/bell shape at the top, widening to a cylindrical shape at the bottom. This was sometimes hollow, which would have greatly reduced the overall weight and cost of the apparatus, whilst also making it more transportable (Jaccottey & Longepierre, 2011, p. 98). The top of the meta had a hole, into which a spindle could be fitted, providing a pivot to facilitate rotary motion.

In-situ examples at Pompeii (figure 2.19) present an example of this mill type at one specific site. It is likely that mills were rotated using donkeys or mules, as shown by the remains of two donkeys still in harness at mills at a

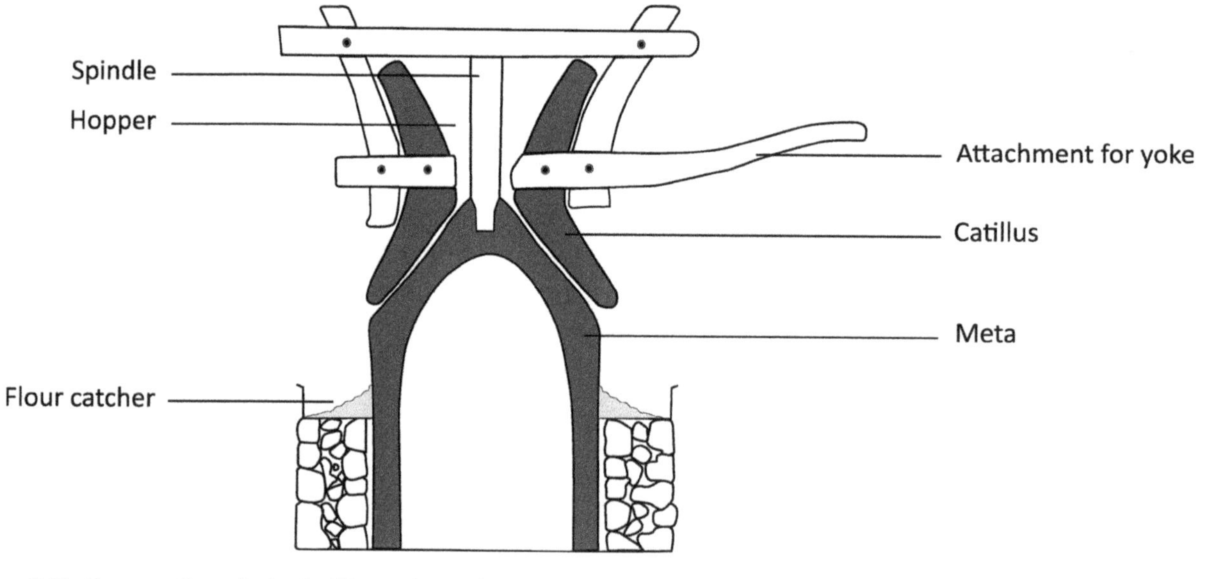

Figure 2.18. Cross-section of a typical Pompeian style mill (illustration by author after Williams & Peacock, 2011, p. 117).

Herculaneum bakery (Deiss, 1966, p. 204). Depictions of donkeys, mules or horses being used in this way have been found on shop signs (figure 2.20) (Wilson & Schorle, 2009, p. 104) and friezes (Hackworth Petersen, 2003) (figure 2.21). Unfortunately, there is less detailed evidence on the operation of such mills in Britain. A reconstruction of a mill using part of a *catillus* recovered from Princess Street, London, currently on display at the Museum of London (figure 1.4), is significantly smaller than those at Pompeii. The examples at Pompeii, therefore, cannot be used in direct comparison with the British material when considering method of operation and use.

Figure 2.19. Bakery with *in-situ* Pompeian-style mills, at Pompeii (Watts & Watts, 2016). These show how the stones fitted together and how they were fixed in position. Trough-like flour catchers (gremia) would have been placed on top of the structure that held the meta in place.

Figure 2.20. Part of a Roman period relief depicting bread-making, found over the oven in the bar of the restaurant 'Romolo' in Rome. The detail shown here shows the Pompeian style mill in use, with a donkey or mule being used to rotate the device (Wilson & Schorle, 2009, p. 104).

Figure 2.21. Drawing of part of the frieze from the tomb of Eurysaces the baker, at the Porta Maggiore. This section of the frieze shows the grinding of grain to flour using Pompeian-style mills powered by donkeys or mules (Wilson & Schorle, 2009, p. 110).

Pompeian mills have been subject to a great deal of research in continental Europe and the Mediterranean region (White, 1963; Williams-Thorpe, 1988a; Peacock, 1989; Williams-Thorpe & Thorpe, 1989; Williams-Thorpe & Thorpe, 1993; Antonelli, et al., 2001; Anderson, et al., 2011; Gallala, et al., 2018). Research into the British material is also extensive despite there being very few examples. However, the fragmentary nature of the evidence relating to this mill type in Britain means that a typology is yet to be produced, though provenance analysis has been used to generate one level of categorisation. In total, there have been: two examples of Pompeian style lava millstones from Dorset, one of lava from the Massif Central area of France and the other, probably non-Roman, of lava from Sardinia (Williams-Thorpe, 1988b); one almost complete *catillus* from Princess Street, London, also of French lava from the Massif Central (*ibid*); several fragments of Pompeian millstones from 1 Poultry, London, suggested by Williams and Peacock (2011) as belonging to three separate mills and provenanced to Mayen; and fragments of at least one example from Silchester, also provenanced to Mayen (Allen, 2012) (see Chapter 3).

The Pompeian type mill has early origins that predate the Claudian invasion of Britain (White, 1963), and could potentially appear in the earliest Romano-British contexts (Shaffrey, 2021, p. 21). However, the evidence for the manufacture of Pompeian style mills of Mayen lava at Andernach is sadly undated and it is not known when this commenced. There is only one securely dated example (11 BC to AD 9–16) provenanced to the Mayen region, and this was recovered from a consumption site at Haltern, Germany (Gluhak, 2010, p. 281). No 'new' Pompeian millstone fragments were recorded as part of this study, and none were examined or identified during data collection. One current published example has been discussed in relation to the urban case study of Silchester.

2.5. Quarrying and Organisation of Labour

Quarry sites are a relatively neglected area of research for a variety of reasons. There is limited information available from ancient sources and quarries are rarely described or explained in any detail. Further to this, the excavation of ancient quarries is an expensive and dangerous undertaking and much of the evidence has often been removed by later activity on the site (Peacock, 2013, pp. 131–132; Russell, 2013, p. 37). As the quern quarries of the Eifel region in Germany are well-known and have been extensively researched, there is an unusually abundant amount of information available.

Mayen lava was sourced from the Bellerberg volcano and is named after the town that is situated close by. The town of Mayen is in modern day Germany (figure 2.22), on the western side of the Rhine and northern bank of the Moselle and belonged to the 'tribal' area of the Treveri during the Iron Age. The territory, which included the modern Eifel region, came under Roman rule after the Gallic war of 58–51 BC and was wholly incorporated into the newly created imperial province of Gallia Belgica in 27 BC during Augustan administrative restructuring (Köstner, 2012, p. 74). The Mayen region would have been substantially affected by the establishment of Roman rule, with further impact possibly caused by subsequent administrative reorganisation phases.

The success of Mayen as a quern and millstone production area is closely correlated to excellent transport links in the form of riverine and road networks. The nearby port of Andernach on the Rhine saw an increase in growth and development during the Late Iron Age and Roman era, indicating that it was probably a key location through which Mayen goods were exported (Deutsche Forschungsgemeinschaft, 2019). Possible lava millstone workshops and unfinished lava millstones found at sites in Andernach also indicate that it may have played an important role in the finishing of lava millstones for export (Mangartz, 2008, p. 76–78). The *vicus* of Mayen was established in the Late Iron Age and was the economic centre of the region (Glauben, 2012; Köstner, 2012; Oesterwind, 2012), growing in size and population during the Roman period, with development of the surrounding landscape visible in the archaeological record (Giljohann, et al., 2017, p. 126–127). Mineralogical analysis of lava querns recovered from military sites on the River Lippe indicate that Mayen lava querns were incorporated into Roman military equipment as early as the Augustan Germanic campaigns (Gluhak, 2010). It is likely that the new role of Mayen as a military supplier initiated widespread social and economic change within the region.

Although details pertaining specifically to Mayen are not known with any certainty, resource management

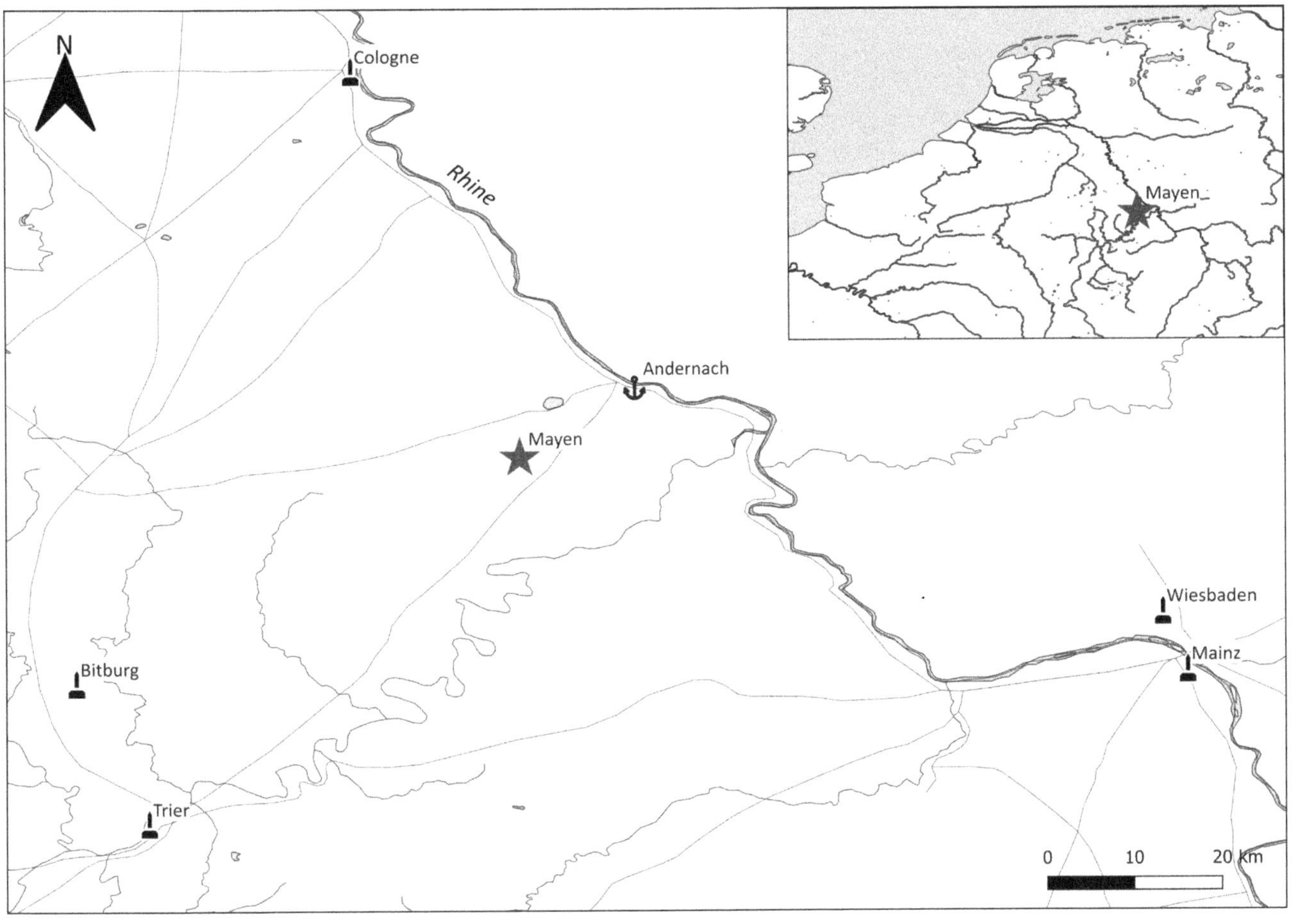

Figure 2.22. Location of Mayen within Germany (illustration by author with background GIS data courtesy of the Ancient World Mapping Center).

strategies from elsewhere in the Empire indicate that state control of quarries was widespread (Köstner, 2012, p. 75–76). Organisation of the Empire's resources would have required a degree of investment in the form of knowledge, technology, and human input; this investment would have guaranteed the outputs of such organization, in the form of production of goods, control of resources and the development of experienced workers (Hirt, 2010, p. 9). Quarries of the Laacher See volcano, used to extract tufa for building material provide evidence, in the form of epigraphy and 'military' material culture at the site of extraction, for direct military involvement in the management of quarrying activity during the early Roman period (*régie directe*) (Giljohann, et al., 2017, p. 127–128). Such evidence is absent at Mayen, and it seems more probable that a lease system was in operation and a local workforce employed (*régie indirecte*). The way that lava was extracted from the Mayen quarries indicates that this is likely as clear divisions exist between different quarry 'plots' in the form of border strips comprising unworked stone (Crawford & Röder, 1955, p. 73) (figure 2.23). If quarry workers were employed as one single taskforce, these features would be absent, as the team would have worked together to extract stone as a single unit whilst minimising waste material.

2.6. Stone extraction: Quarrying process

The extraction of lava at Mayen during the Roman era has left a wide range of evidence in the form of tools (Mangartz, 2012, pp. 4–5; 2008, pp. 58–62), partially worked quarry faces (Mangartz, 2012, p. 6–7) and roughouts (Hörter & Röder, 1951, p. 15). The suitability of the Mayen quarries for quern production lies not only in the properties of the stone itself, but also in the geological formation of the resource as basalt lava columns (figure 2.24). This facilitated extraction of the stone as the natural dimensions of the columns provided the minimum circumference from which workable quernstone blanks could be produced.

The first stage of the quarrying process probably involved some form of preparatory work, as this had the potential to mitigate any potential issues before too much time had been invested in the work (figure 2.25). An experienced quarry worker would be able to identify faults in the lava by 'sounding' the stone or by soaking it with water (Mangartz, 2008, p. 64). During the next stage, measurements were probably taken and marked out to ensure that the minimum size for a quern could be created from the roughout material. In the early Roman period until the second century AD, a groove was incised around

Figure 2.23. Reconstruction of the quarrying process and the division of the quarry face into 'plots' for extraction (Mangartz, 2012, p. 115. fig 39, CC BY-SA 4.0).

Figure 2.24. Photograph of quarry face at Etringer Lay showing the naturally formed lava columns (Weinandt, 2010: CC BY-SA 3.0 DE via Wikimedia commons).

the full circumference of the stone, marking the lower boundary between the roughout and the rest of the lava column. This was used in Mayen prior to the introduction of iron wedges for stone breaking (Mangartz, 2008, p. 64–68). The iron wedges were used to facilitate the breaking of stone by inserting them into small 'pockets' notched into the stone a few centimetres apart around the circumference. An appropriate mallet-type tool would then be employed to strike the wedges in turn to separate the roughout from the lava column (Mangartz, 2008, p. 68–71). Evidence for this process can still be observed at the quarry sites (figure 2.26).

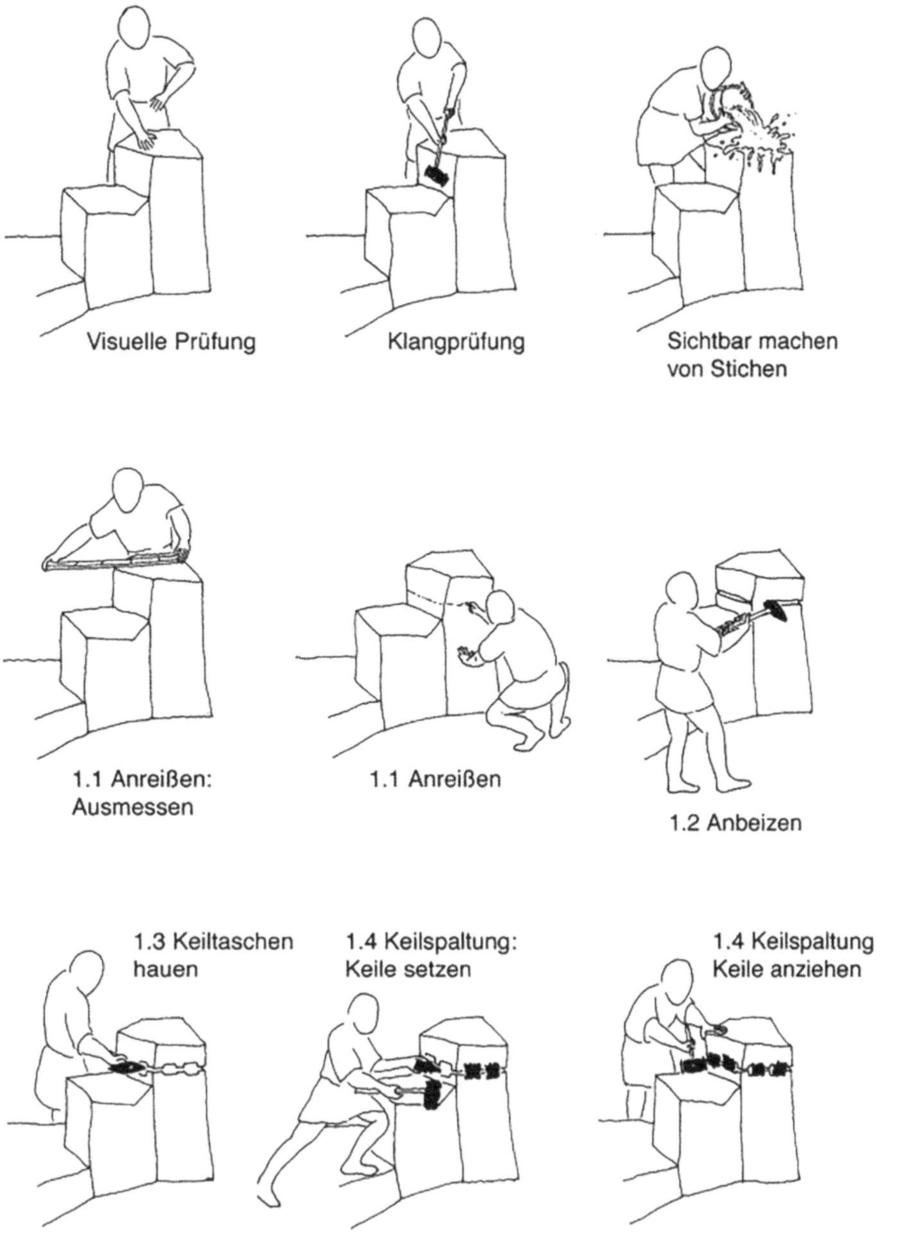

Figure 2.25. Lava quernstone quarrying: Preparatory work (top row)- visual inspection, sounding the stone, checking for fissures. Marking out (second row)- measuring the top, marking the circumference, making a notch. Stone breaking (bottom row)- creation of wedge pockets, adding iron wedges, hitting iron wedges (Mangartz, 2008, pp. 65, fig. 18).

Once the roughout was sufficiently separated from the lava column, a lifting iron or similar cantilever could be used to complete the disconnection and lift or move the block from its original position, though shaping the sides could also have been carried out prior to the complete extraction of the stone block from the lava column (figure 2.27). Iron compasses were probably used to mark out the intended shape of the final product so that excess stone could be removed from the sides more precisely. This initial shaping of the stone would have been relatively roughly completed, allowing for greater blows and the removal of most of the excess stone (Mangartz, 2008, p. 71–72). Once the sides of the roughout had been shaped, the choice of whether the stone was intended to be a *meta* (lower stone) or *catillus* would influence how the process would continue due to the differing shapes and angles needed to provide the appropriate grinding surface. This would have been completed in a rough manner to remove most of the excess stone prior to more detailed work being carried out (figure 2.28). Many of the roughouts retrieved from the Mayen quarries are at this stage in the process indicating that querns were worked in the quarries until at least this stage (Mangartz, 2008, p. 72–73).

The final stage that probably took place at or near to the stone extraction site involved further smoothing the surfaces of the roughout to produce a more well-defined

Figure 2.26. Remnants of stone breaking technique in Mayen quarries to show how the iron wedges were used in the process (Mangartz, 2008, p. 72).

object. Roughouts have been recovered from the quarry at this more advanced stage of working (figure 2.29), but these are less numerous than the rough versions seen from the previous phase of manufacture. It may simply be the case that the rougher stages of the process were more 'dangerous' to the integrity of the quern and more likely to produce discards (Mangartz, 2008, p. 73). Finally, the quern roughouts were moved from the quarry to the nearby *vicus* or *villae* for further working. The stone workshops of the *vicus* were situated on the Nette, and it is likely that the stone was moved via water wherever possible (Mangartz, 2008, p. 73).

2.7. 'Im Winkel': A Possible Quern Finishing Location

Although roughouts left the quarry as identifiable quern stones, their functional and decorative features must have been added in a different location; no waste querns recovered

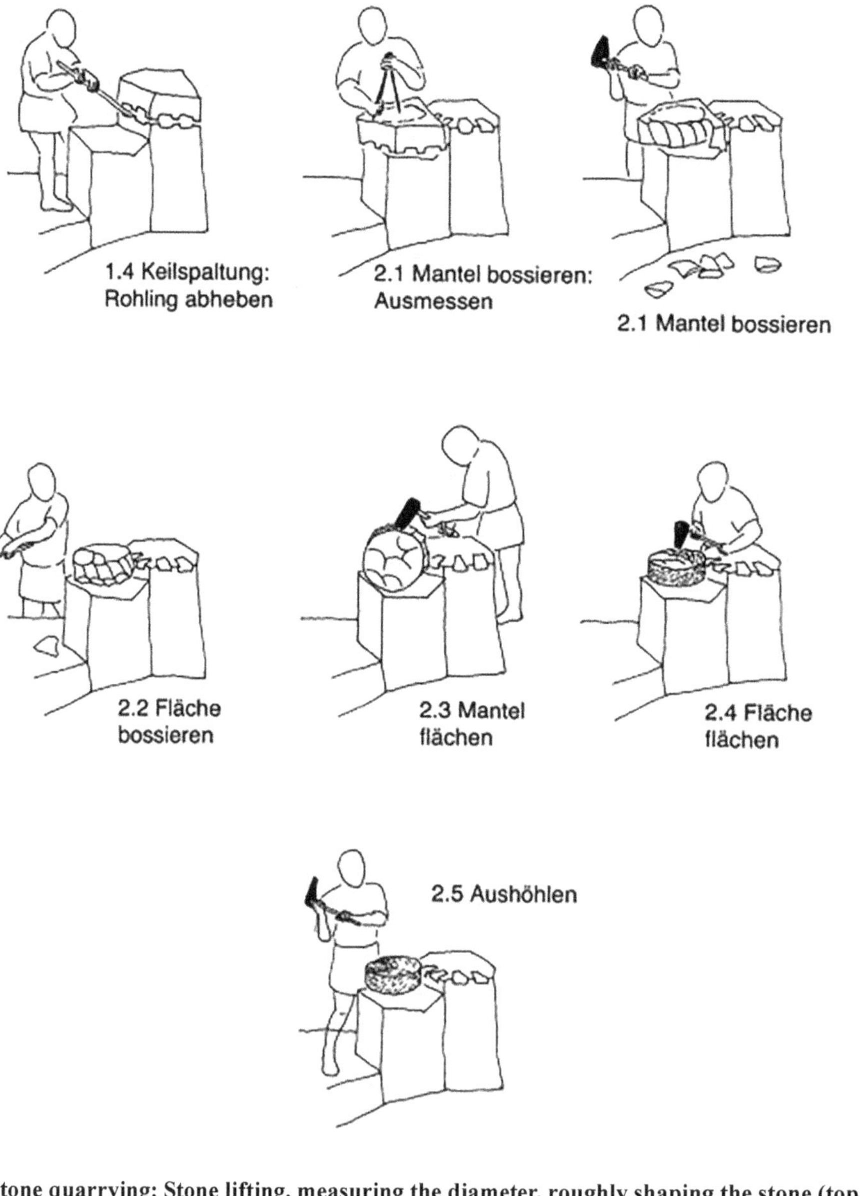

Figure 2.27. Lava quernstone quarrying: Stone lifting, measuring the diameter, roughly shaping the stone (top row). Shaping the surfaces, shaping the sides, smoothing the surfaces (middle row). Hollow out (bottom row) (Mangartz, 2008, pp. 66, fig. 19).

Figure 2.28. Example of a typical roughout recovered from Mayen quarries. All three examples show that the quern stone shape is beginning to emerge, but that the surfaces are still roughly shaped (Mangartz, 2008, Plate 23).

Figure 2.29. Smoothed roughout of a meta retrieved from Mayen quarry site. Tool marks show the direction of working and the smaller size of the tool blade used (Mangartz, 2008, Plate 23).

from the quarry have these characteristics (figure 2.30). A Roman *villa* known as 'Im Winkel' may have been close to a workshop where Mayen quern stones were 'finished', as evidenced by quern roughouts recovered from the nearby stream (Wenzel, 2011, p. 131). Excavations in the basement of the main *villa* building revealed a further 21 quern roughouts (figure 2.31) (Wenzel, 2011, p. 135). The querns were probably deposited in the basement around AD 100 and are of a different production stage from those typically recovered from quarry contexts; some had the beginnings of decorative or functional features present (Wenzel, 2021).

All roughouts from 'Im Winkel' were typically of the Roman style quern in their shapes and emerging features (figure 2.32). The distinctive concave *catilli* and convex *metae* shapes are clearly identifiable, while the presence of a 'kerb' on some of the roughouts is also characteristic of the Roman era lava quern. Most of the roughouts were in the very early stages of processing after having left the quarry and most show deviation from the standardised sizes known to have been imported into Roman Britain, or evidence of probable accidental damage. It is possible that these represent rejected querns that did not reach the required standard.

2.8. Distribution Analysis and Mechanisms of Exchange

The use of distribution analysis to examine ancient economic interactions is one method of investigation that has been used extensively within the discipline and can be helpful in defining the kinds of relations that operated between different groups across space and time. That different forms of economic interaction and mechanisms of distribution existed is a well-established and recognised concept. Polanyi (1968, p. 250) categorised three types of economic transaction: reciprocity, market exchange and redistribution. Reciprocity involves gift or equivalent exchange, used to establish and cement social or political connections as demonstrated by a range of different anthropological studies (Burridge, 1969; Firth, 2012; Malinowski, 1926). As this mechanism of distribution relates strongly to personal connectivity, it is identifiable from the presence of material culture types at locations where there is no specific need or demand. For example, if there are querns of non-local stone types in use in regions where there are local sources of stone and quern production of the same or similar quality/function (see Shaffrey, 2006, p. 69–70). Distribution in these areas would be sporadic and not conform to the 'normal' supply and demand networks.

Market exchange tends to be associated with commercial and profit-making situations, where goods are exchanged for coin or items of greater value. With regards to querns, this form of distribution would largely depend on the cost of transporting such large and heavy objects from the source. If a profit were to be made on these goods, the value of their sale could not be exceeded by the cost of their movement. This will be discussed in greater detail in the next section. Finally, redistribution involves a central agent via whom goods are distributed, such as centralised warehouses or state-controlled resources. This

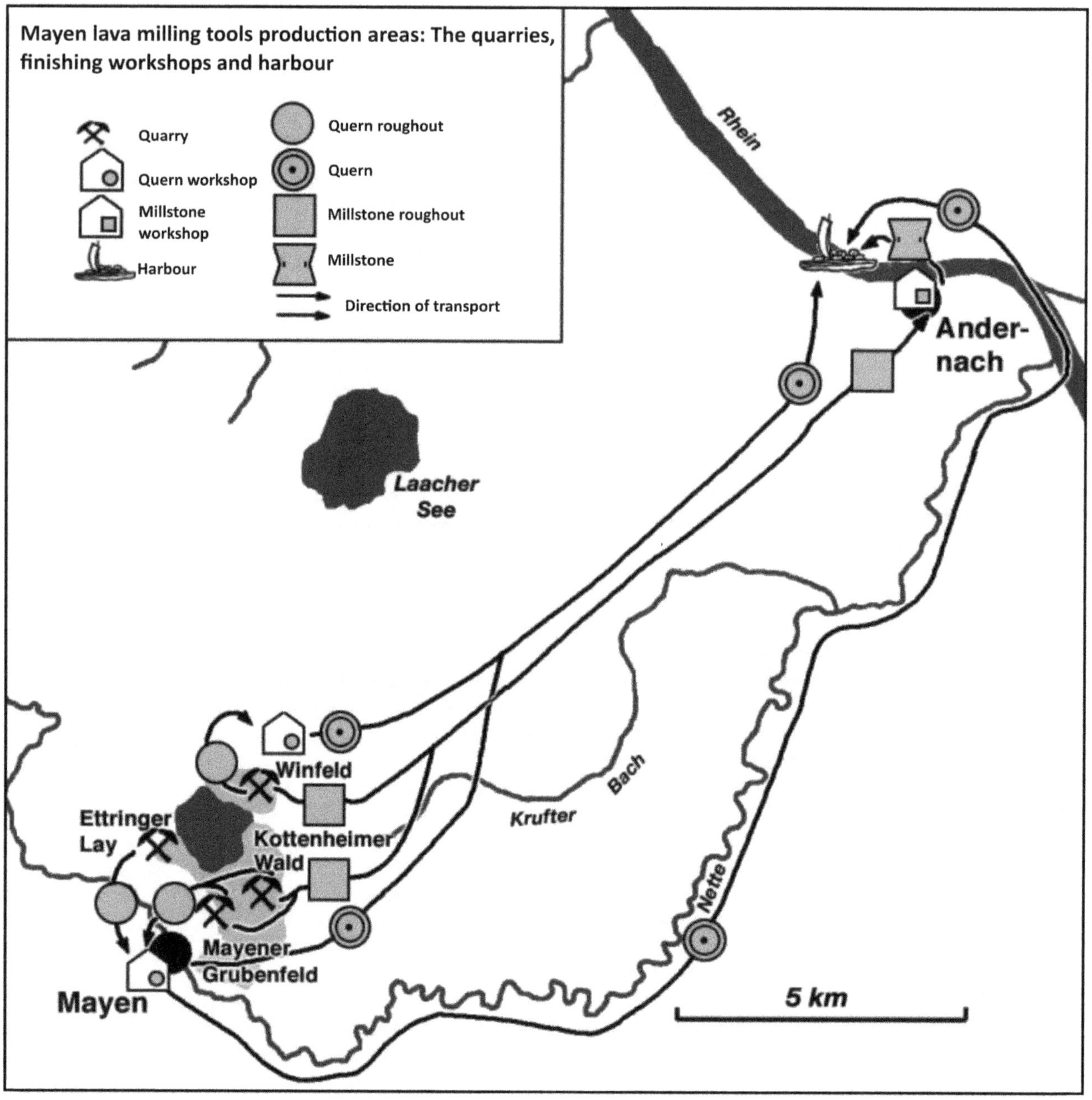

Figure 2.30. Organisation of production and quern finishing workshops (Mangartz, 2012, pp. 11, fig. 4).

is largely controlled by the society and the structure of the community in which it operates and sees a controlling body determining how goods are distributed according to the employment, status, or position of the recipients (Greene, 1986, p. 46–47). Distributions for this form of economic mechanism would show as clusters and concentrations of the material in specific areas, including the source and centre of redistribution; with none, or low volumes, in others.

Identifying these forms of movement using distribution is, however, not straightforward. Some distribution patterns can be interpreted as having been caused by any of these mechanisms, and it can be difficult to pinpoint how or why particular dispersal patterns may have occurred. It should be acknowledged that all three mechanisms may have been involved in the distribution of any one material culture type to varying degrees, and that these may have also changed over time. This adds a level of complexity to interpretations that can only aim to explore possibilities and variations in economic behaviours. Furthermore, these mechanisms only consider the economic 'needs' of the populace, with little consideration for social or cultural 'wants', or how these factors interact. Although reciprocity involves a more personal means of acquiring material culture, our method for justifying such methods of procurement depends largely on economic rationalistic explanations; if there

Figure 2.31. Slot dug through cellar at 'Im Winkel' showing querns in section in the basement fill (Wenzel, 2021).

is no 'need' for an object but it is present, it must have been a gift; though there may have been a social need for the specific object, creating the motivation to obtain it by other means and disrupting the 'rules' regarding how and why goods were distributed.

Interpretation of the demand for goods should be concerned with the balance between 'need' and 'want', as material culture is not simply functional or without meaning. Many objects fulfil both social and functional roles and these both would have been significant factors in terms of demand and distribution. It is not sufficient to explain 'irrational' distribution as being the result of social reciprocal exchange of goods, with 'rational' distributions caused by purely economically motivated exchange or redistribution. The interplay between 'need' and 'want' can greatly disrupt distributions in terms of how we would define 'normal' economic behaviour, and it blurs the distinction of how we identify and differentiate between exchange mechanisms. However, by considering the worldview of participants in the economy and by examining the socio-cultural and functional role of the material culture in question, it is possible to discuss the ways that goods and materials moved around the empire, the interactions involved, and the reason for these behaviours. Therefore, the aim for this research will be less focussed on defining the mechanisms in operation, and more on understanding how and why certain distributions of lava querns came about.

As previously discussed, the socio-political and cultural aspects of Roman period life have been considered by some, such as Finley (1999) and Bang (2008), to have

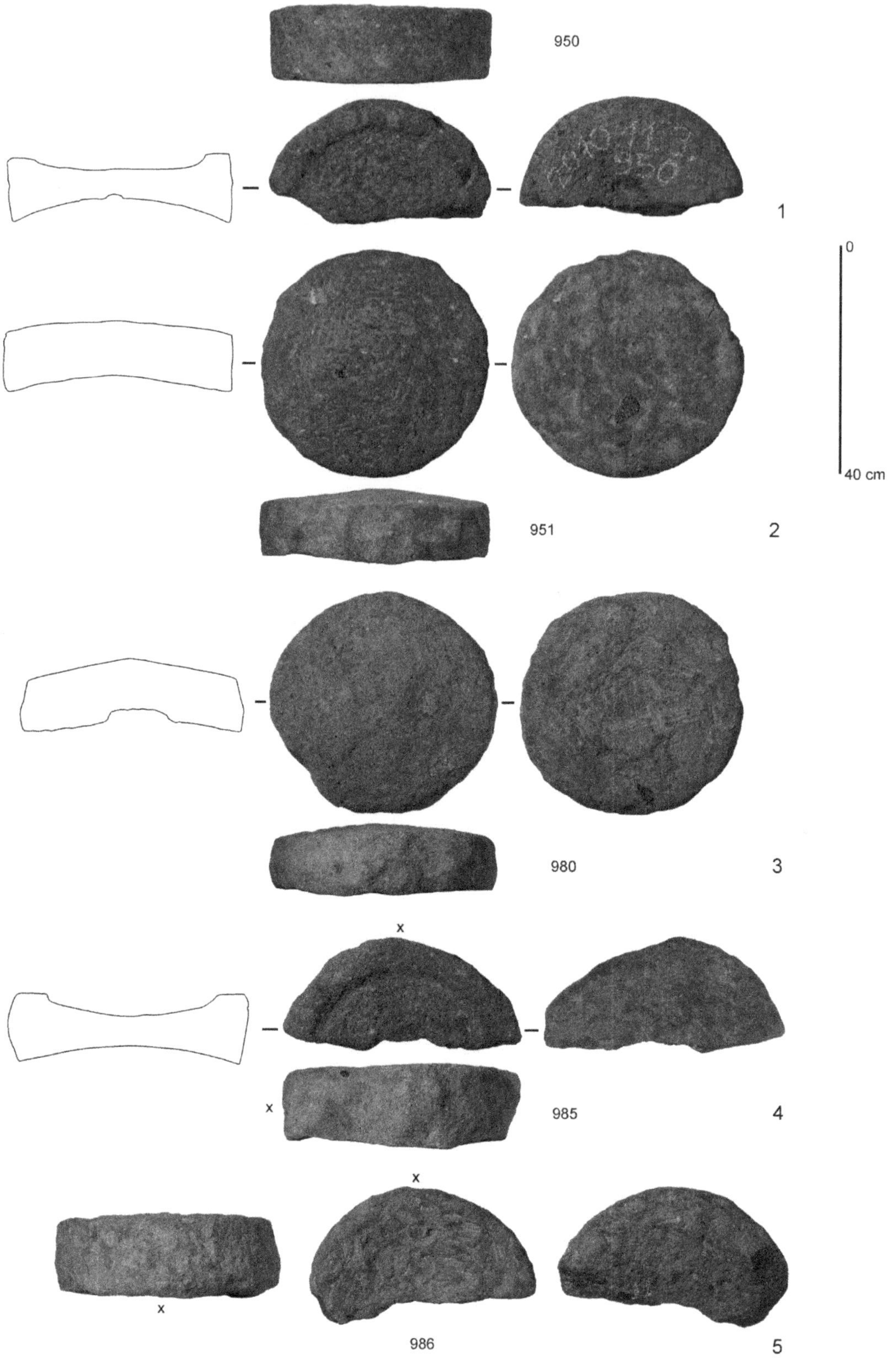

Figure 2.32. Some of the partially finished querns recovered from the site an 'Im Winkel' (Wenzel, 2021).

played a hugely significant role in the economic life of the empire. Political elites had more social power than entrepreneurial endeavours, while state institutions and imperial administrative personnel also held greater economic agency than the rest of the population. 'Directed trade' was a means by which people of entitlement could move goods around the empire without the need for participation in ordinary systems of exchange. Trading activities subjected merchants to high personal cost and risk, factors that had less significance to those who could, and did, use their status, political or institutional affiliation to avoid usual barriers to exchange. Directed trade, to an extent, falls under the 'redistribution' category of Polanyi's three mechanisms of exchange and is thought to have been one means by which the military community was supplied.

Manufacture or provision of certain items may have been solely for the use of the military and taken directly from the production location to the place of consumption, bypassing other status, cultural or social groups within the chain of supply. Alternatively, provisions could be taken as taxation or rent from those producing goods for commercial reasons. These goods would be removed from the 'market' and transported directly to where needed by the military. However, supply of objects or materials did not necessarily require a central redistribution location to operate. It is thought by Evans (2002, p. 482), based on the supply of wine to military sites, that organisation may have been decentralised and conducted on a more localised scale by the fort quartermasters.

Evidence for this method of exchange already exists in the form of studies that demonstrate or consider the social distribution of material culture using archaeological evidence. One example of this includes the distribution of Black Burnished Ware 2, BB2 (figure 2.33) (Tyers, 1996a). This type of pottery is densely distributed in the area surrounding locations of manufacture, on Hadrian's Wall and the Antonine Wall, with very little present in other areas of the province. It appears that though the pottery was consumed locally, there may have also been directed trade between the producers of this pottery and military consumers (*cf* Gillam & Greene, 1981). Coarse ware pottery manufactured in Roman Britain did not often travel far from its production location, meaning that the long-distance movement of BB2 to military sites on the frontier is unusual. As the military were able to avoid many of the usual impediments to free movement of goods around the empire, such as taxation, customs duties, or transport availability (to be covered in more detail in the following section), the risk or cost of these ventures would not prevent exchange so long as there was a genuine need or desire for them. Therefore, goods required by the military saw a greater degree of movement across larger distances, ignoring the economic 'rationality' that capitalistic ventures would ordinarily display in distribution patterns.

Unfortunately, evidence of this kind of movement is limited to the types of material that survive well in the archaeological record, ceramics being the most common. Although pottery is a good archaeological indicator to prove that directed trade existed, the household function of ceramics may not have been the primary reason for their movement, and it is important to recognise what the motivations may have been. Pottery was used to hold and transport imported food, drink, or oil, and can be an indirect form of examining the movement of organic substances, such as amphorae used for wine or olive oil (Peacock & Williams, 1986). Similar but less archaeologically visible necessities, such as cloth, grain, slaves, and other foodstuffs are highly likely to have been used to supply troops under similar arrangements. For example, environmental evidence of non-locally produced foodstuffs has been used as direct evidence of systems of trade and movement of goods (Bastow, 1999), while proxies, such as evidence for pests associated with specific foods (Buckland, 1981), can also benefit analysis. The use of historical written accounts, such as the Vindolanda tablets, can further elaborate on the kinds of goods that were moved through directed trade (for example, Pearce, 2002). However, a great deal more care is needed to generate a better understanding of how directed trade operated within the province and the wider empire, and the scale to which it existed. This requires the inclusion of a wider range of different material culture types alongside that of the ceramic evidence. Lava querns and millstones have excellent potential for this kind of investigation.

2.9. Transporting Goods

When using distribution analysis to investigate the Roman economy, it is important to consider the role of transportation in these distributions. Local and regional economies were heavily influenced by the types of transport that could be used to reach them as this had implications on travelling times and costs. Unlike modern economies which have largely managed to overcome the obstacles of distant trade and interconnectivity through technological innovation, for ancient economies this was potentially a serious barrier. Geography and topography were critical factors in terms of connectivity, though the significance of a location to state interests was also a considerable element. It is generally agreed that goods, and especially bulky objects, such as querns, must have travelled via water as much as was reasonably possible. Bang (2008, p. 134) argues that transport by water was most important when moving products that were low value and high volume, for example grain, as transport cost is more critical to secure a return of investment. There have been several studies that have quantified the cost of transportation using classical texts, and these indicate that it was nearly always much more economical to move goods via water, and that sea transportation was the cheapest option (e.g., Duncan-Jones, 1977, pp. 366–369; Künow, 1980, pp. 21–23). However, greater risks were involved in transport and travel by sea.

There are other less visible costs that should also be taken into consideration, for example, customs duties or harbour taxes. It has been suggested by Steensgaard (1981) that the payment of customs duties would have totalled much higher than the costs associated with transportation. Customs officers were known to extort extra fees under the guise of customs duties and such exploitation of traders has been documented in classical texts (Sijpesteijn, 1987, pp. 90–91; Malmendier, 2002, pp. 98–116).

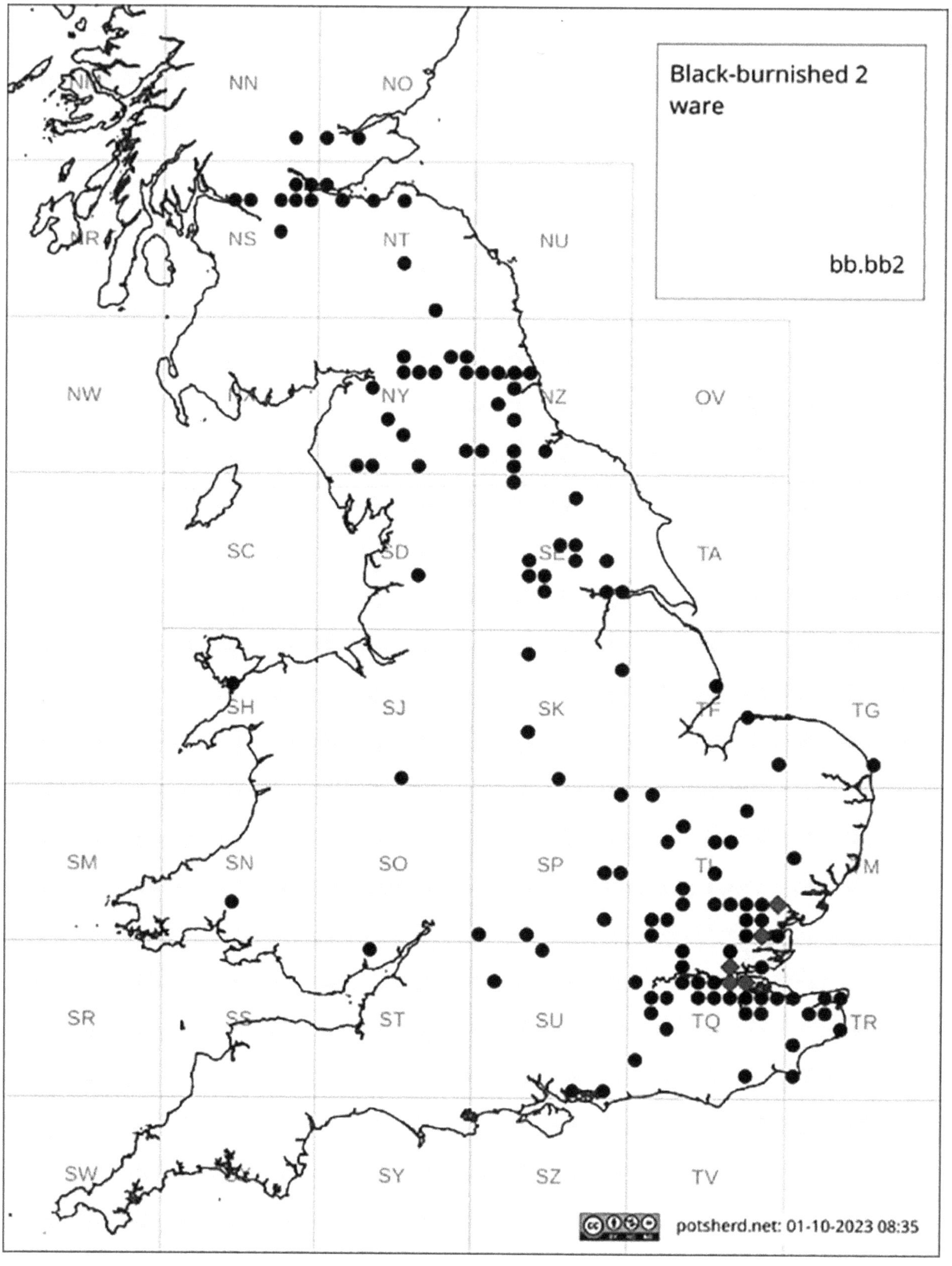

Figure 2.33. Distribution of BB2 pottery in Roman Britain, including production locations (Tyers, 1996b).

Although most of these issues relating to the high-risk venture of sea trade are highly applicable to commercial activity, it is necessary to remember that commercial ventures were not the only means by which goods were moved around the empire. As previously discussed, distribution of various material culture types shows that 'directed trade' can be associated with military and imperial agendas. In such cases, the physical risks of sea transport were just as high as those for any other voyage: bad weather, poor sea conditions, risk of shipwreck. However, unlike commercial ventures, military and imperial authorities had prime access to what they required to fulfil their needs and greater autonomy with regards to risk management. Further to this, harbour taxes, administration fees, customs dues, the cost of 'safe passage' from *pax Romana* and tolls could all be avoided by the military, the political elite or if the correct people were known and the best connections established (Bang, 2008, p. 202–238). All this created greater ease of goods movement and allowed military needs to be fulfilled via whatever means were deemed to be necessary.

The river system was another important means of water transport and may have been less seasonally constrained than that of other transport methods. Goods could easily be moved downstream using currents, or less easily sailed or dragged upstream using draught animals or people. Low water levels in summer may have been an issue, though this is likely to have been mitigated by using flat-bottomed boats/barges, the type of which is well attested for the Roman period. However, when compared to temperamental and potentially dangerous sea conditions, and weather sensitive roads throughout winter, river transport may have presented a reduced risk and a more flexible schedule that extended beyond the summer months.

The alternative to transport by water was to move goods via road. Road travel costs were generally much higher, though the situation would have differed depending on circumstances, and bad weather would have made roads impassable during the winter months. Further to this, the threat of bandits and robbers could potentially jeopardise overland routes (Blumell, 2007). Although the cost of transport should not be overlooked, it would not always have been a barrier to long or short-distance exchange of goods via road, and the significant role of road transport has become increasingly recognised in studies of the Roman economy (see Horden & Purcell, 2000, Chapter 9). Evidence for substantial use of the Roman road network can be related to directed trade arrangements associated with the military. Military site location would have been strategic to local political requirements and not necessarily near transport routes via water. Therefore, transport by road would have been essential in such cases and would not have been an obstacle when connecting with the wider military network.

The many forts situated along Hadrian's Wall that were not connected to navigable river systems demonstrate that the Stanegate and later Military Way must have been fundamental supply routes. The establishment of these, and the many other roads built across the empire, were major construction projects and show a great deal of investment in overland connectivity. Finds and environmental evidence recovered from northern military sites show that there was no lack of access to imported or 'luxury' goods. The Vindolanda tablets especially paint a vivid picture of how military demands for vintage wine and fish sauce were fulfilled to allow 'Roman' standards of living to continue at the frontiers for those who could afford to do so (Bowman, 1974, p. 367). However, the tablets also indicate that there was a dependency on everyday foodstuffs, such as grain, being brought into the site from external sources, suggesting that the movement of goods must have been relatively regular and common. Therefore, it was not only the smaller and high-value goods that were being moved over land to fulfil military needs; the low value high bulk goods were also being moved in this way.

Road construction projects were often carried out by the military (Xeidakis & Varagouli, 1997; Zaharia, 2013), with significant roads created wide enough for two carts to pass in opposite directions (for example, see Castledykes (Maxwell, 1976, p. 96)). Such state involvement indicates that roads were a vital part of the provincial infrastructure, creating viable routes for the movement of goods over land, while also improving communication and connectivity. The presence of the military, therefore, may have acted as a catalyst for other forms of trade to exist and be economically and/or practically viable. Landlocked sites that were not primarily military, such as Silchester, also enjoyed a wide range of imported and exotic goods that must have been brought some distance via road. Stone objects, such as imported lava querns, are also prolific and demonstrate that the significance of a particular site within a wider socio-political and economic landscape would also have determined connectivity and access, and that transport by road was not always an impediment in a variety of circumstances and for a range of different reasons.

'Piggyback' goods also had a significant role to play in a variety of exchange mechanisms and the presence of a particular type of material at a specific site should not be taken as being evidence of prioritised trade of that item. For example, pottery was transported to Byzantium as a secondary exchange item that travelled with the grain tribute (Thomas, 2014, p. 203). Similarly, the exchange of goods for commercial purposes via water and the movement of private possessions may have also been greatly eased by exploiting military contracts. Tax avoidance and cost reductions would have been some the most beneficial aspects of this system (Evans, 2017, p. 159). Distributions of other trade items from the Mayen region, such as 'Mayenware' pottery should be examined to identify if other goods were piggybacking with lava querns, or if the lava querns were piggybacking with these items. It could be possible to distinguish between the desired, and possibly contracted items from the commoditised objects. Geographical and social distributions could reveal how the goods are piggybacking. For example, if Mayenware pottery is found with lava querns at a military vicus, it is likely that these were being transported with military contracts but, perhaps, not as contracted goods.

In terms of connectivity, in general distribution patterns of imported goods show that the south, eastern and south-eastern coastal regions of Roman Britain were more engaged in exchange with other parts of the empire, both in range and volume of imported products. For examples, see distributions of pipeclay figurines (Bémont, et al., 1993, fig. 110), Argonne Ware pottery from northern Gaul (Tyers, 1996b) and Hemmoor buckets from the Rhine provinces and Belgica (Morris, 2010, p. 119). These distributions demonstrate the importance of location with regards to the access to imported goods. It is often argued that pre-AD 43 imports in south-eastern Britain were consequences of the invasions by Caesar and Roman occupation of Gaul. Relationships in terms of exchange, trade, social and cultural interaction are likely to have intensified because of increased contact with the Roman world, while those in the south and south-east of Britain were much more likely to have such interactions due to proximity. Pre-existing political, trading, and social relationships gave these regions a 'head-start' in terms of providing ideal circumstances for the transportation of imported goods via known coastal routes to, possibly, more welcoming coastal regions.

2.10. Trade Routes for Imported Goods: North Sea Trade

Many studies that examine distributions of Romano-British material culture tend to focus solely on movement and use of goods within the province; an approach that is highly informative when considering regional identities, and how and why items manufactured in Britain were moved (for example: nail cleaners, (Crummy & Eckardt, 2003), and copper alloy vessels (Lundock, 2015)). However, lava querns and millstones were imported items and must be considered within a wider context of trade and exchange networks within the empire. Although the provincial distribution will reveal a great deal about how these objects were used and transported, it is vital to explore the routes and methods by which they entered Britain to better interpret such distribution patterns. As these objects are thought to have predominantly been sourced from the Eifel region of the Rhineland, the particulars of the North Sea trade form a vital component to our understanding of how and why these objects found their way to Britain. As the role of lava querns within the North Sea trade is little understood due to the absence of any systematic study of Romano-British distributions, it is important to examine the evidence that we have for the import and export of other goods along these trade routes to understand how lava may fit into the broader picture.

It must be remembered that in terms of exchange with the continent, the movement of goods would have occurred in both directions and journeys were potentially profitable regardless of that direction. From a commercial viewpoint and in terms of economic efficiency relating to market exchange, any saleable or useable cargo would have been preferable to an empty ship. Therefore, the distributions of goods made or grown in Britannia and found on the continent is as valuable as the distributions of imported goods in Roman Britain. Even unladen ships would have required ballast to sail, and material would have travelled across the sea regardless of whether it retained any commercial value or use. Evidence for the use of ballast in the Roman sea trade has been suggested by Peacock *et al.* (2007) using the presence of basalt from Yemen found at the Egyptian Red Sea ports of Berenike and Myos Hormos. The ballast, it is argued, was brought due to ships bearing a lighter cargo of incense on the return journey between Egypt and Yemen (see discussion on Fishbourne Palace, Appendix 3 for mention of ballast at Chichester Harbour).

There is some speculation that Mayen lava, in the form of broken lava querns, travelled to Britain as ballast, though this has not been proven (see discussion of the querns from 1 Poultry, London in Hill & Rowsome (2011)). There are also several publications that state that Mayen lava was brought to Britain unworked and was finished at centralised distribution points, such as London, York, and Colchester. Shipwreck data is often quoted as showing that lava querns were shipped in unfinished form, with evidence of 'roughouts' present in the cargo of sunken ships. It is true that quern roughouts have been found at shipwreck sites, yet none have specifically mentioned Roman period Mayen querns or millstones. One such example involves a wreck dated to 140/150 BC at Isla Pedrosa near Gerona (Catalonia), which was found to contain 59 unfinished sandstone querns (Mangartz, 2008, p. 208).

With regards specifically to Mayen lava querns, the Graveney shipwreck at Whistable, Kent (Fenwick, 1978, p. 131–173) is the most frequently referenced example, though this has been dated to the Medieval period. The evidence at Mayen suggests a different manufacture process, where the finishing stages were carried out away from the quarries at high status villas and in the vicus, meaning that they were likely to have been shipped as finished products. The high level of standardisation is another argument to suggest that centralised production and finishing was occurring at Mayen, as multiple and distant finishing locations would have produced a greater variety of different types and styles. Whether finished or unfinished querns were branded as ballast may have been significant in terms of taxation and harbour fees as this would have affected commercial profitability. Regardless of how they may have been classified, it is more reasonable to consider lava querns and millstones as a combination of saleable ballast and exchangeable goods, as there is no way of fully knowing their role as part of a ship's cargo or the extent to which they were commoditised as potentially taxable imports.

The export of grain to the Rhineland from Britain possibly helps to explain the shipping of lava ballast or finished querns back in the other direction. The invisible nature of grain in the archaeological record makes it difficult to identify whether this may have been the case, and the absence of other recognisably British exported goods of high volume in Germania helps to support such theories. The transport of querns as part of the return journey from the delivery of grain is also speculated for the Mediterranean regions of the empire (Williams-Thorpe, 1988a; Mangartz, 2008, p. 208). Strabo, writing in the pre-conquest period lists other products that have been moved in the opposite direction to the querns, such as cattle hides, hunting dogs, slaves, and metals; none of which are easily identified as British products in German contexts. Similarly, salt and oysters, all of which were produced at scale in Britain, may have been exported to the Rhineland (Morris, 2010, p. 38). Without the physical evidence that goods were exported from Britain to the Rhineland, it becomes difficult to substantiate theories that suggest a market exchange system for querns. This does not mean that such systems of exchange did not exist, but the absence of evidence makes the concept purely speculative and creates difficulties in demonstrating how and why a commercial lava quern trade may have operated. From a more positive perspective, the lack of evidence for imports into the Rhineland does not negate the presence of exports of lava to Britain for military purposes. The transport of querns by the state would not have been governed by the need for profit and considerations of the cost of movement would not have been as significant as that for commercial endeavours.

Regardless of whether there was movement of goods in both directions, it has been established by Morris (2010, pp. 1–2), that there were three different maritime systems that governed the 'catchment areas' for exchange between Britain and the continent during the Roman period (figure 2.34). The system that relates to lava querns is the 'southern North Sea and eastern Channel system', which incorporates the Rhine delta to the Seine, and covers the south-eastern and eastern coast from Portsmouth to the tip of Scotland at Thurso and the Orkney Islands. The crossing associated with this maritime system was focussed on the Strait of Dover, predominantly determined by geography in terms of shorter crossings and coastal routes (Morris, 2010, p. 151). Although exchange via this route is considered by Morris (2010, p. 1) to be 'coherent', he does acknowledge that there is some degree of overlap between the southern North Sea and eastern Channel system and the systems on either side.

In terms of connectivity as part of the southern North Sea system, Morris (2010, pp. 151–154) illustrates that this was not consistent or continuous in all periods of Roman rule and offers detailed discussion of changes seen and possible reasons for these changes. Shifting political and economic circumstances were responsible for periodic changes in the operation of coastal trade systems. As previously discussed, exchange relationships between the south and south-eastern areas of Britain and the continent existed prior to the Claudian invasion, chiefly explained by newly established political affiliations after the invasions of Caesar. These operated via both the Atlantic and the southern North Sea systems. Gallo-Belgic pottery and coins are found in Iron Age contexts of south-eastern areas of Britain alongside evidence of olive oil and wine. The scale of this connectivity increased in the post-conquest period, especially with regards to the southern North Sea system, and saw a huge range of different imported produce enter Britannia, such as wine, fish sauce, samian pottery and lava querns. Any goods assumed to have been sent in the opposite direction are thought to have been largely archaeologically invisible. The movement of goods occurred as a means of fulfilling military supply needs, and because of changing social needs and wants in a 'Romanising' province. Military contracts, state control of goods via taxation and private trade are thought to have been the principal mechanisms of exchange at this time and movement of merchandise was facilitated by the presence of the Roman military in the Rhineland. The presence of the military in both Britannia and Germania would have acted as an economic stimulus, increasing the wealth and consumer culture of the areas in which troops were stationed.

By the late second century and until the mid-third century, exchange along the southern North Sea system decreased dramatically, possibly due the utilisation of goods from more localised sources. Debasement of coinage, changing tastes and the Antonine plague may also have been factors that created the downturn in long-distance trade. From the mid-third century onwards, connectivity becomes more difficult to trace. Coinage is seen to enter the province via the southern North Sea system, which could indicate that there was an increase in the volume of 'invisible' goods, such as grain, being exported from Britain. The third century crisis may have stimulated this change and created a greater dependence on British agricultural production than there had been previously. As soldiers were moved to more critical areas of the continent, reduced troop numbers in Britain may have created a food surplus that needed to be redirected elsewhere in the Empire. Commercial endeavours relating to these changes may have contributed to this influx of cash. Mayenware and Eifelkeramic pottery were also entering Britain from the Rhineland during the third and fourth centuries (Tyers, 1996c), and though the exchange of lava milling tools is more strongly associated with the early Roman period, the possibility for a continuation of lava imports alongside pottery from the Mayen region should be considered.

These conclusions, although plausible, should be tested and explored using the results of this study involving lava querns. These objects were included in Morris' (2011) study of the North Sea trade, though distributions have only included sites in Essex and the north-west Barbaricum (pp. 78 and 122 respectively). More detail is needed to determine whether lava querns and millstones follow the maritime systems outlined by Morris (2011) and whether there are exceptions. How and why these exchanges occurred should be addressed alongside consideration of the changing political, social, and economic circumstances that occurred throughout Roman rule.

2.11. The Economy of Stone Exchange in the Roman Empire

To date there has been no systematic study completed on the role of the lava quern trade in Roman Britain and there is no contemporary written evidence to indicate how it may have functioned. However, there is substantial information relating to the stone trade of the period that may help to

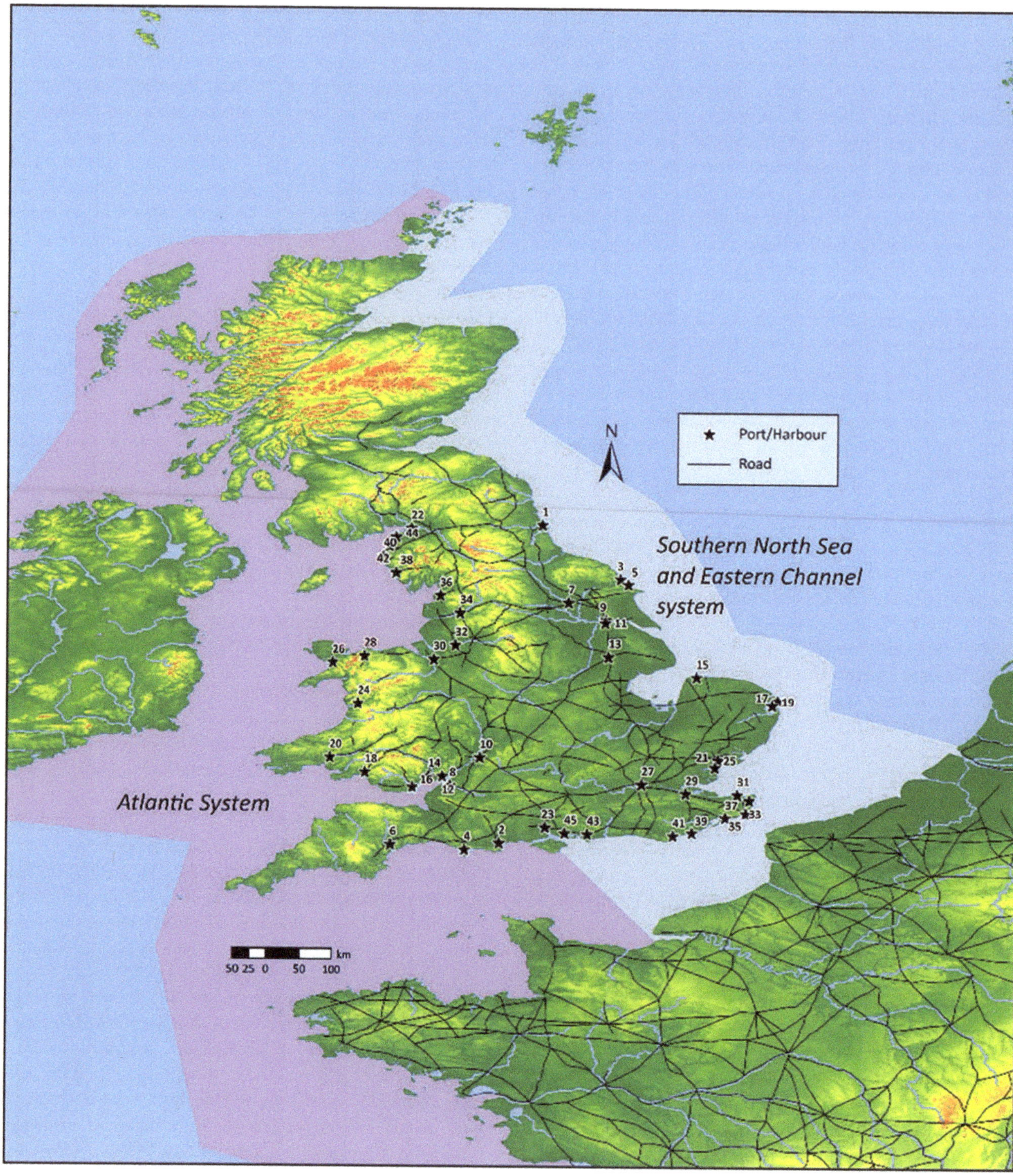

Figure 2.34. Maritime systems in the North Sea and English Channel (illustration by author after Morris, 2010, p. 2).

illustrate how this may have operated in a more general sense. Research in this field has predominantly focused on decorative stone that had a high aesthetic value and often extracted as part of imperial projects (for examples see Ward-Perkins, 1980a; Herz & Wenner, 1981; Ward-Perkins, 1992; Fischer, 1998). However, more recently, there has been a greater focus on utilitarian stone objects and building materials (Allen, et al., 2003; Pearson, 2006; Hayward, 2009; Russell, 2013; Reniere, 2018). Detailed research into more 'mundane' materials has provided a broader sense of more typical economic behaviours associated with stone extraction and movement, as these are the fabrics that were used in greater bulk than any of the decorative varieties.

It should be acknowledged that the function and significance of stone in the Roman world was much more dominant than it had previously been in many regions (Jongman 2007, p. 609). Not only did the demand for good quality stone increase in many territories of the empire, but the value and importance attached to suitable sources was also much higher. In Britain, a province that did not have a strong tradition for building in stone during the Iron Age, this change in demand would have caused fundamental developments in, what had previously been, small-scale exploitation of stone sources. This is not to say that stone was not harnessed prior to the Roman period, or that indigenous Britons were not aware of the best sources for their requirements. Querns, whetstones, and other ground stone tools are well-attested in the archaeological record from pre-Roman contexts and building in stone did also occur, especially in upland areas (Young & Simmonds, 1995; Harding, 2009, pp. 90–91). These examples show careful selection and utilisation of stone to suit a variety of purposes, though most were sourced and used locally. Smaller stone objects and tools travelled considerable distances, for example, querns of Lodsworth stone (Peacock, 1987) and Old Red Sandstone (Shaffrey, 2006, p. 64–65) saw extra-regional distributions.

The scale of exploitation, organisation of labour and transportation of stone must have changed significantly because of an increased demand in the Roman period; both in terms of monumental and decorative stone, but also utilitarian stone required for building projects and ground stone tools. However, most areas with a natural geological deficit of good building stone continued to use traditional building materials and methods, and domestic structures were still predominantly built using timber. Public or military buildings were not automatically built using stone in the first instance, though stone building remains tend to survive much better in the archaeological record and can obscure traces of earlier phases of timber construction (Pearson, 2006, p. 15–16).

Regional differences in use and exploitation would depend heavily on the local presence or absence of a good suitable stone supply, and on the local knowledge of such sources. The regions of Iron Age Britain that had a greater tradition of stone use for building tended to be in upland areas and places where good quality and abundant stone was readily available, and these practices continued into the Roman period (Pearson, 2006, p. 15). In other regions, it may have taken a degree of investigation and prospecting for newly arrived officials and settlers to find suitable materials for their needs at a reasonable distance. This may have been an initial barrier to the use of stone in some areas of the province, and there is some evidence to suggest that some of the earliest stone structures relied more heavily on non-local or imported material as part of state-funded or elite building projects. For example, Fishbourne Palace, used a range of 'exotic' stone brought from the Pyrenees, Italy and various non-local areas of Britain (Peacock, 1998, p. 1), while the triumphal arch at Richborough was built using Ditrupa limestone and Marquise stone from north-west Gaul alongside marble from the Carrara quarries of northern Italy (Pearson, 2006, p. 24).

A deficit in skilled workers may have provided a similar barrier during the earlier post-conquest era, especially considering that different skills are required to work different types of stone. For successful stone-working or quarrying to take place, there is also a requirement for suitable tools and qualified workers to produce and maintain them (see Blagg (1976) for stonemason tools and Mangartz (2012) for those used in quarrying). More intricate work with decorative stone or the production of more complex items with multiple stages of manufacture would also have required more skilled workers. It is unlikely that such craftspeople and managers were extant within the indigenous populations of the newly conquered province and there would have been a reliance on the skills of the colonists and invading forces.

The use of foreign craftspeople in early projects utilising stone is almost certain, and it seems that both skills and stone were initially brought from further afield; for example, the style of construction of the Temple of Sulis Minerva at Bath must have relied on the skill and expertise of workers from outside of Britain (Barry, et al., 1990, p. 171), as did the proto-palace at Fishbourne (Blagg, 1980, p. 28–29) and there is no evidence to suggest that these people remained to train successive stonemasons after these projects were complete. In a province where stone-working was previously virtually absent, these people would have been in great demand and well compensated. Employment opportunities of this kind may have quickly filled with others keen to profit from their craft, and the distinction between military and civilian architecture demonstrates that there must have been further demand for workers that were outside of direct state interest (Blagg, 1976, p. 152). However, the role of the military for providing both a skilled and non-skilled workforce should not be underestimated, and many foreign skilled craftspeople are likely to have been military personnel (Pearson, 2006, p. 21; Hayward, 2009, p. 108).

It seems unlikely that any such deficit in the workforce would have remained the case for long, as there would have been many incentives for the military to develop and train the necessary skills that may otherwise have been absent or in short supply (Blagg, 1980, p. 39; Evans, 1994, pp. 146–147). However, access to this source of labour and skill would have been heavily restricted to those projects beneficial to the state. The level of control and exploitation of stone sources exerted by military and state powers is evidenced at quarries that are known to have been used to build Hadrian's Wall (Bidwell, 1985, p. 4; Dobson & Breeze, 2000, pp. 30–31), and also for the construction of the Saxon shore forts (Pearson, 1999, p. 110–111). These were large-scale projects that were completed in relatively short periods and the requirement for a well-trained, disciplined, skilled workforce could easily have been met using military labour and expertise.

How exactly quarrying and stone extraction operated in Roman Britain is poorly documented, little understood and widely generalised in terms of what is known from other areas of the empire. There is no reason to suppose that there was always a heavy degree of direct state intervention or monopolisation, and it is probable that many quarries were operated in less controlled conditions than those that generated highly prized marble or porphyry. There are no known quarries within Britain that can be considered as comparable to the extent of some of the quarrying operations in other provinces, such as Egypt. Unfortunately, later post-Roman exploitation of good quality stone in Britain has obscured the evidence of Roman activity in many of the areas where larger volumes of stone must have been extracted, such as the Jurassic rocks of Lincolnshire which were widely used (Pearson, 2006, p. 52).

Nonetheless, from the scale of quarries currently known, it seems highly likely that most operations would have been irregular, intermittent, or isolated to occasions when stone was required for specific needs. Extraction may have occurred on a seasonal basis or have employed a special taskforce to complete the necessary work. The timing of quarrying activities could be arranged to maximise the labour of rural workers, without harming agricultural production. An example of quarrying to demand can be found at Vindolanda; a military graffito or graffiti at the nearby sandstone quarry site at Barcombe Down shows that soldiers were employed to carry out extraction work, while a coin hoard dated up to the Hadrianic period suggests when this took place. Stone from the quarry was used to construct several phases of the stone fort (Bidwell, 1985, p. 4), though from the scale of extraction there is no indication that large volumes of material surplus to requirement was ever removed (Pearson, 2006, p. 52).

Stone artefacts saw wider distributions, as these had more of a commodity value than building stone, though heavier and utilitarian stone objects, such as querns, were still relatively restricted in terms of distance and range of distribution. The reason for wider distributions probably relates to the skill and labour input required to make such items, as opposed to the perceived value of the stone itself. Although stone was a precious resource, it was also abundant in many locations and unless specific qualities or properties were required of the stone, there was little reason to incur the expense and time delay that would have come with the transportation of stone from further afield. This was not the case when it came to imperial or military requirements, as it was possible to bypass some of the most restrictive social and economic hurdles relating to the acquisition and transportation of stone.

2.12. The Lava Quern Trade in Continental Europe and Britain

Although this is the first nuanced study on the lava quern trade and how it related to life in Roman Britain, there has been much more work in continental Europe that can be used to assist interpretations within the context of this study. Some of the most extensive research completed so far has included the production stages of the lava quern lifecycle. This is invaluable information, as it highlights ways that lava querns may have played a role in the economy of the Roman world (see section 2.11 on lava sources). Furthermore, social factors relating to extraction and labour organisation can help explore how and why economic conditions and behaviours relating to the lava quern trade may have come about. Provenance analysis has also been used to explore distribution of lava querns sourced from Mayen, and to examine the distinction between distribution regions of Mayen lava and that of similar vesicular lava from the Massif Central area of France. This information will be used to assist interpretations of lava quern distribution in Roman Britain and to understand the reason why certain distributions exist.

2.13. Distribution of Provenanced Lava Milling Tools in Mainland Europe and Britain

The development of geochemical analytical methods used to provenance lava has greatly increased the visibility of the Mayen lava quern trade and has helped illustrate the ways in which this may have operated. To date, most provenance work of this kind has involved querns and millstones recovered from contexts in mainland Europe, though an interest in 'Pompeian style' millstones of lava has resulted in a small number of these undergoing analysis in Britain. The most significant research relating to the provenance analysis of Mayen lava was completed by Gluhak & Hofmeister (2011) who determined that there was a 'catchment area' within which querns from the Eifel region were distributed. A 'competing' trade involving querns manufactured from lava sourced in the Massif Central region, France, had a separate distinct distribution that bordered that of the Mayen querns (figure 2.35). Mayen lava was found in Kempten as the most southern area of distribution, Magdalensberg to the south-east and Mirebeau to the west. Massif Central querns were in the regions of Besançon, Bibracte and Bassou (Gluhak & Hofmeister, 2011, p. 1614). Though this does not provide us with a complete and full picture of distribution and the resolution could be improved, it does provide a new way of considering the specifics of the trade involving Mayen lava and how this related to another similar product. However, as the analysis was completed with the aim of exploring the extent of distribution, there is a lack of information relating to the volumes of Mayen lava at each of the locations. The investigation also does not include querns of other stone types and there should be recognition of the fact that these would also have had a role to play in consumption and distribution.

A similar study that included Mayen lava querns from Roman northern Gaul by Reniere (2018) reached some similar conclusions to that of Gluhak & Hofmeister (2011) in terms of distribution. It was found that 35 of the 37 querns sampled were of Eifel origin and that the study area was well within the catchment area for the trade. The ratios of querns and millstones from each of the Bellerberg lava flows was very similar to that detailed by Gluhak & Hofmeister, though as previously detailed, the timescale of consumption was different for the Ettringer Lay lava and there could possibly be a pattern in the social distribution of use relating to this timescale (see Chapter 3 on previous

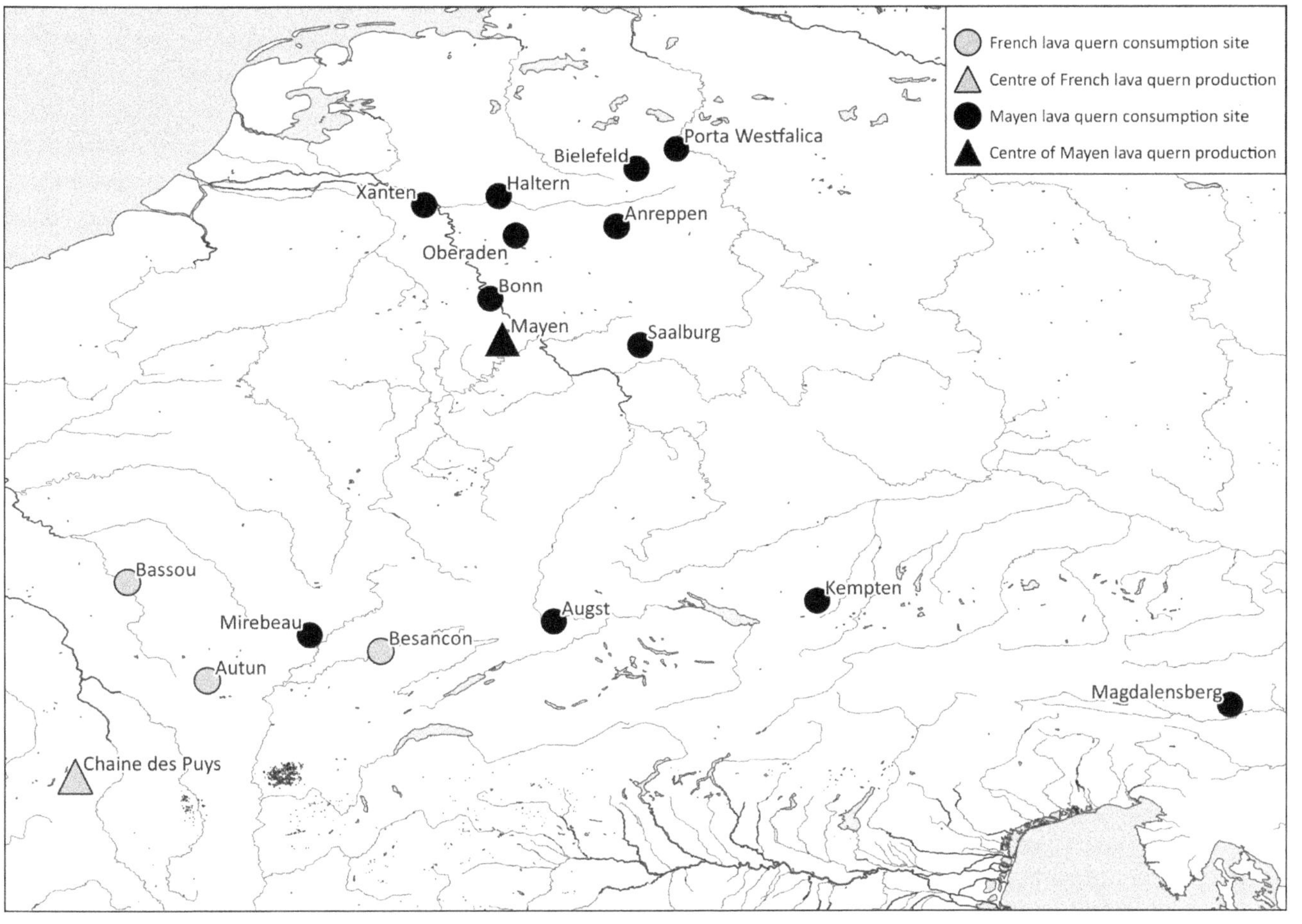

Figure 2.35. Provenanced lava quern distribution (illustration by author after Gluhak & Hofmeister, 2011, p. 1617).

research on the geology of Mayen lava). Furthermore, two querns were found to have possibly been of Hochsimmer origin, which is a lava flow located east of the Bellerberg volcano. Although this source has been previously identified as a Roman extraction area, it had been assumed that this had only been used for building stone. That lava querns were exported from previously unknown quarries in the Eifel area demonstrates the importance of completed geochemical analysis and shows that there are other intricacies to the trade yet to be identified.

2.14. Applying Economic Considerations: Roman Britain and the Lava Quern Trade

It is clear from an examination of the stone trade within the wider empire, that the political elites, military, and imperial administrative powers had a considerable advantage over the rest of the population when it came to economic decisions and agency. In many ways, the empire was very much an economy of status. Trade appears to have been another avenue through which the elite could financially benefit, though only through the risk and hard work of merchants who were at the mercy of the imperial administration. That trade occurred is not under dispute, though the extent to which commercial operations had the ability to expand and grow is debateable.

In considering the lava quern trade in Roman Britain, it seems highly unlikely, at least in the earlier Roman period, that lava querns were traded in Britain as part of 'free trade' for commercial gain. As the Britons already had querns and knew how to make them, there was no financial advantage in transporting large and bulky items over large distances to what was unlikely to have been a particularly receptive market. However, the yet unproven assumption that the first Mayen lava querns were introduced by the military could still prove valid. The earliest Roman Mayen querns along the *limes* have been recovered from military contexts and show that the military had early access to the lava source, or at least to its products. Therefore, Mayen querns were already established within the military and by AD 43, production of these objects was reliable, large-scale, and accessible. As the military were able to bypass all rules, taxes and restrictions that applied to commercial trade, the influx of lava querns post-invasion is likely to have been initiated by the military. This will be investigated and examined further by systematic social distribution analysis to establish whether the earliest uses of lava querns occurred within military contexts.

However, it was not only the military who arrived in Britain after the conquest of AD 43. There would have been an increase in population from other settlers; including military families, administrators, and other colonists. All these people needed to be supplied with the necessities for life, including querns. Although the skills required to produce these more complex tools was clearly present within the population prior to the invasion, as long-lived objects, querns may not have been produced on a large scale. There is no standardisation of Iron Age querns that indicates that organised, high-volume production existed during prior to the Roman era and local types usually had limited distributions. It is also possible that new settlers in the post-conquest period lacked local knowledge with regards to suitable stone sources to produce querns, and that there was an immediate deficit in the number of available milling tools needed to fulfil the growing population's food processing needs. Although speculative, it is worth considering whether there was an initial requirement to import querns in the form of a finished, or nearly finished product as a result. If this was the case, it is unlikely to have remained a permanent or long-term reason for such imports. Other socio-economic aspects of life in the province may have had more of a role to play in the sustained use of imported lava querns, such as a preference for lava and its associated milling properties.

To explore this idea, the chronological change in lava quern volume and social distribution will be a vital form of analysis. Furthermore, it will be worthwhile to incorporate the presence or absence of indigenous type non-lava querns into any considerations of distribution, as these will show the extent to which these were still being produced and used. The Iron Age quern quarries that were still generating products into the Roman era must also be mapped alongside these distributions to see how relative distance from indigenous quern production areas affected the demand for lava and its subsequent distribution.

3

Volcanic Beginnings: Lava Sources

This section discusses the geology, geochemistry, and exploitation of lava at Mayen in the Eifel region of Germany for quern and millstone manufacture. The Eifel region describes a wide geographical area and saw a long chronology of lava exploitation dating from the Neolithic to the present (Crawford & Röder, 1955, p. 71; Belmont & Mangartz, 2006, pp. 25–34; Gluhak & Hofmeister, 2009, p. 1774). It is, therefore, necessary to ensure that the appropriate terminology is applied to address the specific period of quarrying and production areas investigated in this research. Lava used to produce querns sourced from the Eifel region has been given a variety of different names: Andernach, Niedermendig and Mayen. Andernach is the name of a nearby town and harbour, the location of which does not relate to any lava source, whilst Niedermendig quarries only supplied grinding stones in later periods (Crawford & Röder, 1955, p. 68). Mayen lava originated from the Bellerberg volcano at Mayen and may refer to three different lava flows from the same volcano. Mayen itself was a small settlement prior to the Roman period, which later grew as quarrying activity increased. For the purposes of this work, Mayen will be used as the most appropriate name for lava quarried from this region during the Roman period.

3.1. Mayen Lava as a Material

Throughout this publication, reference is continually made to Mayen lava as the dominant source of material used to produce the many lava querns and millstones that were in use in Britain during the Roman period. This conclusion has been reached mainly due to the knowledge that we have about the extensive quarrying and quern production area at Mayen in the Eifel region, plus our understanding of North Sea trade and the connections that Britain had to the Rhineland; proven by looking at the distributions of a variety of products that were imported via this route. However, we do know that other sources of lava were used in the province, such as French lava from the Volvic area of France (Massif Central) that was used to make 'Pompeian' millstones present in Romano-British contexts (see Chapter 2). Though the potential for different lava sources has never been systematically studied for Roman Britain, it has been explored more extensively in continental Europe and several small-scale studies have been completed for the British material. This section aims to examine what the geological background and identification of Mayen lava is and how lava has been studied with regards to geological provenance. There will then be a short discussion on what should be done in the future to identify and understand the different lava sources that may have been in use in Roman Britain.

3.2. Geological Background

The East Eifel region, where Mayen and the Bellerberg volcano are located, can be found in western Germany and is part of the Cenozoic Central European Volcanic Province (CCEVP), which includes several regions in mainland Europe. The Eifel is at the centre of this region, which consists of a wide arc of volcanic areas north of the Alps, comprising tertiary and quaternary volcanic fields. It stretches from the Auvergne to the Carpathian Mountains and is related to the tectonics of Central Europe, specifically the folding of the Alps, the opening of the North Atlantic and the formation of extensive fault systems such as the Upper Rhine Rift Valley and the Lower Rhine Bay (Mangartz, 2008, p. 3–4). As a relatively young volcanic region, volcanic activity began in the East Eifel around 0.65–0.45 million years ago, when the first lava flows and tephra deposits were formed (Smithsonian Institution, 2021). The Bellerberg volcano from which the Mayen lava querns and millstones were formed, consists of multiple lava flows within one volcanic group. The 'cinder cone' is known to have been formed as recently as around 200,000 years ago, created during an eruption of low-gas basaltic magma from a great depth below the Earth's surface. This was cooled by the air as it emerged, forming loose cinder walls around the location of the eruption. A cinder cone eruption generally loses momentum over time and magma then emerges with a lower gas content and thrust, creating a lava flow. This process can occur multiple times during the active life of a volcano group, as it did in this case. The different lava flows of the Bellerberg, thus, have different names, which correspond to the quarries from which lava was sourced; Ettringer Lay, Mayener Grubenfeld and Kottenheimer Winfeld (Gluhak & Hofmeister, 2009).

3.3. Geological Description and Properties

Mayen lava from the lava flows of the Bellerberg volcano in the Eifel region of Germany is of a rare type and can be defined as a nepheline tephrite (Peacock, 1980, p. 49); a crystalline, basic igneous rock, though not a 'true' basalt. During formation of the lava stone, gasses within the magma cause the liquid rock to solidify around bubbles when cooling to create vesicles (holes) of a few mm in size within its structure (figure 3.1). The presence of these vesicles is one reason why the material is suitable for milling, as these allow the grain to be 'cut' instead of crushed, creating a more efficient means of flour production and a better grade of flour. They also helped to reduce the density of the stone as these 'bubbles' have the potential to make up around 25% of the volume of the material, making it up to 25% lighter than comparably dense rocks (Mangartz, 2008, p. 9). The very fine crystalline structure

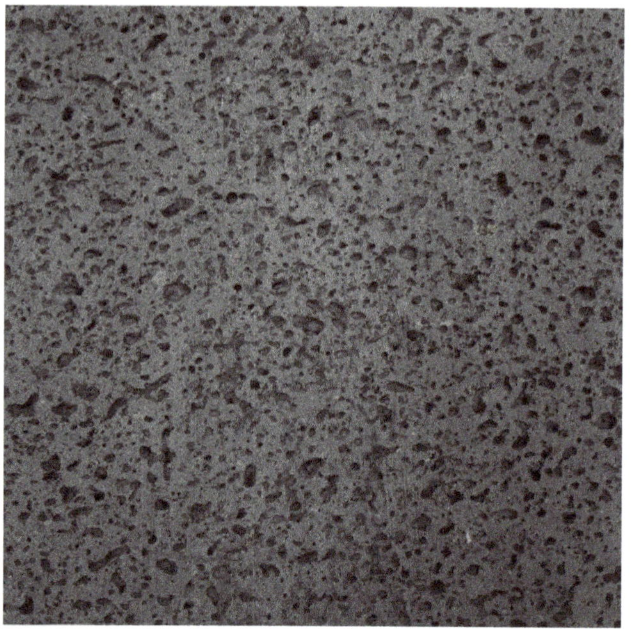

Figure 3.1. High resolution image of Mayen lava to show the vesicles (holes) that are an identifying feature of the material. It is also possible to see some of the sanidine phenocrysts, visible as white crystalline material (Roll-stone, 2009: Creative Commons License CC BY-SA 3.0 DE via Wikimedia).

Figure 3.2. Photomicrograph of Mayen lava in thin section with cross polarised light (Roland Dreesen © Gallo-Romeins Museum Tongeren in Gluhak, et al., 2015, fig 32).

of the rock is another feature that assisted grinding, and lava milling tools did not need to be dressed to function well as they were 'self-sharpening' (Fitzpatrick, 2017, p. 196). The structure described was formed during rapid solidification of the lava flow, meaning that the mineral matrix of the rock hardly had time to form crystals. This process results in fewer abrasives within the fabric, such as quartz or quartzite, which are often found in the sandstones used for other quern or mill types. These abrasives can be damaging to teeth if they get into the flour (which they must have often done), providing lava with some clear benefits for use in milling (Mangartz, 2008, p. 9). Some of the physical properties of Mayen lava can be found in Table 3.1.

The physical appearance of Mayen lava can be described as a fine-grained pale to medium grey vesicular rock with frequent felsic xenolith inclusions, these being foreign rocks within the lava matrix. These inclusions comprise

Table 3.1. Physical properties of Mayen lava (Mangartz, 2008, p. 24)

Physical property type	Physical property
Bulk density	2.3 g/cm^3
Porosity	15–25 vol.%
Compressive strength	165 N/mm^2
Flexural strength	6.1 N/mm^2
Abrasion resistance	12 cm^3/50cm^2
Water absorption	8 vol.–%
Thermal expansion	0.9 mm/mK
Elastic modulus	5800–10300 N/mm^2

small clinopyroxene phenocrysts, seen as dark coloured crystals, and occasional larger sanidine phenocrysts, seen as white crystals, of up to 10mm. In thin section, it can be described as having a fine-grained matrix composed predominantly of nepheline, sanidine, clinopyroxene crystals, plagioclase and opaque ores with minor grains of leucite and apatite (Reniere, et al., 2016, p. 408–409) (figure 3.2).

3.4. Petrographic and Geochemical Analysis

As scientific methods of geological analysis have developed over the years, the ability of archaeologists to determine the provenance of stone building materials and artefacts has improved considerably. New methodologies have created new ways of answering questions relating to the use and distribution of different ancient stone objects, such as millstones and querns, and these have been well utilised within the discipline for certain artefact categories and stone types. Lava as a fabric has been well-researched, especially with regards to trade and exchange of Roman period millstones in the Mediterranean region of Europe and North Africa (Williams-Thorpe, et al., 1991; Williams-Thorpe & Thorpe, 1993; Renzulli, et al., 2002). Many of these studies have made use of XRF geochemical analysis, thin section analysis, or a combination of both approaches to compare the types of lava used to make millstones with reference samples taken from potential millstone quarry sites.

More recently, a similar methodology has also been applied to lava quarried in regions of mainland Europe north of the Alps. Reference samples of lava have been taken from known lava quarry locations in use during the Roman period and throughout the geographical reaches of the empire. These have been geochemically analysed using XRF to produce data for comparison with lava querns or millstones of unknown origin. At Mayen, several quarry locations were in use during the

Roman period (see Chapter 2) and reference examples exist of individual lava flows of the Bellerberg volcano, which each have unique geochemical signatures (Gluhak & Hofmeister, 2009). This data has been used to further investigate the provenance of lava querns and millstones recovered from Roman period sites north of the Alps and to study their distributions relative to provenance (Gluhak & Hofmeister, 2011; Dreesen, et al., 2014; Reniere, 2018). Of forty-four Eifel lava artefacts in the study, all had been manufactured using basaltic lava from the Bellerberg volcano and were, thus, of East Eifel and Mayen origin. It was found that most of the querns and mills analysed could be provenanced to Mayener Grubenfield, with some from Kottenheimer Winfeld and significantly fewer from Ettringer Lay. Most of the Eifel lava from the querns and millstones tested was found to be from the Mayener Grubenfeld quarry, with fewer sourced from the other two locations. None were produced using basanite lava from other volcanic areas in the East Eifel, which appear to have been predominately used for the manufacture of building blocks. There is, therefore, the indication that individual quarries provisioned specific manufacturing industries, and that some quarries were used to supply different regions. The lack of any querns or millstones from the West Eifel region has been used to support Mangartz's (2008) discussion that lava from there was used only for localised consumption of querns.

Exploitation of the different lava flows also showed some variation in terms of when each of these were more extensively used for quern production. Although the querns from Ettringer Lay were a minority, totalling only 4, these provide some of the earliest examples, dated to 7–5 BC/9–6 AD and 11 BC/ 8–7 AD. Of these four, three were recovered from military contexts, suggesting that the quarry at Ettringer Lay was a significant supplier to the military along the *limes* in the earliest phases of Roman lava exploitation. A study by Reniere (2018) on Eifel querns recovered from northern Gaul has added some detail to this. Although the ratios of querns from Mayener Grubenfeld and Kottenheim were very close to that determined by Gluhak & Hofmeister (2011), with the majority coming from Mayener Grubenfeld, the Ettringer Lay querns were dated to the second and third century. This could indicate that there was a change in the way that lava from Ettringer Lay was exchanged and that it may not have had such an important role to the military in later phases, but the low numbers of querns from Ettringer Lay do not allow for absolute conclusions to be drawn. Further provenance analysis of a range of lava querns from military sites of different periods, both along the *limes* and further afield, could help to further refine the chronology of military consumption of lava querns from specific lava sources.

Distribution analysis also demonstrated that there was another competing lava quern manufacturing area, probably located in the Massif Central region of France, and that the Mayen and French lava querns had separate 'catchment' areas of consumption (Gluhak & Hofmeister, 2011, p. 1617–1618) (see Chapter 2). On a macroscopic scale, the French lava looks very similar to the Mayen lava. Thin section analysis can be used to differentiate between the two (figure 3.3), but as lava is inconsistent in the way that it forms, the specific characteristics that separate French from German lava can often be absent in any given sample. Geochemical analysis has provided a more uniform approach for identifying the material, as the two are chemically distinct.

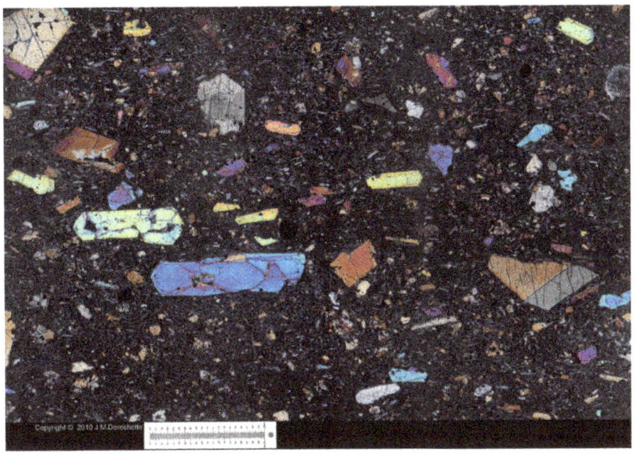

Figure 3.3. Photomicrograph of basalt lava sample from Banne d'ordanche, Monts d'Ore in the Massif Central region of France (Derochette, 2010).

There have been smaller scale geochemical and thin section analyses carried out on lava from British contexts and these have been very revealing. Most of these have been carried out on the iconic 'Pompeian style' millstones (Williams-Thorpe, 1988b; Williams & Peacock, 2011; Allen, 2012) (see previous section, this chapter). The results from these separate studies have shown that both French and Mayen lava was present in Roman Britain in the form of animal-powered mills. Though an important revelation, the relative rarity of these objects in Britain means that these results present us with exceptional examples of unusual objects that are unlikely to tell us much about normal trade and exchange in the province. It is possible that the French sourced example is a rare case of a specially commissioned tool. Thus far, there is no data to suggest that French lava is present in Britain in the form of hand operated querns, but systematic geochemical analysis of these objects has never been carried out for Romano-British material, and even small-scale provenance analysis is currently rare. Shaffrey and Gluhak (2020) have demonstrated how this can be done using material taken from a single site. Results of this study have shown that querns manufactured from the Kottenheimer Winfeld lava quarries were exported to Kent in the early post-conquest period. Though significant, this needs to be put into wider context with the inclusion of more data.

3.5. Future Research

As has already been indicated, a great deal more work remains to be done with regards to examining the provenance of lava querns and millstones in use during the

Roman period in Britain. The importance of understanding the provenance of lava quern stones is becoming increasingly recognised as a gap in our current knowledge of the Roman economy and society. The geochemical data of the reference lava samples taken from the Bellerberg quarries is now widely available, as is the methodology used as part of German and Belgian research. It appears very possible that there may have been more than one source of lava being used to produce these objects, which were then exported to Britain from the continent. Geochemical and thin section analysis of 'Pompeian style' millstones has already demonstrated that French lava is present in the province. The different distributions of Mayen and French lava handmills in continental Europe suggest that distributions of lava querns in Britain may also be more nuanced when provenance is considered. A systematic scheme of sampling and analysis of the British material is needed to include lava from both millstones and quern stones of different chronologies, social and geographical distributions. In doing this, it will be possible to further clarify the different trade routes that were in place between Britain and mainland Europe, how these changed over time and how they impacted different communities. Further to this, it will help us to gain a better understanding of distribution networks that existed within the province. For example, connectivity between different groups and settlements and the identification of well-travelled or lesser travelled exchange routes. All this information will provide a much richer vision of the Roman past.

4

Methodology

This study investigates relationships between lava milling tools and people through their economic behaviours and social or cultural practices. This includes behaviours and interactions associated with imperial institutions and the state, different socio-economic groups, within different regions and chronological periods. It is an evidence-based examination of the types of economic activity that were present, while considering how and why they may have existed. The approach does not assume one particular economic paradigm, but utilises a range of different ideas and economic models to make comparisons and as tools for contextualisation. Large-scale data has been used to do this, quantifying the presence, absence and volumes of lava querns and millstones from around the province. This has helped determine whether regionality, geographical or topographical location was a factor in the supply and use of lava querns within the province. In terms of Morris's (2010) martime systems of exchange, it has been possible to test the North Sea exchange model of distribution for imported goods. Furthermore, the study includes an investigation into the presence, absence and volume of lava querns in terms of social distributions and site type to examine how these objects may have 'spread' between supply routes and been moved around Britain under different personal, commercial or political motivations (cf. Eckardt, 2005). Relationships between forts and their hinterlands, between urban centres and rural locations, between ports and surrounding settlements and towns can now be visualised using mapping software to contextulise and conceptualise connectivity, including both social and economic interactions.

In terms of food production, distribution analysis has the ability to show us how food processing activities may have been organised on a regional and provincial scale. The idea that urban centres were reliant on the rural hinterland for food production does not necessarily relate to how food processing was structured. The inter-relationship between food production, processing and consumption provide interesting insight into the broader workings of the Roman economy and illustrate the role of social status and location in determining a person's involvement in food systems. How lava querns functioned within these systems is also of significance. As imported objects, not all people would have access to, or wanted to have a lava quern. Often classed as being technologically better than querns of indigenous stone types, there is an assumption that these were desireable and more efficient tools. They may have been used by individual familial groups due to being the most preferred or most accessible means of daily food processing, or used in mass food processing activities to supply flour to larger towns, settlement or communites. Therefore, how and why these objects were distributed in particular ways could potentially unveil the relationships that people had with lava querns and how they fit into everyday life for people living in the past.

Evidence for lava querns in Roman Britain allows for conclusions to be drawn in terms of specific patterns of economic activity or behaviour relating to the material. However, these are not necessarily representative of the economy of the empire as a whole and a range of different economic behaviours are present across different regions, social groups and periods. The reason for this relates to the relatively unique fact that stone, in the form of lava querns, was transported over large distances, resulting in widespread distributions in Roman Britain; a situation that is unparalleled on this scale with any other utilitarian stone object. The lava quern serves to demonstrate the variability that ancient economies had, adding to a growing body of evidence that can be used to create better models and ways of understanding the ancient economy. These different behaviours are just as significant when it comes to understanding how economies operated during the era and can contribute to broader insight into this area of research.

4.1. Practical Aspects of Data Collection, Recording and Analysis

This section aims to set out the more practical aspects of how the dataset was generated with an overview of the tools, both digital and physical, that were used to collect, store, and analyse the data for this study. This is so future work can replicate and reassess the results and interpretations generated here. Both lava and non-lava milling tools were recorded as part of the research, with lava milling tools recorded as individual objects alongside associated site information, such as location of site, deposition context and context date. Sites where milling tools were present, but none of lava were also recorded. For these sites, the number of other milling tools present were recorded alongside location and site type data. For some case study sites, individual non-lava querns were also recorded, but these were kept separate from the lava dataset. The number of objects recorded, and the method of their recording has been summarised in Chapter 5.

4.1.1. Quantification

Quantifying lava milling tools is a tricky process as soil and deposition conditions can result in extreme fragmentation and degradation of the material. A method for quantification had to be established prior to any recording taking place to ensure consistency in the analysis and results. As with pottery, it is possible to quantify the number of querns using quern edge fragments (Orton, 1989). A larger form of a pottery rim chart can be utilised to determine the percentage of quern present by measuring the proportion of edge remaining, thus allowing for a minimum number of querns to be established. However, as the survival of lava quern fragments is dependent on soil conditions and deposition practice, this does not provide a consistent method of quantification across all sites. The fact that the dataset was generated using a large volume of published information and grey literature reports has also meant that quantification methods have relied heavily on the level of description and quantification used by the stone specialist or person writing the report. It is very rare for querns to be measured in this way, and so this could not be fed into my own database.

During in-person data collection, fragmented milling tools were first checked to see if any could refit. If so, some fragments could be grouped and counted as a single object. Where this was not possible, fragments were grouped and counted by archaeological feature. A similar approach was taken for published material, though there was less opportunity to assess whether fragments would refit. Counting groups of fragments by archaeological feature provided the most consistent approach for quantification, and this was information that was most readily available in quern reports. In the data analysis chapters, small fragments recovered from one feature are referred to as 'lava groups', and these groups are counted individually. The method works under the assumption that most pieces found in the same pit/ditch/dump probably belonged to the same object originally. This is not a perfect system, and there are bound to be some issues relating to the degree of fragmentation and spread of material at any one given site compared to another. However, as the quantification of querns and measurements provided in relation to them is so variable across quern reports, it is not possible to refine the process until issues relating to how fragments are recorded has been dealt with. Specific problems relating to the recording of lava milling tools have been discussed further in Chapter 5, while suggestions for future recording methods have been outlined in Chapter 9.

4.1.2. In-Person Quern Recording

Quern dimensions were measured using rulers and callipers where appropriate, while an enlarged pottery 'rim chart' was utilised to estimate quern diameters for fragmented examples that had an identifiable edge and sufficient curvature for this to be practical. Weights were measured using scales, while a hand lens and torch was employed to examine stone surfaces for signs of working, inscription, or dressing. A quern recording sheet was used to manually record the dimensions and descriptions of the milling tools when this was necessary.

4.1.3. Digitising the Data and Completing the Analysis

Data for all milling tools were recorded in an Access database designed for this purpose. Each lava milling tool was recorded with a unique catalogue number, which are used as a point of reference throughout this study. Where individual non-lava milling tools were recorded for case study sites, these were also given catalogue numbers, but these begin with the prefix 'X' to differentiate them from the lava. Similarly, individual site data corresponds to a unique site code, which for the non-lava sites has been recorded with the prefix 'X'. Data was organised in a way that would enable queries to be made across data tables to answer specific research questions. Analysis of the location data was completed using QGIS, while statistical analysis was completed using both Excel and Minitab.

4.2. Object Biographies

This study utilises an object biography, or object lifecycle approach, where the different events in the life of objects, typically from manufacture to discard, are explored to understand the histories, or biographies, of objects during the various processes that they undergo. As rooted in social anthropology, objects are an important part of the lived experiences of people, they should not be considered as static but rather that different points in their 'lives' need to be examined and understood, looking beyond original purpose and function to how past people may have related to them (Appadurai 1986; Kopytoff 1986). Even though object biography has been dismissed by some as a theoretical approach itself, it is generally agreed that it provides an effective way of thinking about archaeological material culture, so long as its use is theoretically informed (Burström, 2014, p. 68–70). Object biography may be better defined as an object-based approach that examines synchronic past processes of practice and behaviour relating to those objects (Salisbury & Rebay-Salisbury, 2017, p. 26).

Most studies that apply an object biography approach consider the different stages of an object's 'journey' alongside associated changes in meaning that are renegotiated from manufacture to deposition (Smith, 2016b). The significance of these journeys is emphasised by Joyce and Gillespie (2015), who discuss the importance of detailing the movements of objects, their circulation, exchange, and itineraries. By doing this, the whereabouts of objects can detail aspects of economic, ritual, or personal exchange (see also Gallardo, et al., 1999; Hamilakis, 1999; Peers, 1999; Saunders, 1999; Seip, 1999). Mode of exchange and specific origins or movements can help form perspectives relating to the meaning of objects and their value. However, object biographies should not be thought of as a linear concept according to Joy (2009), as there is a tendency to focus on the chronological changes that occur in the biography of material culture, which fails to explore the intricate web of human-object relations that can occur. Gosden and Marshall (1999) maintain that it is these connections that furnish an archaeological object with its meaning, importance, and role in present-day interpretations. The complexity of such relations has been presented by Swift (2012, p. 202), who examined the reuse or recycling of Roman period bracelets as rings in the Late to Post-Roman era in Britain. This process of repurposing is described as a 'moment of object transformation' creating 'divergent life histories' for objects, extending object use-lives in different forms with different meanings; transcribed, correctly or incorrectly, from the meanings of the objects they have been formed from.

Relations between people and objects and the creation of an object biography are one way that an object can become 'loaded' with agency. The creation of meaning through human-object interaction can affect people and result in altered behaviours or meanings associated with that object (for example, see Hamilakis, 1999) (Burström, 2014, p. 66–67). This approach is often associated with 'entanglement theory'; the idea that people are reliant on 'things' and that 'things' are reliant on people for reasons that go beyond subsistence and innovation. Social relations, embodiments and meaning are generated by and generate human relations with 'things', resulting in human-thing entanglement caused by a network of dependency that drives social change (Hodder, 2012, p. 177). Alternatively, there have been arguments in favour of object agency derived from the materiality of an object, as opposed to its biography; its physical properties, material characteristics, and effect on its environment (Burström, 2014, p. 67). Though analysis may begin as an empirical exercise to examine technological innovation, fabric, form and evidence of wear, interpretations focus on social relations, time, space, and materiality to understand how objects help shape the human experience (Meskell, 2005, p. 1–3).

In some analyses, biographies of objects post-discard have also been considered, such as when they have been identified as heritage objects or reused in a modern-day context (Brien, 2019, p. 37–38). Similarly, Herva (2005) discussed the concept of 'invisible object afterlives', whereby the breakage and deposition of an object does not necessarily signify the end of its use-life, while Joy (2009) considered the concept of object reincarnation; when objects continue to exist outside of the cultures or civilisations that once understood their meanings and functions and they take on new identities (see also MacGregor, 1999; Eckardt & Williams, 2003).

Many earlier works concerning object biography for Roman period material culture have focused on items of a personal or commoditised nature (for example, Joy, 2009; Swift, 2017). This is probably because evidence for human interaction with such objects is more apparent and more relatable, allowing for theoretical discussion to be placed in the foreground. However, it is possible to apply an object biography approach to a wide range of different archaeological material types and remains, and more recent studies have done so using utilitarian and 'mundane' objects or assemblages (for example, Sands & Marlière, 2020). Milling tools have not previously been considered in such a way, despite having a prominent role in the lives of past peoples. As long-lived objects that were vital for daily food processing needs and often at the heart of domestic life, querns must have held complex and varied relations with people and been agents of socio-cultural practice and change. Often recycled, modified, and reused, both querns and millstones probably experienced multiple reincarnations and reinterpretations throughout their biographical journeys and shaped the human experience in manifold ways.

The potential for object biography approaches to promote more creative ways of thinking through interpretations of archaeological material is one reason why it has become a popular application within the discipline. It is most successful when used analytically and interpretatively and to relate object biographical narratives to wider phenomena. This has been achieved in this study through the production and analysis of empirical data to explore the materiality of lava milling tools recovered from Romano-British contexts. Discussion and interpretation of the dataset has been carried out in the form of site-based case studies to explore the human-object relationships that must have existed to generate the data seen. This has then been examined in terms of wider provincial behaviours to contextualise individual and site-wide biographies. Although information concerning specific objects can be fragmentary, gaps in data used to generate narratives can be addressed by examining the trajectory of other similar objects (Burström, 2014, p. 69). The large dataset generated as part of this project has been used to fill in some interpretive gaps to maintain biographical narratives where appropriate, but these tend not to 'highlight the exceptional or unusual features' of the lifecycle, which are key to the production of a good biography (Dannehl, 2009, p. 124). However, knowing what is unexceptional and ordinary is key to identifying those events and the generic lifecycle of lava milling tools must be understood first.

4.3. Imported Lava Querns and Millstones: Applying an Object Biography Approach

Understanding the lifecycle of lava milling tools is one of the objectives for this research and has been explored thematically according to site type (rural, urban, or military) with reference to relevant case studies. Five main stages of analysis have been identified that describe significant and illustrative events or processes in the life cycle that might produce unique indicators or general patterns of evidence for human-object interaction. These comprise manufacture, distribution, primary use, reuse/modification, and deposition. Although this has been presented in what appears to be a reductive linear chronological format, it must be noted that a logical structure for analysis and communication of results must exist to provide a means to contextualise and conceptualise the outcomes. However, it is recognised that stages of the quern or millstone lifecycle can occur out of this implied sequence or happen simultaneously, and that interactions are not constrained by the way that they have been presented in the analysis. For example, a quern stone might be manufactured, exported to Britain, distributed, and then modified prior to its primary use. It could then have undergone various stages of remodification to extend its primary use life prior to deposition. The exact order of each biography 'stage' can be difficult to reconcile with the material evidence, and it is recognised that organising the analysis in stages may result in detailed and complex biographies becoming simplified and descriptive only. However, discussion of each biographical stage has been completed with the aim of exploring the human-object relationship, interactions between people and milling tools, the agency of the milling tools and the role that they may have played in shaping human experience.

Although the focus is predominantly oriented around the primary use, re-use/modification and deposition stages of a quern's lifecycle, there has also been the potential to explore aspects relating to technological innovation, typological changes and production technique or design that relate to the initial manufacture of the object, and to modification. The methodology for examining lava querns and millstones within each of the lifecycle stages is detailed here, including the recording methods, observations and measurements used to investigate specific research questions relating to the material and its interactions with past peoples.

4.3.1. Manufacture

Mayen querns imported into Britain were highly standardised as they were centrally produced (see Chapter 2), but it is possible that variations of the commonly identified type exist. These may have occurred due to chronological development, technological innovation, or been the product of different quern manufacture or finishing regions. A clear form of standardisation at Mayen means that differences in finishing are unlikely to be very prominent, though subtle variation may still be visible (figure 4.1). Chapman (unpublished) identified two basic types of lava quern that were distinguishable by diameter and the way in which the rynd was fitted (see Buckley & Major, 1983 for mention of this work). Therefore, the following measurements/ observations were made for each object:

- Diameter of the top and bottom stones
- Presence or absence of kerbs
- Width of kerb
- Description of original handle hole

Figure 4.1. Upper stones of two lava querns recovered from Vindolanda. Although differences are subtle, the kerbs are slightly different widths, the crosshatch design is inscribed at different angles, and the central eyes may have had different shapes (CAT 0048 right, 0014, left) (Image ©Vindolanda Trust)

- Central perforation shape and size
- Rynd fitting type for upper stones.
- Presence or absence of the crosshatch design on upper stones.
- Chronological change.

As grinding surfaces become worn and are redressed during the life of a milling tool, the dressing on the grinding surfaces are unlikely to reveal details pertaining to the manufacture process.

4.3.2. Distribution

Small-scale spatial distribution analysis of finds on an individual site level can reveal aspects of social practice regarding deposition querns once they could no longer serve their primary function. Deposition location is unlikely to relate to location of use in the case of milling tools, unless they were used in a fixed position that was preserved until the time of their discovery. Therefore, lava querns and millstones are more likely to be recovered from locations far removed from the use and social context to which they originally belonged. Yet as querns are unlikely to have been the subject of casual loss in the same way that smaller objects are, a degree of intentionality can often be associated with their deposition or redeposition locations. Redeposition, reuse and residuality can be difficult to differentiate from initial deposition or initial use, but where identified these stages of the lifecycle can reveal interesting and complex biographies. For example, querns were often recycled as bulk building material. At some sites where there is a high volume of this type of reuse, spatial distribution of quernstones relates to locations of refurbishment at specific times when building work was being carried out. Therefore the value of examining the distribution of lava milling tools within a single site lies in the ability to visualise behaviours associated with depositional practices and those that involve the reinterpretation of lava milling tools when used to fulfil new roles as reused or redeposited objects.

With regards to the national distribution of lava querns, GIS was utilised to produce spatial distribution maps to visualise the data, with the inclusion of chronological and social distributions of these objects relative to location. Sites with querns present, but none made of lava were also included in the analysis. This was done to create meaningful distributions that relate to how lava querns circulated, when, and for what reasons. For example, if lava querns were part of a century or legion's military equipment, there is a strong possibility that they were taken on campaign and were transported during troop movements. At times when the military were highly mobile, such as the earlier period of Roman Britain, lava querns may have been disposed of in a variety of locations where their use may not have held any long-term meaning. Temporary camps would have seen a high influx of military personnel remaining for a short period of time. Deposition of querns at these locations would not relate to domestic occupation, and it is unlikely that soldiers would form a deep cultural connection to temporary surroundings for ritual deposition to be common. This is in direct contrast to what might be expected of a 'domestic' quern, which would have strong associations with social context, place, and culture.

As an overview, the following aspects of distribution have been examined:

- Volume- Using spatial data on smaller site-based scale where available for case studies and national distribution on the macro scale, volumes were investigated by location. Volume was recorded by the number of individual quern stones present. Fragments that are not identifiable as belonging to one object were recorded as a single quern stone if recovered from the same feature. Sites that have querns present but none of lava were included to show where there is also an absence recorded.
- Chronology- This was recorded as precisely as possible from the excavation report to allow for the data to be manipulated differently if necessary. Date categories were allocated as: very early Roman, AD 43–100 (VER); early Roman, AD 43–150 (ER); mid-Roman, AD 151–250 (MR); late Roman, AD 251–410 (LR); very late Roman, AD 351–410 (VLR). Spatial data was used to see how distributions changed over time. There is some overlap with the categorisation of periods, but the very early and very late Roman groupings are to allow for the differentiation of objects with narrow date ranges from those with wider ranges at the earliest and latest parts of the Roman period. These may help with interpretations relating to the start and end of lava milling tool imports in Roman Britain.
- Site type- Site types were assigned on a general scale: rural, urban, military, military/urban, unknown, other.

4.3.3. Primary Use

Use-wear analysis can be used to identify how people interacted with querns, how their bodies were used as an extension of the milling tools and the human-object relationship. Tools are usually created in a form to suit intended use and the features of its configuration allow us to assume certain movements based on the way the human body is able to engage with it. Experimental work can help to reveal how wear patterns may have been achieved, though it is necessary to appreciate that an object can be used in many ways and hold multiple meanings or functions. The user of the object does not have to conform to the intentions of the manufacturer and can use it in very different ways from those that it was designed for (Swift, 2017). Furthermore, ethnographic studies show the range of possibilities that exist in terms of the range of movements applied to tool use, and how these can be socially and culturally determined. The same tool can be used very differently by separate groups of people, and it should not be assumed that there is a standard practice that applied across all periods and locations.

One way of looking at quern usage and use-wear traces is by examining the degree and extent of wear on the grinding surface and this can be correlated with handle placement. There is a suggestion that some rotary-type querns in Britain and Ireland were, in fact, oscillated (Heslop, 2008, pp. 53–55; Shaffrey, 2019, pp. 394–395) (figure 4.2), and this has also been claimed in association with the use of lava querns (Röder, 1953). If a quern was oscillated or rotated, operated using the left or right hand, used in a fixed or unfixed position, traces of wear on the upper and lower stones would vary. A summary of these variables and their outcomes on wear traces can be seen in Table 4.1. However, this situation would be further complicated if several people used the same quern on different occasions and did not use the same motion as each other. Further to this, mending of a quern, such as to replace a worn handle, or modification to enable it to be used differently to its original design would have consequences for the wear traces present on a quern stone.

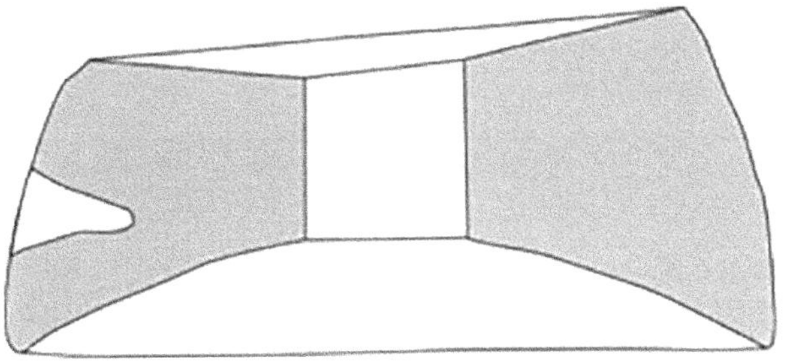

Figure 4.2. Typical indigenous pre-Roman rotary quern top stone with uneven wear, possibly from oscillating motion. This is often found to occur on both the top and the bottom stones. It is worthwhile to note that the heavy wear has occurred anticlockwise from the handle slot (image by author after Curwen, 1937).

The stones that can reveal the most evidence for method of use are those that are recovered as pairs, as this allows for the wear traces of both top and bottom stones to be examined in correlation with each other. However, finding paired stones is very unusual and only one possible pair is present in the dataset. Furthermore, the fragmentary nature of the evidence means that it is difficult to obtain a full picture of wear patterns on any one piece of quern. Therefore, there is no simple systematic approach for examining how querns were used that can be applied to all material remains. Nonetheless, the lava querns have been examined to identify what the wear traces can reveal about their use.

A simpler measure can be taken to examine the extent of use for a lava quern or millstone, and this is of the thickness of the stone. This was taken at the edge of the stone to ensure consistency, as different parts of the stone can wear differently, and was important to be able to compare measurements across datasets. Similarly, upper and lower stones were compared separately, as the edge is the least worn part of an upper stone, but the most worn for a lower stone. This has provided a means by which the supply and perceived value of lava querns or millstones has been explored between different site types, chronologies, or regions.

Table 4.1. Summary of wear traces and possible ranges of movements that they might reveal. Note that 'before the handle' refers to wear being clockwise from the location of the handle and 'after the handle' refers to wear being anticlockwise from the handle location

	Direction of movement	Top stone wear	Bottom stone wear
Unfixed quern position	Oscillated	Both before and after the handle	Even wear
	Rotated both ways	Both before and after the handle	Even wear
	Clockwise rotation	Before the handle	Even wear
	Anticlockwise rotation	After the handle	Even wear
Fixed quern position	Oscillated	Before and after the handle	Uneven with specific heavy area of wear
	Rotated both ways	Both before and after the handle	Uneven with specific heavy area of wear
	Clockwise rotation	Before the handle	Uneven with specific heavy area of wear
	Anticlockwise rotation	After the handle	Uneven with specific heavy area of wear

The final aspect of primary examined was the prevalence and type of decoration or personalisation visible on the querns, both lava and non-lava. Although an uncommon feature, the presence of such added details have yielded fascinating insight into the relationship between people and their milling tools. In Collingwood and Wright (1992) 'The Roman Inscriptions of Britain, Vol 2, Fascicule 4', a total of twenty inscribed querns are recorded, twelve of which are made of lava. These were recovered from the following locations: Binchester, Vindolanda, Holt, Carlisle, Balmuildy, Chester, Carvoran, Castleford and Ebchester; all of which housed forts, except Holt, which served as an industrial area where there is strong evidence for a military presence. These inscriptions have been interpreted as being military in nature due to the context of their recovery. They tend to be read as cohort or century numbers and names, which most identifiably are. Others have been subject to some level of assumption as to their meaning. There is one example, from Carvoran, that appears to be the end of a name in the singular genitive case, and which may relate to personal ownership (Collingwood & Wright, 1992, p. 98).

In summary, the following variables have been examined wherever possible:

- Identify the method of use where lava querns occur in pairs using use-wear analysis. For individual top and bottom stones, explore the possibilities of the range of movements that they may have been subjected to by examining wear traces before or after the handle slot/s.
- Compare the measurements for upper and lower stone thickness between different site types, locations, and chronologies.
- Investigate lava quern inscription- the type of inscription and message being conveyed, the placement location of the text on the object and the social context, in terms of site type, in which the quern was deposited.

4.3.4. Re-use/Modification

There are a variety of different ways that querns can be reused and some correlate with the final stages of the quern lifecycle, deposition, and disposal. For example, if a quern is incorporated into a structure, flooring or into earthworks, this can be classified as belonging to both the reuse and final deposition stages. For the purposes of analysis and interpretation, these forms of recycling have been investigated within the discussion on deposition, though the blurring between stages of the biography is recognised here. Instead, reuse has focused on the material signs on querns or millstones themselves for evidence of how they may have been reused. For example, differential wear patterns that indicate that a stone was utilised for sharpening or processing activities different to that of its primary function as a milling tool. This could constitute a purely economic interaction between person and object, with the form of reuse determined by the material properties of the lava. Such reuse may also refer to the socio-cultural meanings associated with the original object or material and its utilitarian function. However, it is more difficult to determine other forms of reuse that may have had more symbolic or non-tangible roles that leave no physical traces. The exploration of reuse is, therefore, limited to that which is identifiable from the evidence present in the archaeological record, though other uses should not be ruled out as having possibly existed.

Therefore, it was necessary to examine wear traces other than those associated with primary use to see if signs of secondary wear could be isolated from these. As primary use tends to make use of relatively consistent movements to provide the grinding process necessary for quern function, it is possible to identify secondary use wear. Wear on surfaces other than the grinding surface are also indicative of alternative use, though these may have occurred while the original function of the quern was being performed, or once the object had ceased to function as a quern. The properties of lava make it an excellent stone type for grinding, processing, or sharpening activities and many wear traces can be attributed to this type of occupation. However, it is difficult to determine precisely what type of pursuit was being undertaken to produce such an effect as any one of these might produce very similar kinds of wear. Nonetheless, if querns are often reused in such a way, it can reveal details relating to the kinds of activities that were taking place at the site and the extent to which they were carried out. An industrial site or area would probably have a much higher level of reuse for this type of endeavour, though this would also be the case for a region that was not well-supplied with local stone suited for such purposes. For example, at Silchester, where there is a deficit of natural stone sources and considerable industrial activity, there is a high level of stone reuse for sharpening and processing; more specifically, roof tiles of Brownstone or Pennant Sandstone (Allen, 2018, p. 156–158).

Modification, which has been included within the same stage of the object biography as reuse, can occur at any point during the quern or millstone lifecycle, and does not have to conform to any specific sequence. Modification here is defined as any alteration to the design, form or specification of the originally manufactured quern or millstone. Due to the high levels of standardisation for lava querns, it is relatively clear where modification involved alteration to the stone. For millstones, this may be less obvious as there is no one standardised type with which to make comparisons. As there is currently no typology for the British lava querns (see Chapter 2), care was taken not to confuse different typological characteristics with features related to modification. It has also been possible to identify a sequence of changes to a quern stone, and isolate changes that must have developed post-manufacture. For example, if the kerb has been cut through to attach a different type of fitting, the fitting must have occurred after the addition of the kerb.

Processes and interactions that involve modification have been considered in terms of the interactions that may have occurred between people and the quern/millstone. Was the modification added to improve the primary use of the milling tool, or to change it completely? Was the aim to extend the use-life associated with the original function? Modification can completely alter the human-object relationship through investment in materials, time, and skill, or through changing function, value or meaning. These ideas have been discussed in relation to the physical evidence observed.

To summarise, the following aspects were explored in terms of reuse/remodification:

- Traces of wear aside from that on the grinding surface.
- Location and extent of alternate signs of wear- can a specific type of repurposing be identified or speculated?
- Level of reuse in consideration of local geology, location, and social context (military, urban, rural etc.)
- Identification of possible signs of modification
- Discussion of the possible reasons for modification according to socio-economic context.

4.3.5. Deposition/Disposal

Two factors were considered in relation to the final disposal or deposition of lava querns as part of this study; where they were finally placed (findspot) and the condition of the object when this process occurred. Location of deposition can be investigated on multiple scales at an intra-site level; there is the immediate context of the find, such as within a pit, ditch, or wall; the placement within that feature, such as the top or the bottom; and the social context of that feature, such as domestic, ritual or military. There may also be particular types of find associated with the quern; if recovered from a pit, for example, with articulated animal skeletons or complete pots, which might suggest ritual deposition as a possible interpretation. Unfortunately, the precise placement of an object within a feature is rarely recorded. It is, however, sometimes possible to extract this information if the sequence of context data, such as a Harris matrix, is available within the excavation report, and if multiple contexts were recorded for a feature allowing for there to be differentiation between top, middle and bottom. Similarly, consistent recording in terms of social context is not always present, as there is a degree of complexity involved in forming such interpretations. Furthermore, re-deposition and relocation of objects should always be a consideration, both with regards to the physical location of the find as well as its social context. Nonetheless, where possible location and social context have been explored.

Reuse of milling tools in construction has been included within the stage categorised as deposition. This includes deposition within walls and foundations, incorporated into earthworks or as paving and flooring, both internal and external. This type of reuse may simply have been the recycling of good quality and readily available stone, as is also seen with other stone types during the Roman era. Whether or not there was a socio-cultural dimension to this particular method of reuse is difficult to determine, though its prevalence and the ways in which stone is incorporated might introduce the probabilities of this having been the case, as has been identified in studies of stone reuse in other periods (Eaton, 2000). Again, context information has provided some indication of whether this occurred.

The condition of querns when they were deposited or disposed of is also important and can be related to considerations of ‘value’. As querns are very unlikely to have been subject to accidental loss, disposal in most cases would have been a deliberate action. The extent of wear can, therefore, reveal perceived value attached to the quern; heavy wear prior to disposal might show great significance attached to its primary role as a grinding tool. This significance can relate to both economic and social value and can help us explore possible supply systems and tool choices made by those who specifically used lava querns to complete milling activities. It is worthwhile to consider whether querns of non-lava types had a similar degree of wear when they were disposed of and what this might tell us about different meanings and values attached to these objects.

Overall, the categories that were recorded for the benefit of examining deposition have included the following:

- Type of feature quern recovered from- pit, ditch, structural etc.
- Social context from where quern was recovered. This can be examined using deposition context- type of structure or role of associated feature- e.g., boundary, domestic, religious etc.
- Extent of wear of quern on the grinding surface.
- Percentage of quern present.
- Whether any quern pieces refit with others from the same context/area/site.

5

Making Flour the German Way: Analysis of Lava Quern and Millstone Use Across the Province

Prior to examining the lava quern and millstone data from rural, urban, and military sites, this chapter explores the full dataset to identify province-wide patterns. This can be achieved using the same approach that applied in the thematic chapters that follow, by exploring object biography (see Chapter 4). However, as this has been done using the data produced for all sites, individual objects and individual object biographies are not the focus of this discussion. The aim is to produce a general picture of lava quern and millstone types, distributions, use and discard across the whole province and to scrutinise what this can tell us about the lifecycle of lava milling tools and, consequently, how this relates to life in Roman Britain. One of the current issues relating to the ways that lava milling tools are dealt with in excavation reports is that interpretations relating to quern and millstone use are rarely correlated with data from other sites, which limits our view of lava use on a local, regional, or provincial scale. The data from an individual site needs to be put into context with what is happening elsewhere if we are to identify socio-cultural meaning or economic significance, as this can highlight similarities and differences between varying site types and geographical locations. This chapter does not only provide a means of comparison for the thematically generated datasets, but it also provides the data that will allow future assemblages to be subjected to similar comparisons to produce more nuanced interpretations of lava milling tool use in the province.

5.1. The Dataset

The full dataset consists of 2,707 lava milling tools from 564 sites in Roman Britain. A total of 1192 objects were identified as quern stones; 299 of which were lower or probable lower stones, 551 were upper or probable upper stones and 342 unidentified. Millstone numbers were significantly lower at 79; upper or probable upper stones totalled 35, with 16 lower stones, and 28 unidentified (figure 5.1 and table 5.1). Added to this, 601 sites were recorded where milling tools were present, but none of lava, making a total of 1,165 sites in the dataset.

Data was collated from a variety of sources:

- In-person data collection using archived material was conducted for the bulk of the case studies, including 18 sites: Chichester and 11 sites in its hinterland, Verulamium, Vindolanda, Corbridge, Housesteads, Binchester and Chesters. The querns recorded for these sites comprise 301 lava and 127 non-lava examples.
- Excavation monographs, county journals, and 'grey literature' reports from developer funded excavations, to include 910 sites. Much of this data was compiled with the help of the Rural Settlement of Roman Britain database.
- Data extracted from Ruth Shaffrey's personal database collated over her career in commercial and research archaeology. Of the full dataset, 232 sites were from this database.

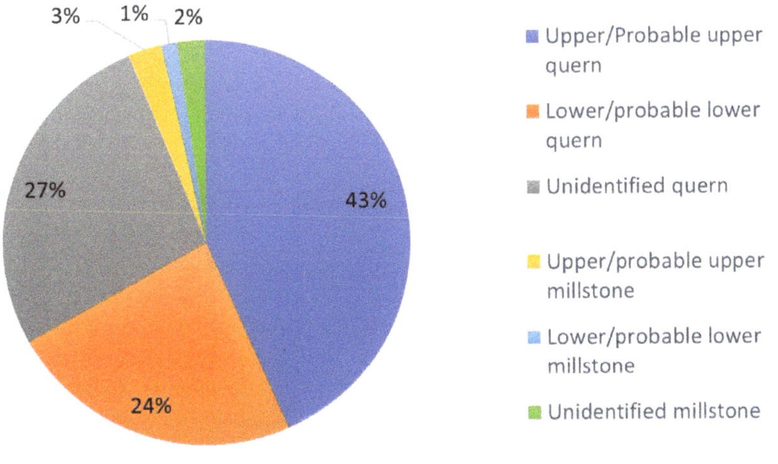

Figure 5.1. Chart showing the breakdown of querns, millstones, and unidentified lava in the assemblage according to whether they are upper or lower stones (T= 1271).

Table 5.1. Data for figure 5.1, showing the breakdown of querns, millstones and unidentified lava in the assemblage according to whether they are upper or lower stones

Type of stone	No Examples
Upper/Probable upper quern	552
Lower/probable lower quern	299
Unidentified quern	342
Upper/probable upper millstone	34
Lower/probable lower millstone	16
Unidentified millstone	28
Total	**2171**

- A small volume of data was taken from online collections from museum archives with detailed records and images of their querns. Five sites were recorded this way.
- Data was recorded, stored, and analysed using an Access database whereby all lava milling tools were recorded using unique catalogue numbers. These have been used as a way of cross-referencing the dataset.

As with all large datasets, the way that this dataset has been compiled must be understood prior to making interpretations using it. One of the most significant issues relates to the regional density of information collected. The regions shown here are arbitrary and are used here to highlight the potential issues in the data. As a large part of the data was compiled using Ruth Shaffrey's database, milling tools from the south-east of England have been extensively recorded as this is the region where she predominantly works. This detail is not replicated in other parts of Britain and the sparsity of querns in other regions (figure 5.2) cannot be taken to signify a lesser presence of milling tools overall. Further to this, as developer funded excavation reports were used to form the bulk of the dataset, areas of more intensive construction and development are more heavily represented in the data, which includes much of the south of Britain and London. Access to excavation data has also created some challenges in terms of regional representation. As much of the data collection took place during Covid lockdowns, some less accessible county journals have not been checked, which has led to sparse data in those counties. Similarly, stone type for querns is not always recorded and lava is not always quantified. It was found that the method of recording and the detail presented often related to the commercial unit completing the work, which created further regional bias in terms of where usable data existed. This has been mitigated as much as possible by recording sites with milling tools, but none of lava, as this highlights where gaps in the data exist.

The dominance of rural sites in the dataset is a similar issue when considering the way that data was collated. Although rural sites are prevalent across Roman Britain, these have received more attention in the data collection phases of this research than urban or military sites. This is simply due to the availability of data during Covid lockdowns and the online accessibility of rural data due to the work of the Rural Settlement of Roman Britain project. At a time when museum, archive and even library visits were not possible, data collection was able to proceed for rural sites, but this has created a bias to the dataset. Of the 1165 sites recorded, 942 were identified as rural sites, 78 as urban, 73 as military, 6 as military/urban, 3 as other and 63 as unknown (figure 5.3 and table 5.2). Even post-lockdown, urban and military sites have not been examined in the same level of detail as the rural sites in the database. For example, it was not possible to get timely access to the collections at Colchester and London, where the most prevalent use of lava is likely to have occurred. These are significant locations that should form the focus of any future work into the urban use of lava milling tools.

The final potential problem in creating the dataset has involved the quantification of lava, which is rarely conducted systematically. It is likely that some error has been introduced to the data and that this is skewed to show

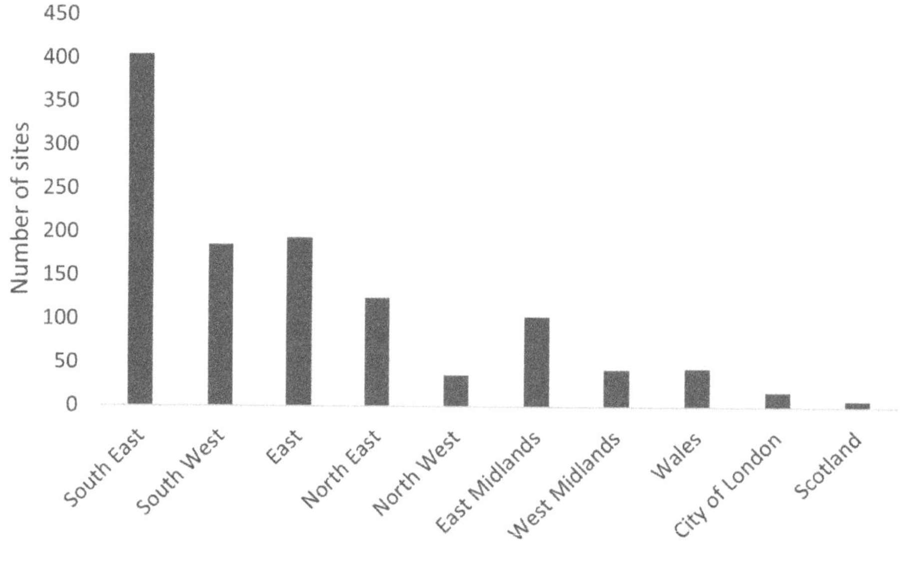

Figure 5.2. Chart showing the number of sites recorded within each region in Britain (T=1165).

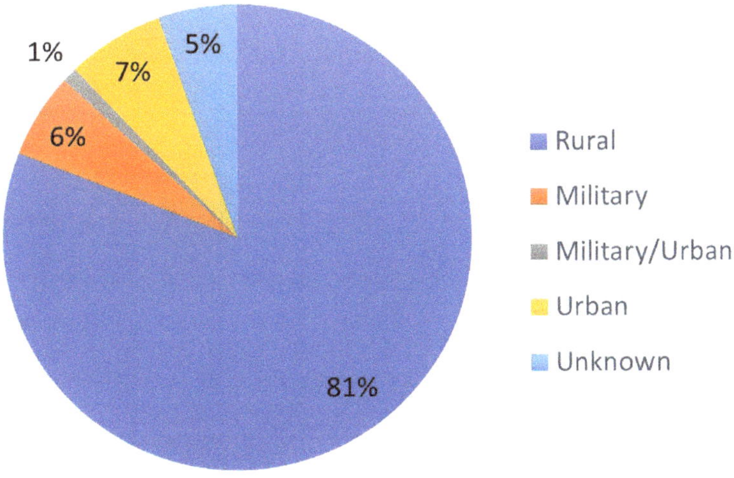

Figure 5.3. Chart showing the breakdown of the site types included within the dataset (T= 1167).

Table 5.2. Data for figure 5.3, the breakdown of the site types included within the dataset

Site type	No included in dataset
Rural	942
Military	69
Military/Urban	12
Urban	78
Other	3
Unknown	63
Total	**1167**

a greater prevalence of lava at sites where fragmentation was common, or reuse that involved spreading stone across multiple contexts. More detail pertaining to this issue has been discussed in the methodology (Chapter 4), but it is necessary to mention here that the highest incidence of fragmentation and reuse appear to have occurred in the east and south-eastern parts of Britain. Bearing all these considerations in mind, the next section examines the data in five steps: manufacture, distribution, primary use/reuse, and deposition.

5.2. Manufacture

Elements of manufacture that can be explored for such a large dataset must be those that are regularly captured in excavation reports. Although some inaccuracies and inconsistencies must be expected when comparing data that has been gathered from numerous sources, the volume of data involved should mitigate much of this. Features that have been included that relate to manufacture include:

- Diameter of the stone,
- Width of kerb,
- Handle types,
- Rynd chase types.

It has not been possible to investigate all the common features of a lava quern, such as central hole shape and size. This is because there are many typical characteristics of a quern that can be removed due to wear, reuse or from mending and this has been observed with several examples that are likely to have begun their lives as standardised lava querns. Good examples of this can be found in the Vindolanda assemblage (Chapter 8). These altered features do not relate to actual typological characteristics or the outputs of the manufacture process. Care has been taken, therefore, to select common features that are less likely to have been altered and that can be compared easily.

5.2.1. Diameter of Stone

The diameter of the stone was the most common feature of a lava milling tool to be recorded across the dataset, with 607 objects complete enough for this measurement to be taken. The smallest was a miniature quern from the A14 road scheme excavations in Cambridgeshire measuring 210mm (CAT 1483), while the largest was a millstone measuring 910mm from Syndale Bottom in Kent (CAT 0856). The range of sizes represented is, therefore, broad. However, most of the data show clear standardisation and similar sizes. The few examples that deviate from this standardisation stand out clearly from the bulk of the data (figure 5.4). At the upper end of the scale, this occurs at 500mm, which provides the clearest point where a division can be drawn between hand operated querns and mechanically driven millstones. There is more ambiguity for those milling tools with diameters below 500mm that are larger than the standard quern. These can only be defined as millstones or quern stones if evidence relating to the mode of operation exists on the stone (Shaffrey, 2015, p. 73–78).

If probable millstones are removed from the chart, we can leave known quern stones and the more ambiguous milling tools that have not been identified as either a quern or millstone (figure 5.5). Again, this emphasises the extent of standardisation that is present in lava querns from Romano-British contexts, with 50% of all querns and unidentified milling tools falling into a diameter range of

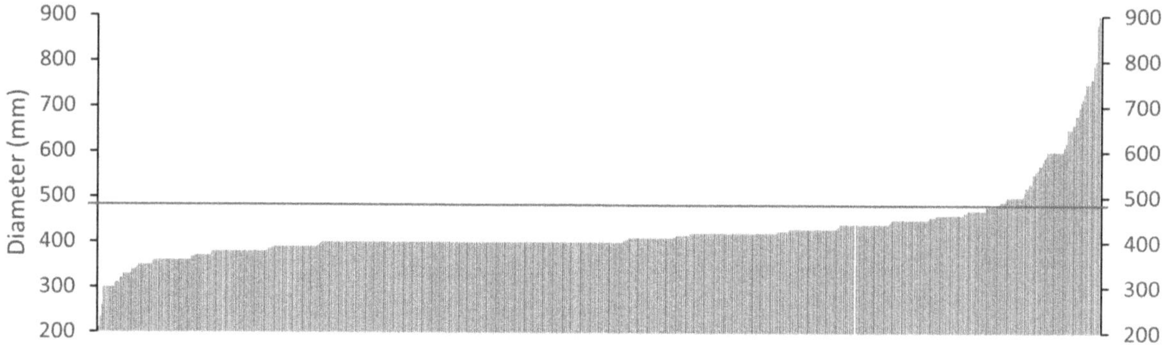

Figure 5.4. Chart showing the diameters for all lava milling tools for which this measurement has been taken, with a line marking where the difference between a quern and a millstone is most clear. (T= 607).

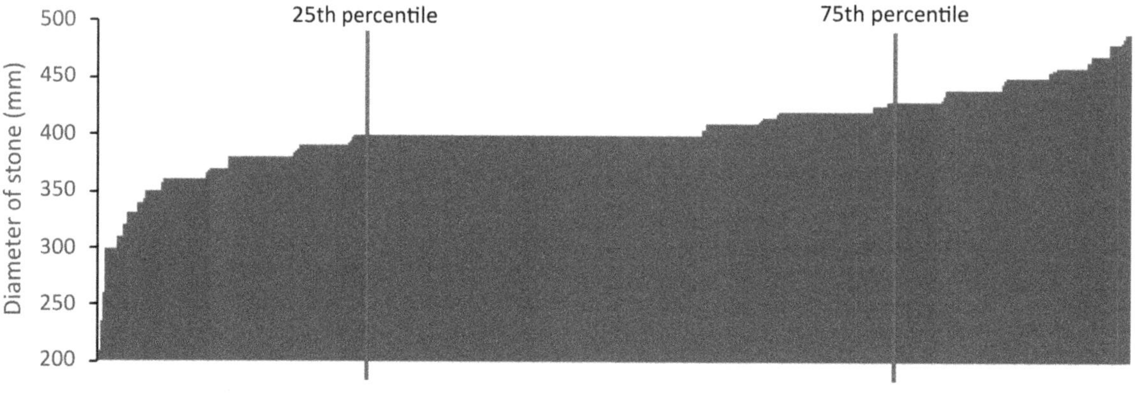

Figure 5.5. Chart showing the diameters of all lava querns and unidentified milling tools from Britain. The 25th and 75th percentiles have been indicated with lines to demonstrate the extent of standardisation in the dataset (T= 553).

400–420mm, shown within the lines on the chart. Any lava milling tools that fall within this range can confidently be referred to as a quern. Those in the upper and lower 25th percentiles for diameter sizes and appearing to the right of the rightmost line and to the left of the leftmost line, show less conformity to this standard. Though some room must be given for inaccurate and estimated measurements, it is likely that those in the more extreme ends of these groups were specialist tools. Some of the larger examples may have been smaller millstones or larger sized querns, made to order for specific requirements. The possibility for some of these to be lava milling tools not produced in the Mayen region must also be considered. Miniature querns or millstones (see Chapter 7) may have also been manufactured specially or created from waste material in Britain to fulfil requirements that were not met by the quern producers in Mayen, or by other lava milling tool production areas that may have been supplying Britain.

5.2.2. Width of Kerb

The raised annular kerb that is found on the top surface of upper lava querns and some lava millstones manufactured at the Mayen quarries (see Chapter 2) is a feature that is often remarked upon and measured in excavation reports. With the high level of standardisation for lava querns, it would be expected to be relatively consistent in its dimensions. However, observations have shown that the kerb is sometimes very prominent, and sometimes almost indistinguishable, with a variable width (figure 5.6). The role of the kerb appears to have been to strengthen the sides of the stone where the elbow handle socket was fitted. By this logic, one would expect for there to be a correlation between the width of the kerb and the size of the stone, but no such relationship can be discerned from the data (figure 5.7).

Any relationship between the handle fitting and the kerb width is difficult to identify as evidence for handles is often lacking. From the typology by Röder (1955), kerbs were added to querns manufactured at Mayen throughout the Roman era despite changes in the way that the handle was fitted. Later lava quern types did not actually require the extra strength that a kerb would have offered due to these changes, but kerbs continued to persist as a feature (Chapter 2). It is possible that as the related purpose of the kerb ceased, its presence became more of a suggestion than having any functional value. In the British assemblage, only elbow socket handle fittings have been observed, and this has led to the assumption that this may have been the only type imported into Britain. However, later types with other handle fittings may also have entered the province, which could explain the variation seen in the kerb data and the observed differences in how pronounced kerbs

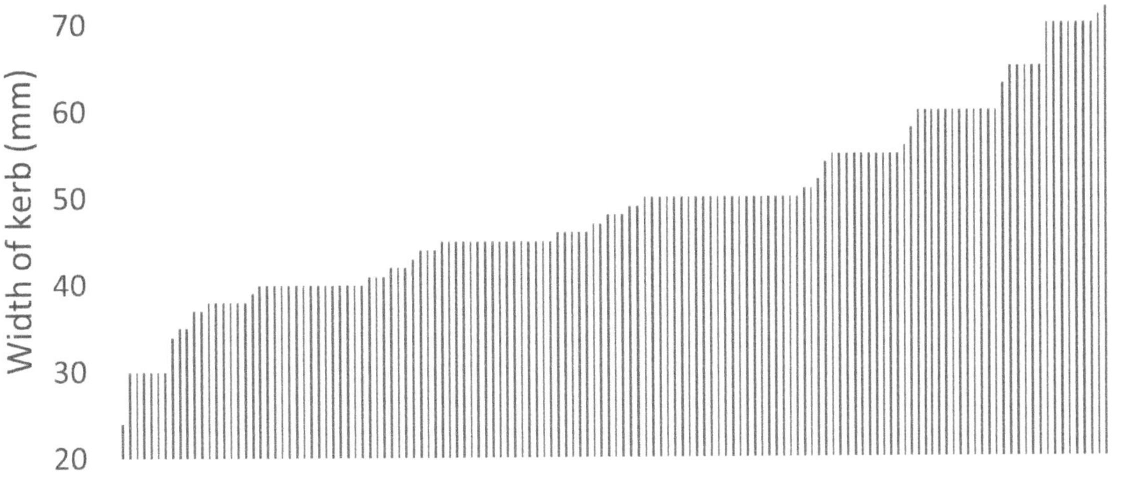

Figure 5.6. Chart showing the variation of kerb widths for all lava milling tools from Roman Britain (T= 137).

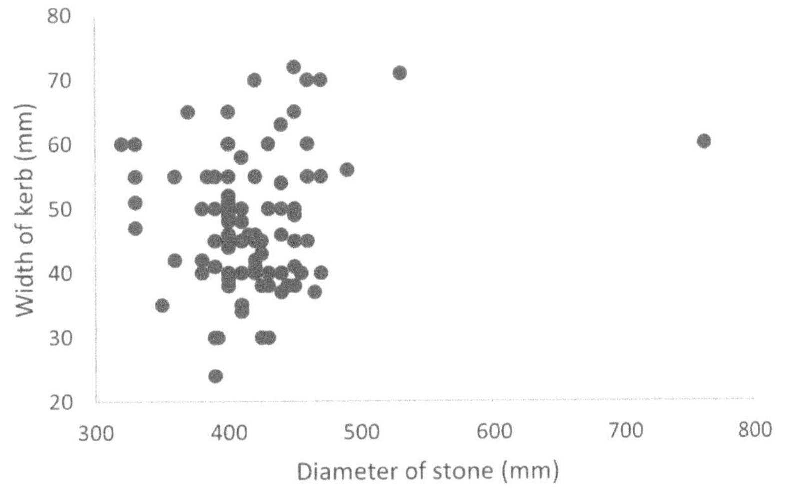

Figure 5.7. Chart showing the relationship between lava milling tool diameter and kerb width for the British assemblage (T= 103).

appear. Preliminary examination of the data does not support a chronological change in kerb width (figure 5.8), but chronological information for querns is rarely detailed enough for this type of investigation. It is worth noting that the number of querns that displayed the presence of a kerb does appear to have decreased over time. This may be an issue of preservation as late Roman querns tend to be more fragmented, yet it is a pattern that should be explored further.

Another possibility might relate to the quarry of manufacture, as there were three sources of lava that were exploited to produce Mayen querns. Each of these quarries may have produced querns that, though generally standardised, had slight differences. Any correlation would have to be investigated through systematic geochemical analysis and comparison with different features alongside consideration of chronological change. Future measurements of the kerb should be completed as part of standard quern recording practice to help generate a more refined typology of Romano-British lava quern stones.

5.2.3. Handle Fittings

Handle fittings only exist on milling tools that were manufactured or adjusted to be operated by hand. Within the dataset they are rare as they form the most vulnerable part of the stone where breakage is most likely to occur. Breakage across the handle socket makes it less visible and harder to define, meaning that this feature is less likely to be identified. Sites that had good levels of lava preservation were more likely to reveal handle features and this is likely to have skewed the data as conditions at the northern military sites, especially those at Hadrian's Wall have yielded many well-preserved lava querns that can reveal handle fittings. Sites in the south, especially in and around Kent and Essex, generally produced only poorly preserved fragmented lava. This might relate to a higher incidence of recycling for the lead or iron used for the handle fittings, as fragmentation would provide one way of accessing these fittings. In total, 100 lava querns were recorded that had evidence for a handle fitting, and these were derived from only 33 sites. In particular, Vindolanda yielded 14

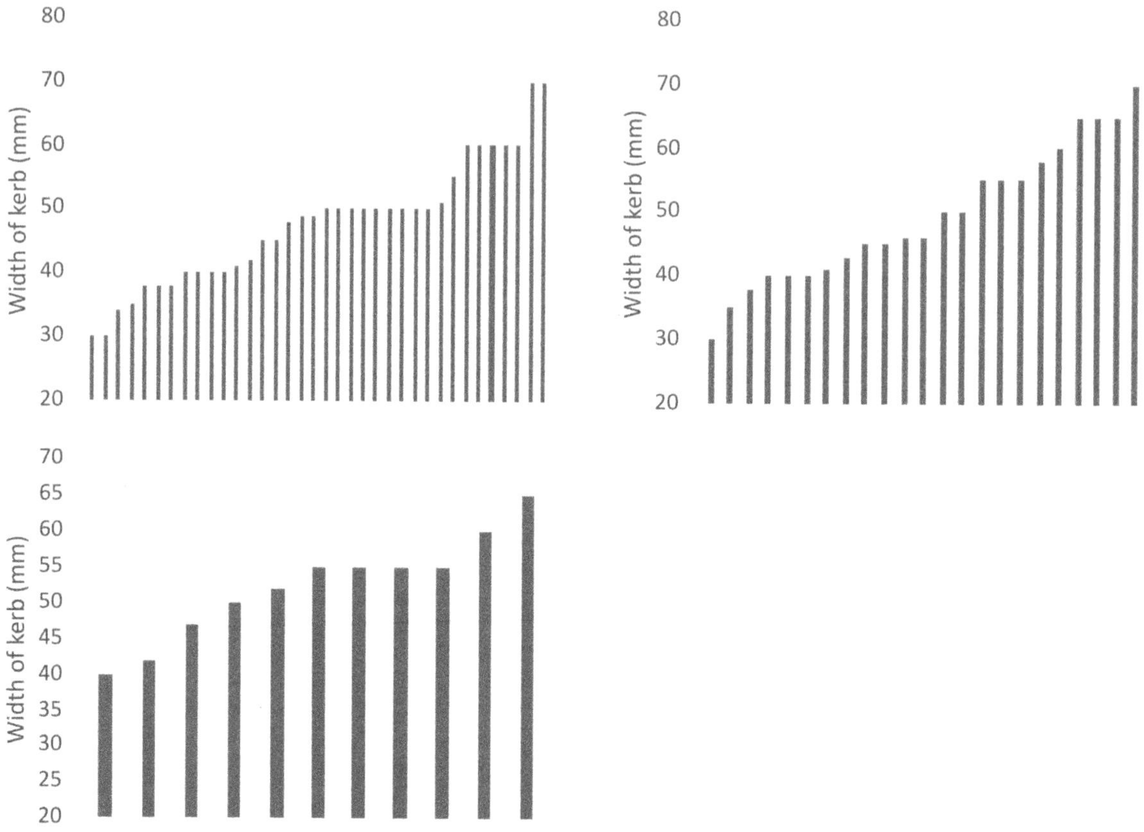

Figure 5.8. Top left, kerb width from lava milling tools of early Roman date (T= 36). Top right, kerb width from lava milling tools of mid-Roman date (T= 23). Bottom left, kerb widths for lava milling tools of late Roman date (T= 11).

examples and the site at 1 Poultry, London produced 34. This presents some obvious problems, especially with regards to the Poultry querns, which are likely to have all been discarded within a brief period of time. There is, therefore, much less variety in the sources of the data than is desirable and this must be considered when interpreting the results.

Overall, elbow type handle fittings were the most dominant type, with very few examples of other varieties (figure 5.9). One possible other type is that of an 'iron band,' which was observed on two lava querns (CAT 2284, 0032), and were probably employed on three others that lack any evidence for a handle fitting, despite being complete stones and showing signs of wear (CAT 0002, 2175, 2179). A further example from Usk (CAT 2347) is described as having traces of iron residue in a band around the circumference of the stone, which may also suggest such a fitting (Welfare & Campbell, 1995). Unfortunately, there is no way of proving that an iron band was used to fix a handle in all cases other than those with the iron band remaining *in-situ* (figure 5.10), but their use seems likely. A reconstructed non-lava quern in the Clayton Collection at Chesters Roman Fort has been put together to show

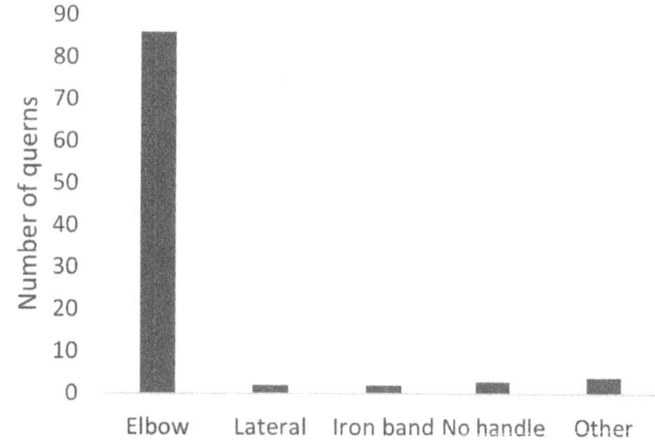

Figure 5.9. Presentation of different handle types in the whole British lava assemblage (T= 100).

Figure 5.10. Fragments of CAT 0032 showing the pieces of iron band adhered to the outer circumference (photo by the author ©Vindolanda Trust).

this format of handle fitting, though how the decision was made to present it in this form should be thoroughly explored to examine the extent to which this reproduction is based on evidence. An iron band was also found adhered to the circumferences of a lava millstone from the Hornet 1994–6 excavations in Chichester (CAT 2511), yet this example cannot be defined as necessarily Roman in date. It is possible to argue that the presence of an iron band is not related to mending the stone, which could provide one interpretation. The absence of other handle fittings on the complete querns in this group suggest that the iron band fitting was intended at the point of manufacture.

The two lateral handles were both present on querns retrieved from 1 Poultry (CAT 1919 and 1926). The rest of the large assemblage from this site only had elbow handle fittings, and the lateral handle querns can be dated to the same phase of deposition as these. From this, there is a suggestion that distinct types were in circulation at the same time, which would help to explain the variation seen in other features, such as the kerbs, which do not currently seem to correlate with chronological change. The strangest examples from across the assemblage include three complete upper stones that lacked any sign of a handle at all, with no evidence for how they may have been rotated (CAT 0002, 2179, 2300), all three of which were recovered from northern military sites. These are discussed in more detail in Chapter 8, but it should be noted here that all three showed traces of wear. 'Other' handle types comprise objects where a handle is mentioned in the literature, but is not described or illustrated (CAT 2417, 0414, 0101).

5.2.4. Rynd Chase Types

As with other mechanical features of a lava quern or millstone, the rynd chase (see Chapter 2) tends to be worn away or broken to the extent that it becomes unidentifiable. Even if evidence of one is present, the grinding surface tends to have been worn right through to the rynd chase, making it impossible to determine if the rynd was fitted in an over or underdrift fashion (figure 5.11).

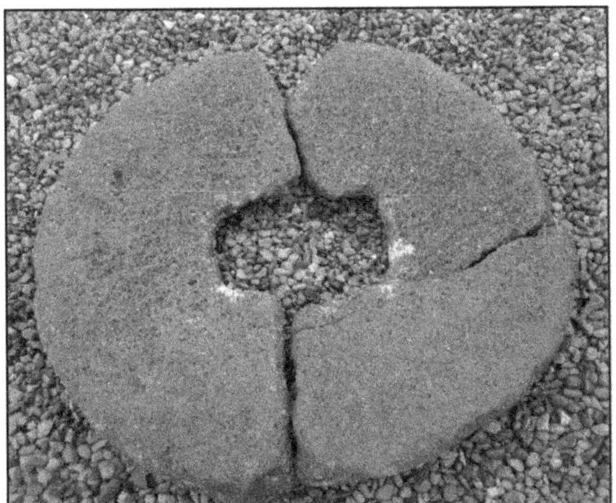

Figure 5.11. Upper quern stone from Vindolanda showing a worn rynd chase that cannot be identified as having either an under or overdrift fitting (photo by the author ©Vindolanda Trust) (CAT 0030).

In addition, particularly in comparison with the data collected for handle sockets, descriptions for the rynd chase and images of this feature are not regularly captured in excavation report data. Consequently, only 45 examples are present in the dataset, most of which were found to be overdrift types, with the iron bar recessed into the top surface of the upper stone. Three occasions were recorded where underdrift rynds were fitted, one of which was identified as a millstone because of this feature, despite having no estimated diameter (Shaffrey, 2015, p. 67–70) (CAT 0940). The other two underdrift querns have been identified as miniature millstones due to their small sizes and the presence of features that ordinarily correspond to millstones (CAT 1483, 2679, see Chapter 7). As evidence from Mayen shows that lava querns were manufactured as overdrift types (see Chapter 2), it seems highly probable that the bulk of the dataset comprises this type.

5.3. Distribution

Distribution was examined spatially, chronologically, and with regard to site type to explore social distribution. The results of the analysis are presented below.

5.3.1. Overall Spatial Distribution

Distribution of lava across the whole province can be seen in figure 5.12, which also shows the distribution of non-lava querns at sites that do not have lava present. It should be noted, the sites that did have lava present often also had non-lava querns, but these are not presented on these distribution maps. This is because lava is the focus of the study and the non-lava distributions are simply to demonstrate which sites and regions have been checked. This way, it is possible to note where gaps in the data exist and where a genuine lava absence can be suggested. The number of non-lava querns at sites with lava has been partially captured in the database produced as part of this study. This can be used to produce more complex distributions of lava in comparison to that of non-lava for future research, but as an incomplete dataset, has not been presented here. Distributions in Scotland have also not been presented since data from this region was less accessible during the Covid lockdowns and the few sites identified had the potential to provide a misleading view of lava distributions. Some quern data in Scotland was captured, but the few examples recorded were not positively identified as lava within the excavation reports. Descriptions of the quern fabric are likely to relate to Mayen lava, but this could not be confirmed without in-person examination. Further work is needed to explore Roman period lava distributions and that of non-lava milling tools in this region.

From the distribution, it is possible to see that there are some strong regional and probable site type associations to the concentrations of lava. Lava is most common in the south-east and eastern parts of Britain, with the largest clusters centred around London and Colchester. Focal areas west of these areas include the Thames Valley and the main urban centres, such as Silchester, Cirencester and Gloucester, with decreasing volumes away from these. Other concentrations are observable around Hadrian's

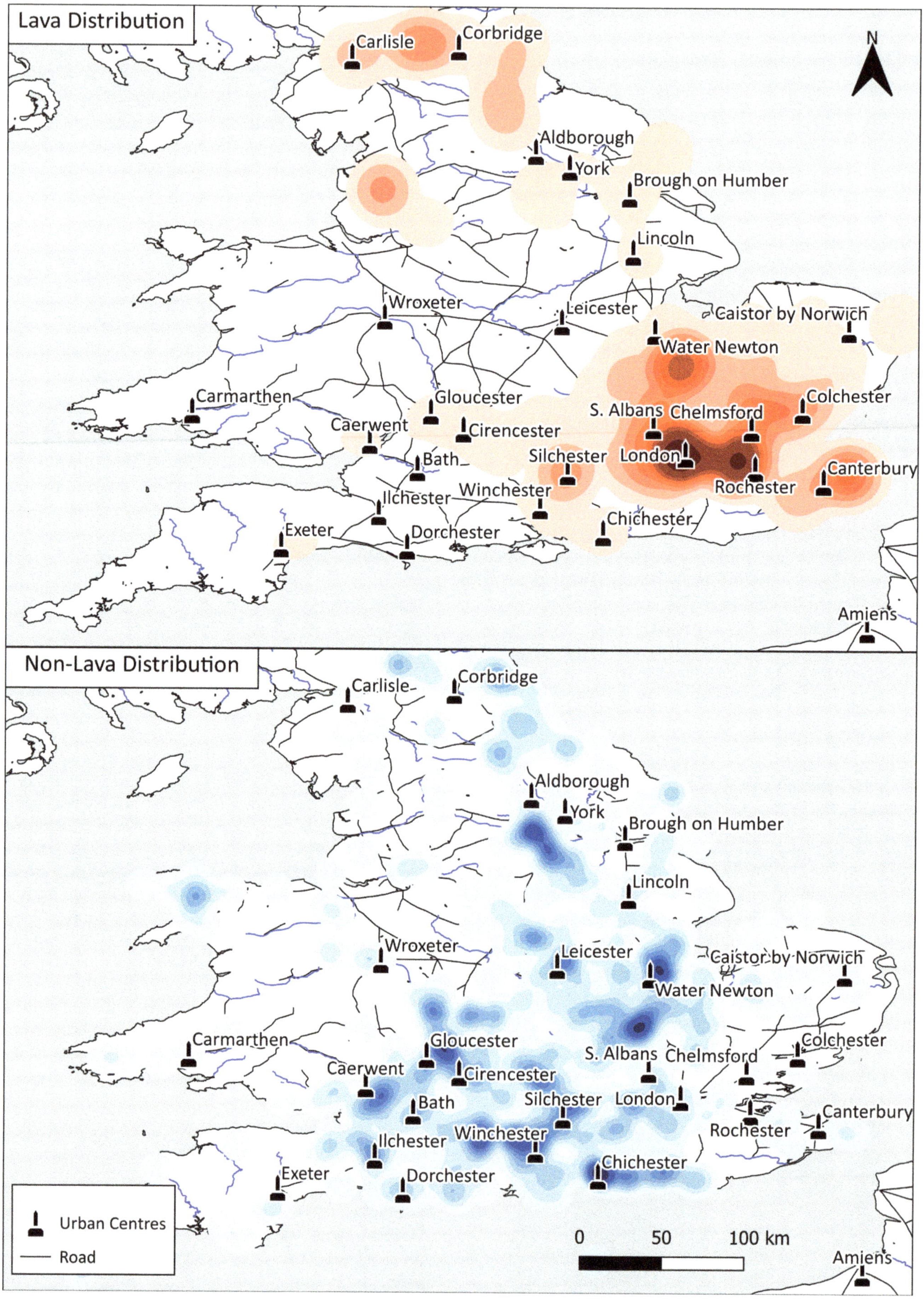

Figure 5.12. Distributions of lava (top) and sites with milling tools but none of lava (bottom) based on the whole dataset (Background GIS data courtesy of the Ancient World Mapping Center).

Wall, with some continuation along the line of Dere Street, which is more likely to correlate with the intensity of military activity in these areas. Similarly, in Wales, smaller clusters are visible in the south of the country where fort sites are prevalent. This pattern has been investigated more fully in Chapter 8.

First impressions indicate that non-lava milling tools have a much wider distribution. This is expected, as milling tools were highly significant objects that were used in everyday life and would have played a vital role in food preparation activities. This certainly is the case in certain parts of the province, such as the south-west. However, scrutiny hints at a more complex picture. One example includes the region around Hadrian's Wall and Dere Street where clusters of both lava and non-lava can be observed. The centre of these distributions is marginally different, with lava centred on the fort locations and non-lava in the hinterland areas around the forts. This is an important indicator that lava was being consumed on military sites but was not necessarily reaching the surrounding rural areas that were only using indigenous stone.

Similar patterns can be seen around Silchester, Gloucester and Cirencester, though the density of indigenous stone is much higher and 'outcompetes' the lava in terms of how far it extends. As with the military sites, lava is most concentrated in the urban centres within these areas, with a mixture of lithologies used in the urban hinterland regions. This is interesting as it suggests that urban sites were the focal areas for lava distribution, which allowed for lava to filter into the surrounding rural areas, where take-up did not always occur.

5.3.2. Distribution by Site Type

The addition of the urban centres and fort locations on the map showing spatial distribution has given some indication that the site type may have been a key factor in how and why lava milling tools circulated in some areas and not in others. Therefore, a separate map showing the distribution by site type of sites with lava and only non-lava milling tools has been produced, with a specific emphasis on whether the sites can be classed as military, military/urban, rural, or urban. The site designation has been identified for each location where milling tools were recorded, alongside the number of objects (figure 5.13).

Initial observations suggest that lava was distributed across a variety of different site types, while a more constrained social distribution occurs for sites with milling tools of indigenous stone types only, with rural sites prevailing over other site types. Though non-lava sites show some distribution at urban and military locations, these are predominantly on the western side of Britain and only small numbers of milling tools are recorded at these places. However, as data is sparse in these areas, it is difficult to know how representative this might be of quern use in general. A much greater volume and density of lava exists at military sites in the north and in Wales, as would be expected from what we understand about the role of lava milling tools as part of military equipment. As was suggested in the spatial distribution, rural sites in the region of military sites in the north and at Hadrian's Wall were highly likely to have milling tools, but none of lava. Redistribution of lava via military sites did not occur in these regions. This may suggest that greater social differentiation may have existed in the frontier region at the time when lava milling tool use was most prevalent and that the rural hinterland of military regions did not have the same access to imported goods as the military community.

A vastly different pattern of distribution is seen to occur in the south-east, where urban sites appear to have been the vectors of redistribution. Volumes of lava are high at urban sites which are the focal areas of lava use for surrounding urban hinterland areas and rural sites. Distribution of lava may have been 'piggybacking' on military supply routes emanating from London and Colchester. This does not appear to be replicated in the south-west or along the south coast. Although lava does still occur at urban sites in these regions, volumes are much lower and do not spread out to rural sites to the same extent as seen in the south-east. This implies that the south-eastern urban centres were more involved in the redistribution of lava milling tools and that this did not extend to urban centres in a general sense. The eastern coast and its connection to the North Sea trade must have been more significant in terms of access to imported lava products than the site type.

5.3.3. Chronological Change in Distribution

The chronology of lava use in Roman Britain is an area of research that has never been examined on a provincial scale. Although observations have been made in the past regarding the initial introduction of lava milling tools to Britain, and assumptions made on the decline of this use, this chapter presents the holistic examination of the data. The maps in figure 5.14 show the changes by century in the density of lava according to location. It should be noted that some objects were dated across more than one century interval, in which case they were included in more than one map. However, objects were only selected for use in this analysis if their context date fell into a maximum of two date groups. For example, a date of AD 150–250 was included in both the second and third century maps, but a lava quern with a data from AD 150–310 was not included. This was to provide some clarity in the visualisation of chronological change, as this becomes blurred if less concise dating evidence is used. Unfortunately, many examples have wide date ranges that could not be included, which is often due to the way that milling tools are deposited. Added to this, the reuse and redeposition of querns, including lava, creates a time lag from the moment of first use and final deposition, creating a delay of 50 years or more between these dates. Added to this, it is not always possible to determine the extent to which the chronology of lava deposition depicted here relates to residual material and some care needs to be taken in the interpretation of change in deposition over time. There is no obvious way

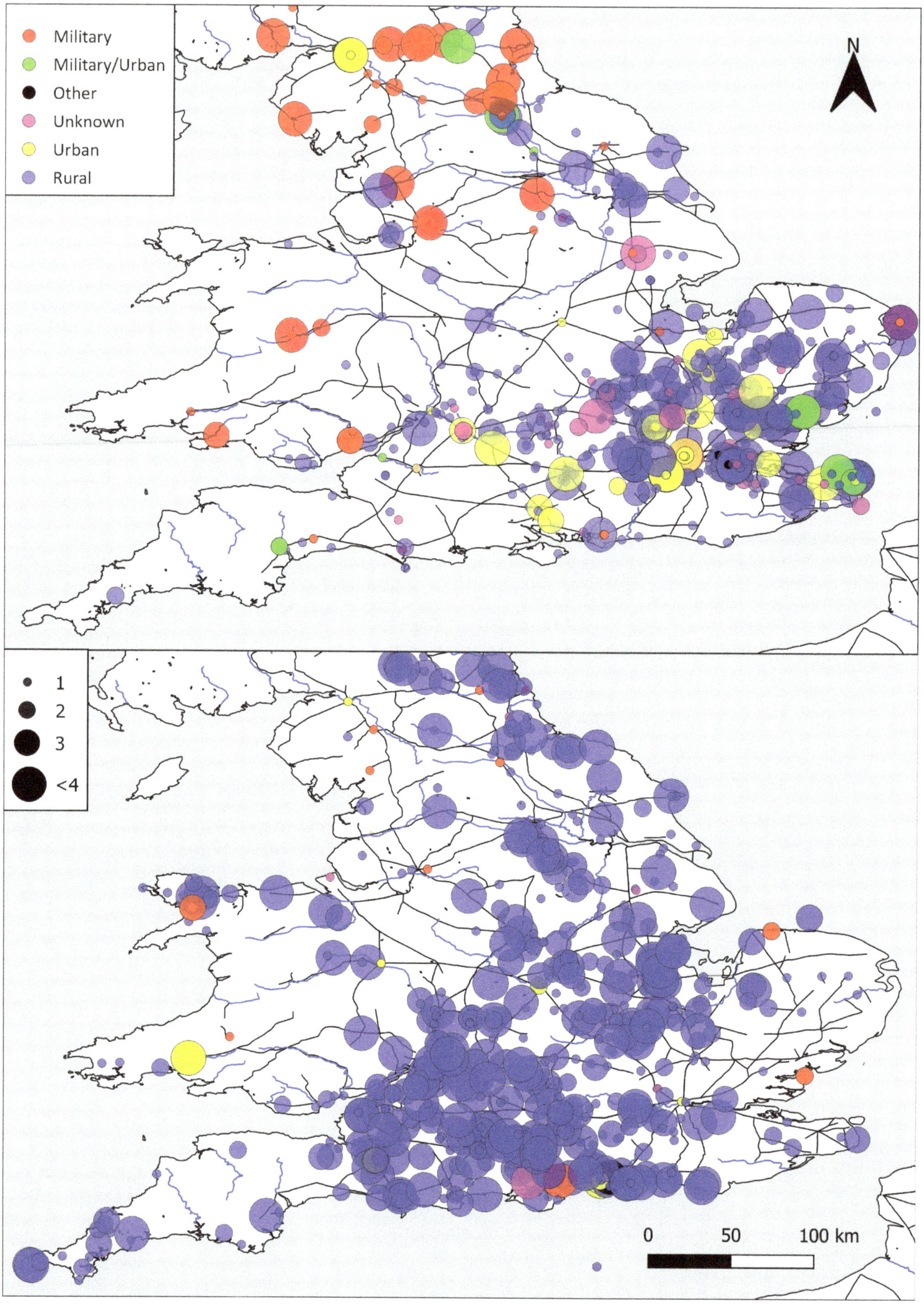

Figure 5.13. Map showing the social distribution at sites with lava (top) and only non-lava (bottom) milling tools, alongside the number of objects (Background GIS data courtesy of the Ancient World Mapping Center).

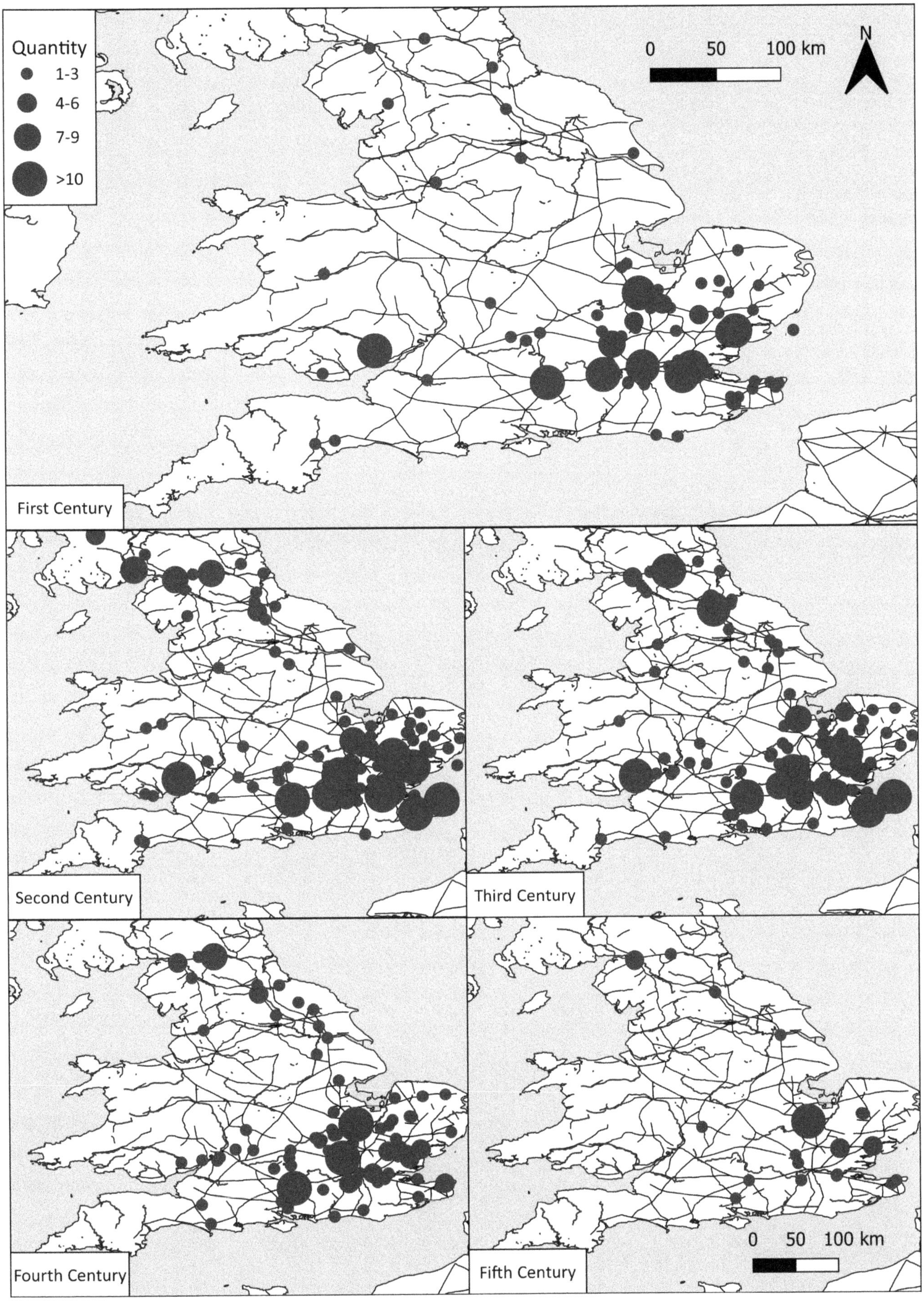

Figure 5.14. Chronological change in lava milling tool distributions from the first to fifth centuries in Roman Britain (Background GIS data courtesy of the Ancient World Mapping Center).

to define the extent to which residuality has affected the results, but a time lag is to be assumed in all cases.

The results show that lava was introduced initially mainly in the south-eastern and eastern regions, with the greatest density being around London and Colchester, as might be expected due to our knowledge of Roman activity in these areas during the conquest period. Urban centres in the south and, to a lesser extent, the north also show clusters of lava. Boudiccan fire deposits at Verulamium, Colchester and London and other southern areas affected by the Boudiccan rebellion, are also more likely to have produced lava from securely dated first-century contexts. Lava at the military sites follows what we understand of military occupation and movements in the first century, with clusters appearing in Wales and at southern military sites. This and other evidence related to lava distributions confirm a link between lava introduction into Roman Britain and the military. The first century lava distributions cannot, however, be identified as being solely military as there is no way of clearly separating the social context of use in urban centres that also had military populations present or moving through. A strong military association does, nonetheless, appear to have existed and this may have initiated civilian and other uses for these tools.

When and how such a transition may have occurred is also not clear, but rural use in the eastern region of the province does appear to have happened in the early phases of lava introduction, as seen in the first century distribution map. This may have been the result of the proximity of these sites to Colchester, where wider distribution of lava is likely to have occurred. This situation and the deficit of suitable sources of stone for querns in the east, may have been the driving factor behind this early uptake. Volumes of lava increase from the second and third centuries, when deposition was at its peak. The appearance of lava milling tools at Hadrian's Wall and around the Antonine Wall in the second century again mirror the known movements of the military and emphasise the important connection that lava had to this section of society. Expansion of use saw greater concentrations and density spreading west of London across the second century, with a slight contraction observable by the third century when volumes in Wales also appear to decrease. This contraction continues into the fourth century, resulting in no lava present in Wales during this period, with a rapid decline in the fifth, when the distribution is more related to redeposited material.

5.4. Primary Use

Measuring and comparing the primary use of a quern for a large dataset has only been possible using measurements of the stone thickness. As explained in the methodology (Chapter 3), this has been completed for upper and lower querns stones separately, with measurements taken at the edge for both stone parts to ensure consistency and the unused querns from the site at 1 Poultry have been excluded from the analysis. Charts have been produced for the whole British dataset, which has been used to compare the urban, rural, and military case study datasets in the following thematic chapters. However, here analysis has also been undertaken on all military, urban and rural sites collectively, as well as looking at regional and chronological differences in extent of use.

5.4.1. Stone Thickness for the Whole Dataset

Upper and lower stone thickness for the whole dataset are shown in figure 5.15 and 5.16 respectively, with the 25th and 75th percentiles indicated with lines. From this, it is noticeable that the range of thickness apparent on upper stones is much narrower than that of the lower stones. As more upper stones were identified than lower, this might suggest greater resolution in the data, though it is also possible that lower stones were used in a less consistent way than upper stones. Possible evidence for

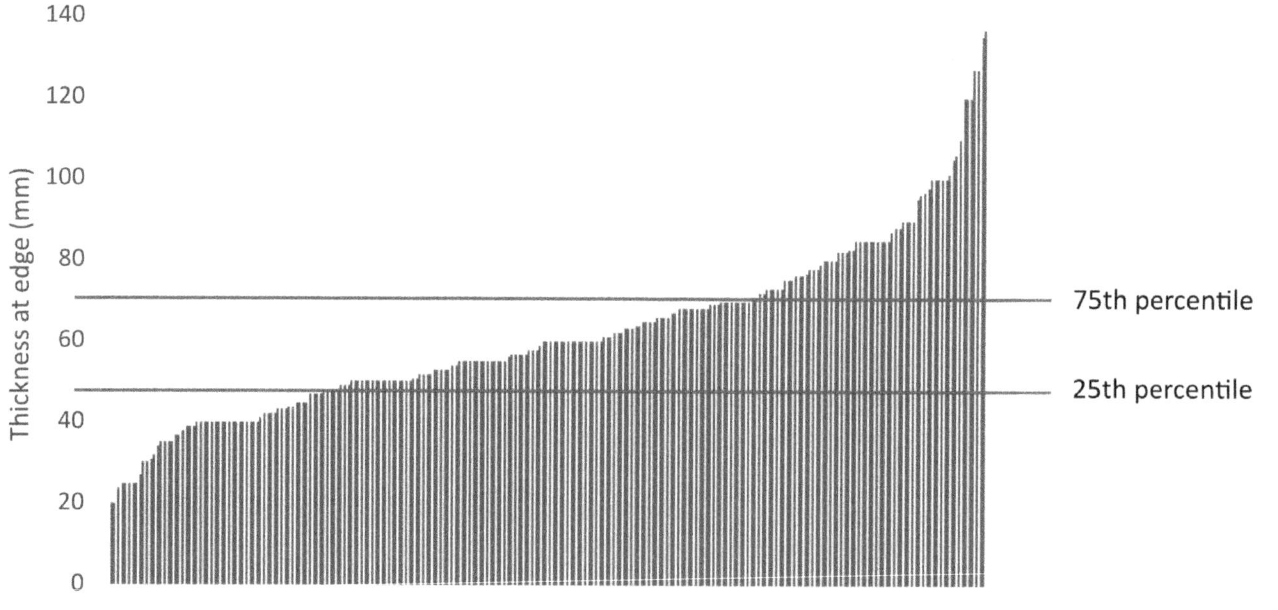

Figure 5.15. Upper stone thickness for whole dataset with 25th and 75th percentiles shown as lines (T= 248).

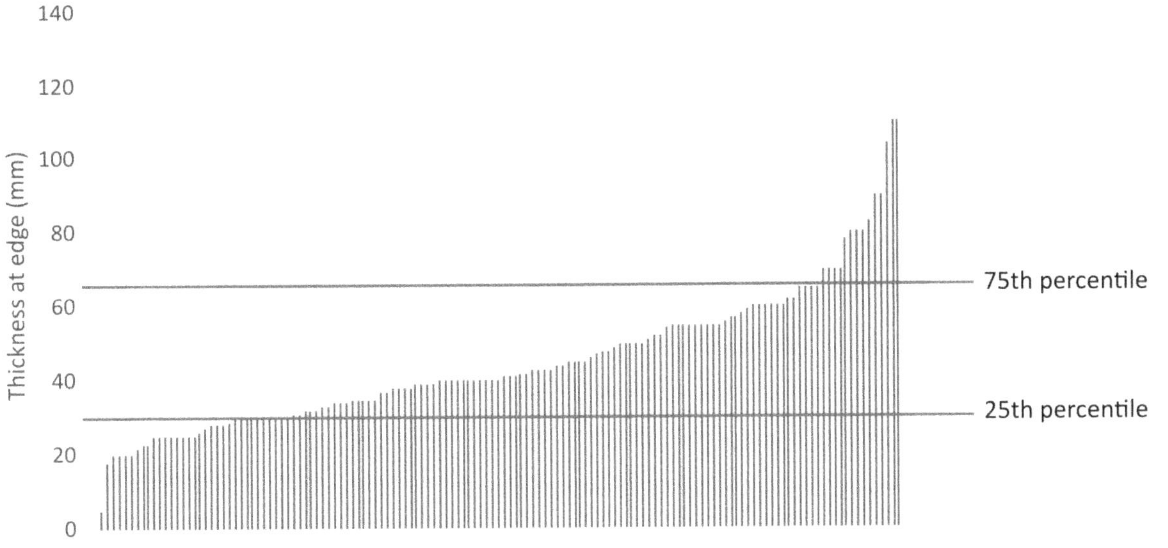

Figure 5.16. Lower stone thickness for whole dataset with 25th and 75th percentiles shown as lines (T= 138).

fixed position querns set into the floor or other surface at Verulamium, Vindolanda and Silchester (see Chapter 7 and 8) suggests that lower stones may have been used and, thus, worn differently according to their placement, which may have varied according to context of use. However, it should also be noted that upper stones that accommodate most of the mechanical parts for a rotary quern are more vulnerable to damage due to wear. At a specific point of wear, irreparable damage would occur whereby the quern ceased to function. Though mending might allow use to continue past this point, some consistency in thickness should be expected. Alternatively, lower stones would have been disposed of before full wear was reached due to the faster wear and damage of its paired upper stone or may have been paired again with an alternative upper stone. This would create a wider range of thicknesses depending on the options explored by the quern user.

5.4.2. Stone Thickness by Region

Stone thickness relates to intensity of use, and regional difference in stone thickness and wear for lower stones are shown in figure 5.17. Insufficient data on lava quern thickness was available for the south-west, East and West Midlands regions. For the West and East Midlands, too few sites are represented to generate suitable comparison, while the south-west saw a low level of lava quern use (see discussion on distribution for this chapter). Interestingly, there does appear to be some regional variation in the data. In both the north-east, Wales and the south-east, a greater range of lava quern thickness can be observed, with this being more pronounced in the north-east. This can also be seen in table 5.3, which provides the 25th and 75th percentiles for thickness for each of the regional datasets, alongside the range between these two values. Although the dataset for Wales is small and thus questionable, interpretations of this range for the north-east and south-east might consider the possibility that lava querns were used, and thus worn, in wider variety of ways than that of other regions. It may also relate to greater divisions in society in terms of the access that people had to imported goods in these areas and, thus, the intrinsic value attributed to these items. Some people may have repaired their lower quern stones or repaired them with new upper stones to allow for continued use to the point of exhaustion. Alternatively, others must have disposed of their querns well before this point had been reached, perhaps because they were able to replace them more easily. The wide range shows that behaviour in this regard was not consistent, and that choices varied within these regions.

However, the upper and lower values for the 25th and 75th percentiles in the south-east and the north-east regions vary considerably. Lower stones were much thicker when they were deposited in the north-east when compared with the south-east. The north-west and Scotland were similarly thicker, while Wales may also be tentatively categorised within this group. As regions that were more heavily militarised, it is likely that the pattern of stone thickness relates to the prevalence of forts and the use of lava milling tools by the army. Military sites contributed to the bulk of the data for all these regions (see following section), and this has influenced the data significantly. In contrast, the south-east and the east regions where rural sites provided the bulk of the data, both the lower and upper percentiles for lower stone thickness is much smaller.

The data for upper stone thickness was more extensive overall, allowing for the inclusion of the east midlands and the south-west for comparison (figure 5.18). However, it should be noted that these regions, with the inclusion of Scotland, Wales and the south-west produced very few measurements for comparison. These may be considered alongside the regions that had more data available, though interpretations must be generated with caution. For upper stones, the greatest range of thicknesses is observed for Scotland, Wales, the south-west, the west midlands and, tentatively, the north-east. It should be noted that most of this group comprises small datasets, except for the north-east, meaning that extreme values are more likely to skew

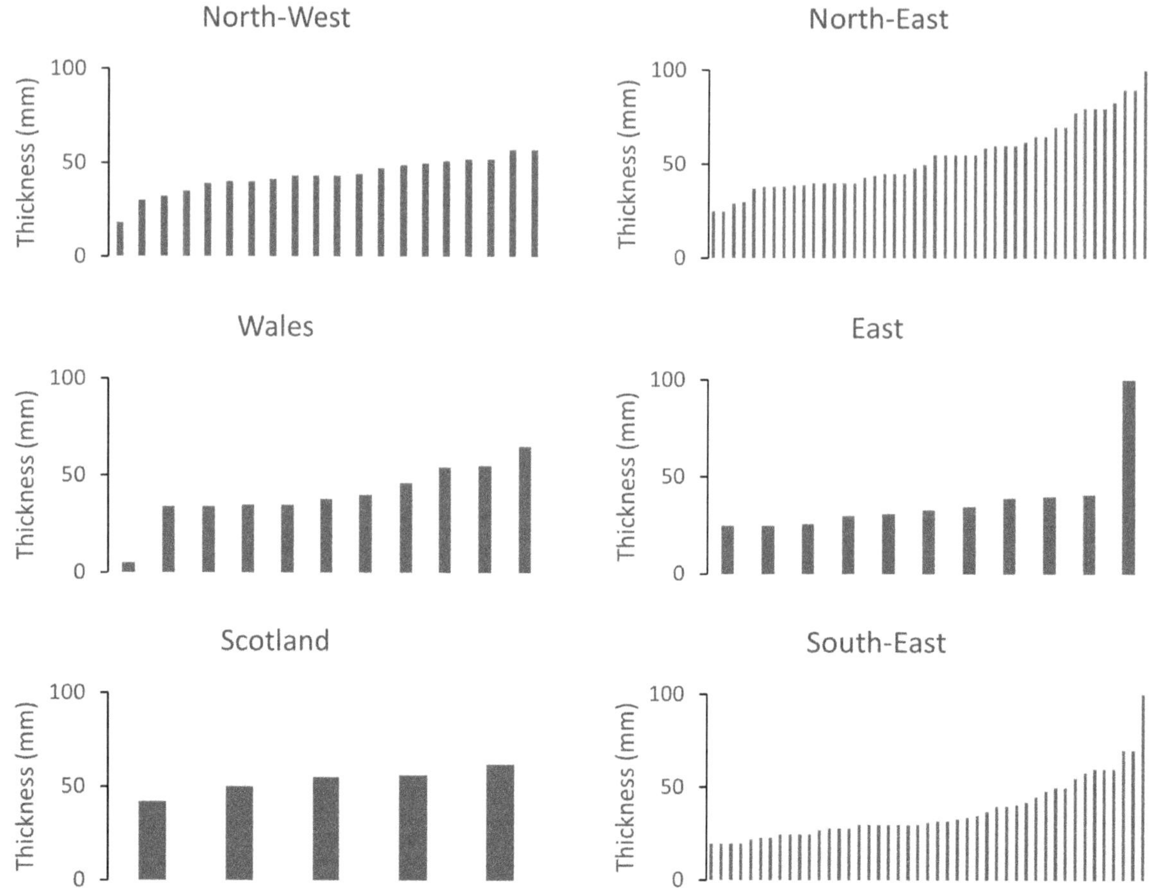

Figure 5.17. Charts showing the relative thickness of lower stones by region. The south-west, east and west midlands regions are excluded from the analysis due to insufficient available data. East T=10, north-east T= 44, north-west T= 20, Scotland T= 5, south-east T= 45, Wales T= 11.

Table 5.3. The 25th and 75th percentiles of lower stone thickness by region, with the range between the two values

	East	North-east	North-west	Scotland	South-east	Wales
25%	26	40	39	46	26	34
75%	40	65	51	59	47	54
Range	14	25	12	13	21	20

the results and generate an inflated range. It is, however, interesting to note that the north-east region shows a wide range for upper stone thickness, as was also seen for lower stones, while the south-east region does not. Upper stones were used and worn in a more consistent way in the south-east compared to the north-east, while greater disparity existed for the use and wear of lower stones in both regions.

As with the lower stones, the south-east and east regions present similar values for the 25th and 75th percentiles for thickness (figure 5.19 and table 5.4), with the possible addition of the east midlands in this group. This shows that similar behaviour relating to lava quern use and deposition may have existed across these regions. The thickness measurements also show a greater degree of wear in these areas, suggesting that upper lava quern stones were extensively used prior to deposition. However, as with the lower stones, the dataset is dominated by quern stones recovered from rural sites in these regions, meaning that this may be a factor that is also related to the site type, as opposed to it simply being a regional variation. Nonetheless, the fact that all three of these regions are located to the east of Britain suggests that regionality did have a role to play. The north-east, north-west and Wales show similar figures, as was similarly observed for lower stones and is likely to also be related to the location of military sites in these areas.

5.4.3. Stone Thickness by Site Type

The results of the examination of stone thickness by region has suggested that thickness of stone type may also be related to the site type, and that this may have been a more significant factor contributing to the extent of wear prior to deposition. This has, therefore, also been explored and the results of the analysis can be seen in figure 5.20 and table 5.5 for lower stones and figure 5.21 and table 5.6 for upper stones. These confirm that there is a clear differentiation in levels of wear on lava querns recovered from military (including military/urban) and urban sites when compared with rural, and more slight deviation observable between military and urban. Both upper and lower stones present similar trends; at military sites, both

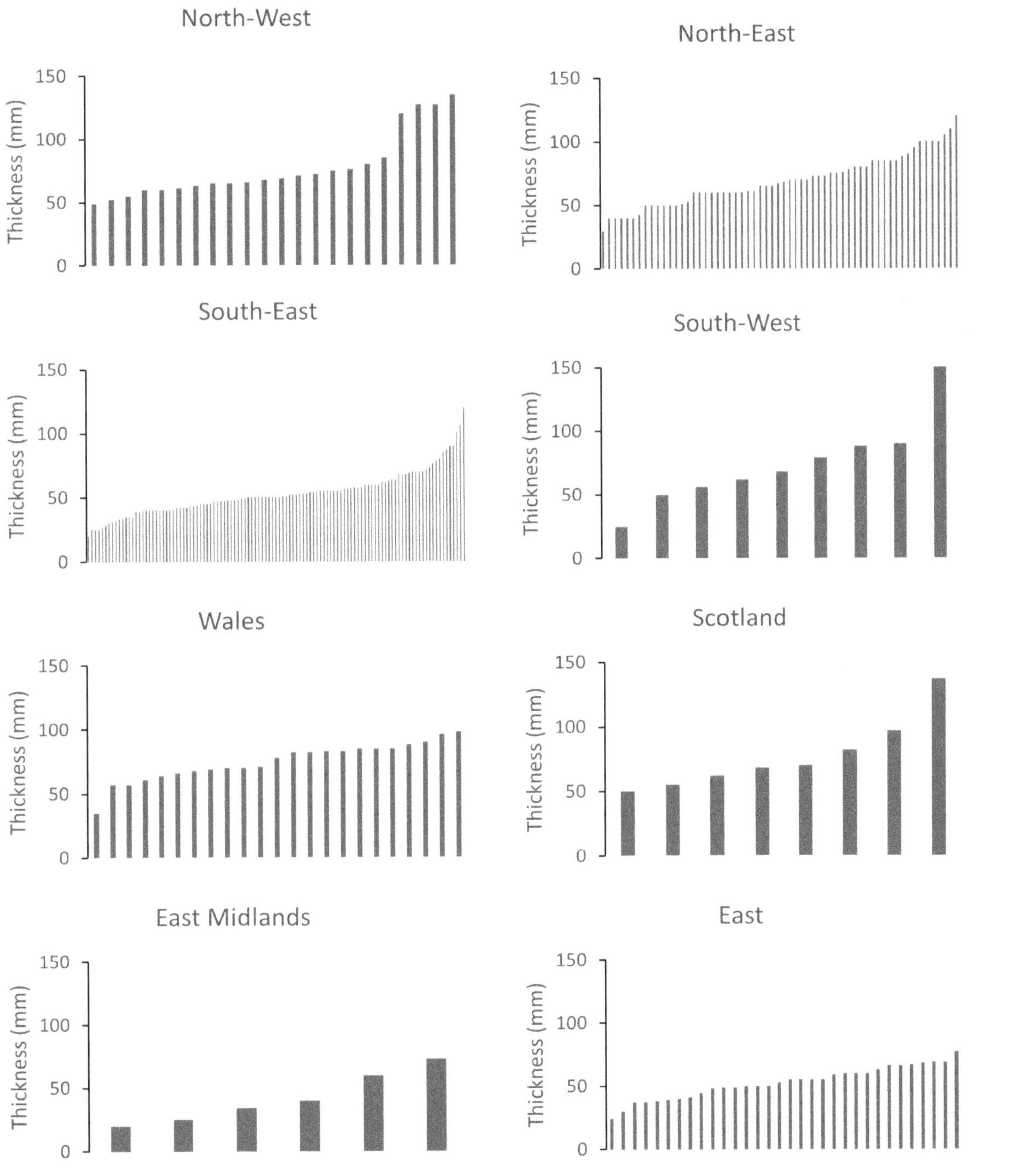

Figure 5.18. Charts showing regional difference in upper stone thickness. North-east T= 60, north-west T= 22, south-east T= 111, south-west T= 9, Wales T= 23, Scotland T= 8, East Midlands T= 6, East T= 33.

the 25th and 75th percentile measurements for thickness are significantly larger than that for rural sites, and slightly higher than for urban sites.

This relates to the privileged access that the military community had to imported goods through military contacts. The range of thicknesses seen at military sites is also greater than for urban and rural, showing that there was less conformity in the extent of wear prior to deposition in military contexts. The reason for this remains unclear. Interpretations should consider whether ritual deposition may have had a role to play, as querns may have been selected for deposition that were still usable and, thus, thicker. These would contrast with querns that were discarded after their original use lives had ended. Military supply, which would have been more reliable and extensive is another factor, as good supply would promote disposal of milling tools at an earlier stage when they were heavily worn, with less anxiety about its replacement. There is also the possibility that the redeployment of troops to new locations may have resulted in the abandonment of heavy milling tools before full use had occurred. Military sites often saw gaps in occupation and on withdrawal from a fort, mass destruction and dumping of material appears to have occurred, which is likely to have included querns. Examples of such events include the large dump of nails at Inchtuthil (Cool, 2003, p. 1), and the Vindolanda tablets (Bowman & Thomas, 1975, p. 465–466). However, this

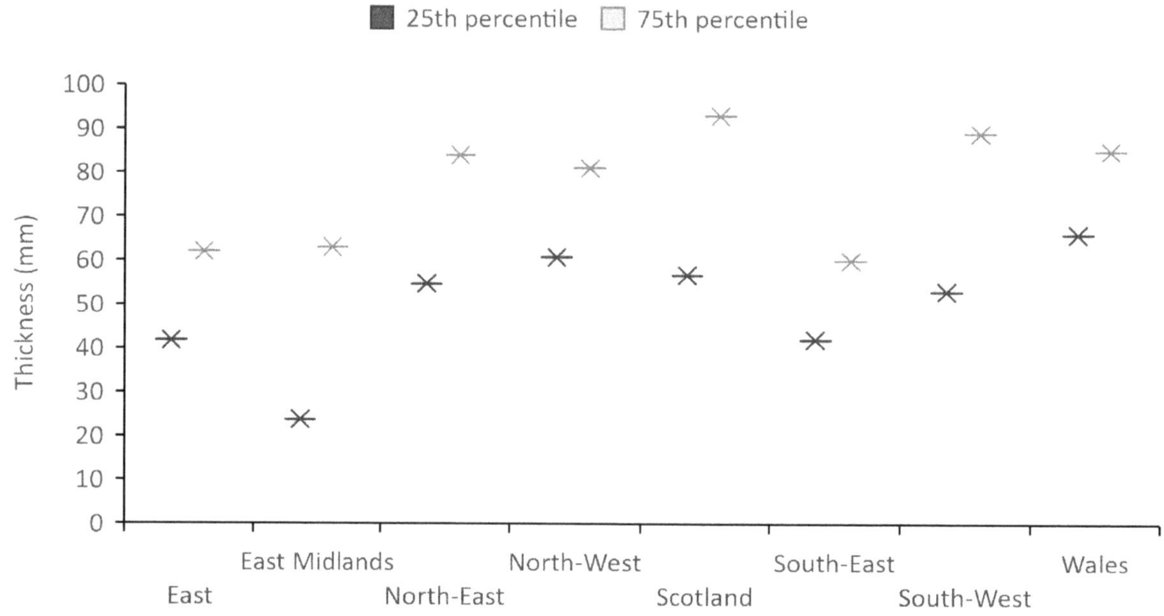

Figure 5.19. Chart showing the 25th and 75th percentiles by region for upper stone thickness.

Table 5.4. The 25th and 75th percentiles for upper stone thickness by region, alongside the range for these values

	East	East Midlands	North-East	North-West	Scotland	South-East	South-West	Wales
25%	42	24	55	61	57	42	53	66
75%	62	63	84	81	93	60	89	85
Range	20	39	29	20	36	18	36	19

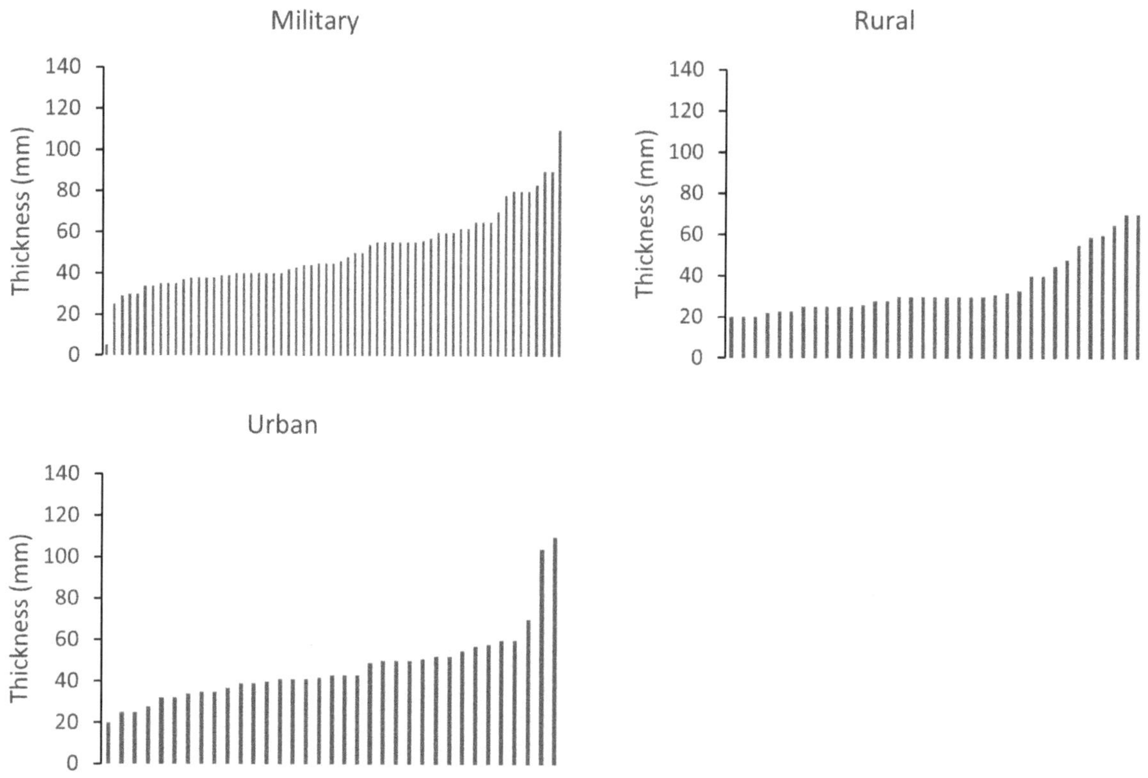

Figure 5.20. Analysis of the thickness of lower stones by site type. Military T= 62, rural T= 35, urban T= 35.

Table 5.5. Data for the thickness of lower stones by site type

	Military	Urban	Rural
25%	39	35	25
75%	60	52	40
Range	21	17	15

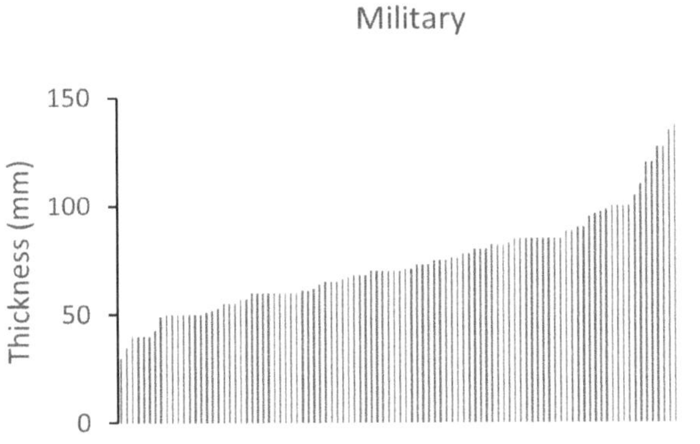

Figure 5.21. Analysis of the thickness of upper stones by site type. Military T= 98, rural T= 83, urban T= 52

Table 5.6. Data for the thickness of upper stones by site type

	Military	Urban	Rural
25%	60	55	40
75%	85	72	58
Range	25	17	18

pattern may also relate to excavation methodologies and finds retention policies at sites that were subject to earlier and antiquarian investigations. Larger fragments are more likely to have been kept in these cases, and if later excavations followed where smaller pieces were then also retained, it might exaggerate the range between thicker and thinner stones.

Urban sites show values for quern stone thickness that lie between military and rural site values. The 25th and 75th percentile values show that querns were deposited when still relatively thick when compared to rural sites but saw a greater degree of wear to that of military sites. This fits with common perceptions of how the economy worked in Roman Britain; the military received priority access to anything required, urban populations benefitted from good access to infrastructure and exchange systems, while the people living and working in rural areas may have had less access to, or desire for, imported goods. The extent to which this was true for the rural sites is clearly visible in the data; the value of the 75th percentile for quern thickness is equivalent to that of the 25th percentile at military sites. Some of the thickest querns deposited at rural sites were, therefore, almost the same thickness as the most worn examples at the military sites. Significant degrees of repair must have been necessary to allow for continued use in these cases. Despite the variation in thickness between urban and rural sites, the range of wear is similar. This presents the possibility that there was a socially acceptable extent of wear for a quern prior to its deposition in both rural and urban contexts. Once a quern stone had reached a particular extent of wear, it would be rendered unusable and cease to function. However, this was different between site types, indicating that this was not a purely economic concern as an element of choice must have been involved.

Overall, the data analysis presented here does not tackle the intersectional nature of the variability involved. Some patterns have been identified for site type and region, but consideration has not been given for chronology or the more nuanced aspects of site type, such as the differentiation between rural villas, defended settlements, farmsteads, roadside settlements etc., because the respective datasets would become too small. Future work might use multivariate analysis to explore these themes, though more data would need to be collated first.

5.5. Reuse

Reuse is a feature of the object biography that was rarely identified with any confidence or recorded in site reports. Most of the examples presented here were either personally recorded, taken from Shaffrey's dataset or from instances of more detailed quern reports. It seems therefore, that many more cases of reuse exist within the dataset than can be presented here and it is hoped that more effort will be invested in acknowledging reuse of querns in future work. Inhabitants of sites with poor access to local or imported stone suitable for ground stone tools often reused quern and building stone for other purposes (for example, see Allen, 2022). Reuse, therefore, can reveal a great deal about the economic and social circumstances of a particular site and changes in connectivity with other parts of the province. This section only deals with the reuse that is perceptible on the stone itself, as reuse in the form of deposition has been explored in the following section.

Table 5.7. Sites, locations, and site types where reuse of lava querns has been observed and recorded

Site Name	No Querns	Site Region	Site Type
Vindolanda	1	NE	Military
Colchester	1	E	Military/Urban
Great Dunmow	1	SE	Rural
Carlisle	1	NW	Military/Urban
Mucking	1	SE	Rural
Silchester	4	SE	Urban
Church Field	1	SE	Unknown
Wanborough	1	SE	Urban
A14 Cambridge	4	E	Unknown
Carvossa	2	SW	Rural
Billingley Drive	1	NE	Rural
Corbridge	2	NE	Military/Urban
Usk	1	Wales	Military
Verulamium	5	SE	Urban

Overall, 26 examples of reuse were recorded from 14 sites (table 5.7). Very few of these had identifiable traces that would indicate a specific type of reuse, and many interpretations simply mention wear traces that do not correspond to original use as a quern. Of the 14 sites with reuse observed on lava querns, six are in the south-east and these contributed to the bulk of the dataset. However, as three of these sites were recorded personally or by Shaffrey, there are clear issues in interpreting this pattern. No clear trend is apparent in terms of reuse according to site type. Although rural sites appear to be more prevalent in their reuse of lava, there are many more rural sites than other site types in the dataset and this must be compensated for. Generally, exploring the full dataset for reuse does not appear to be an effective approach for identifying trends in behaviour. However, on an individual site basis, this has been more successful, as seen in the following chapters. Overall, reuse tended to be associated with other processing activities, such as for whetting and sharpening of blades, but specific reuse is hard to pinpoint from wear traces alone. However, there would have been a continued value associated with lava for reuse purposes, especially in regions where suitable stone for processing was absent or inaccessible. The extent to which lava querns are recovered as fragments in the south and southeast of Britain may relate to the value of broken-up querns to supply this need, and the extent of fragmentation may not be solely explained by taphonomy or deposition practices.

5.6. Deposition

Unlike smaller objects, the find spots of querns and millstones cannot be directly correlated with use contexts as they are not subject to casual loss. Deposition was more likely to be intentional, either through disposal, ritual practice or reuse/redeposition. The wide variety of deposition and redeposition contexts is presented in figure 5.22, which shows the most common Romano-British archaeological features from which lava querns and millstones were recovered and their relative volumes.

Although 1241 lava milling tools were associated with a specific deposition context, as shown in the chart, many more were from unspecified contexts (1201), unstratified (83), or were not associated with a particular archaeological feature (177). These have not been included in the chart as their presence would dwarf the other data and obscure any patterns. However, it does emphasise gaps in the data.

From the chart, deposition in pits occurred most frequently, with floors/roads/levelling and ditch deposition also common. This should not be surprising considering how often pits and ditches occur on archaeological sites and given that material culture is typically found in cut features. The extent to which these may relate to ritual deposition, practical reuse or the dumping of rubbish is a complex matter that should be considered according to context and has been explored further in the thematic chapters and individual case studies. The frequency at which deposition into roads, floors and for levelling occurred might be the consequence of conflating three different feature types. Though these contexts may seem similar in functional terms, meaning associated with deposition into a road surface may have been vastly different to that for levelling or for floors. There are issues associated with interpretation on a site level for this category, and it was not always possible to distinguish between the three possible feature types. Therefore, they have been grouped together during the data recording process, as have ovens, corn dryers and hearths as similar issues exist for these features. It should also be noted that the site at 1 Poultry, London, has also inflated numbers, as the large lava quern assemblage was incorporated into a cobbled floor surface, comprising 213 such examples (Hill & Rowsome, 2011).

After the removal of the 1 Poultry querns and after further defining the type of deposition context, it is possible to see that there is little to distinguish between the occurrence of deposition in road surfaces, floors (internal to a building/structure), surfaces (external to a building/structure), and use for levelling (figure 5.23). However, these do

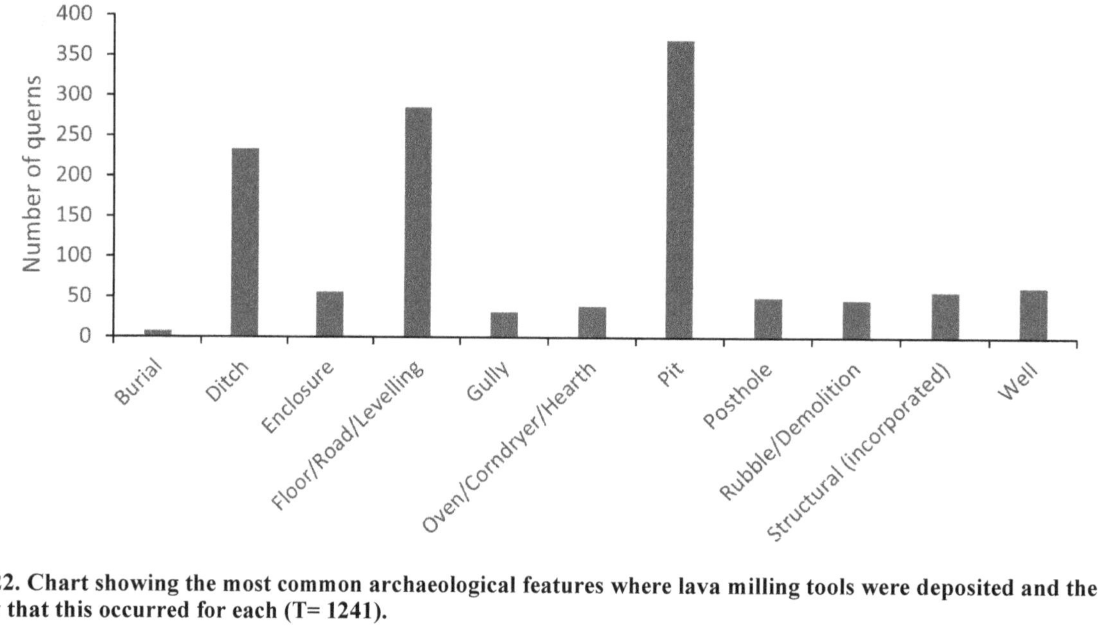

Figure 5.22. Chart showing the most common archaeological features where lava milling tools were deposited and the frequency that this occurred for each (T= 1241).

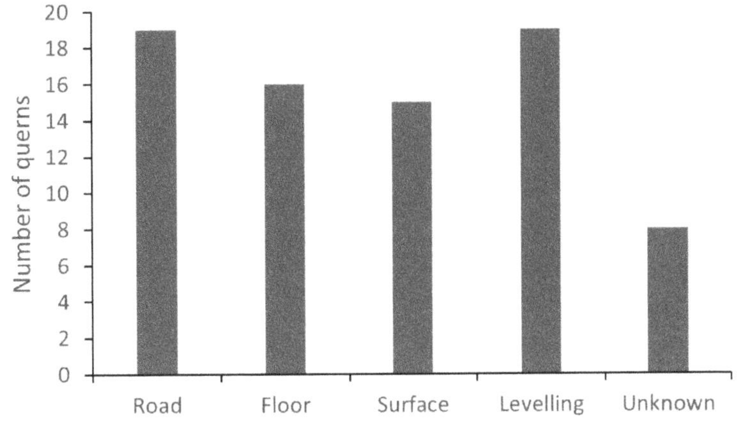

Figure 5.23. Chart showing the frequency that lava querns were recovered from road, floor, surface, or levelling contexts, and those from an unspecified surface (unknown) (T=77). The data excludes the querns recovered from 1 Poultry.

appear to occur differently depending on the site type, though this is difficult to reconcile with the variable extent to which certain site types and feature types occur. Therefore, to explore this further, correspondence analysis was carried out to explore whether specific site types saw more frequent deposition of lava within floor, road, surface, or levelling contexts. The results, shown in figure 5.24 indicate that urban sites were more likely to yield lava querns from levelling contexts, while military sites saw greater frequency of deposition in road surfaces. Rural sites had less of a strong correlation to specific deposit types, but clustered most closely with floors and surface contexts.

Again, it is worth considering whether this demonstrates different deposition practices across different site types. It should be noted that of the 17 recorded lava deposits in urban levelling contexts, ten were from Silchester. This might be more of a site-specific phenomenon or may be due to better preservation of the stratigraphy where levelling occurred,

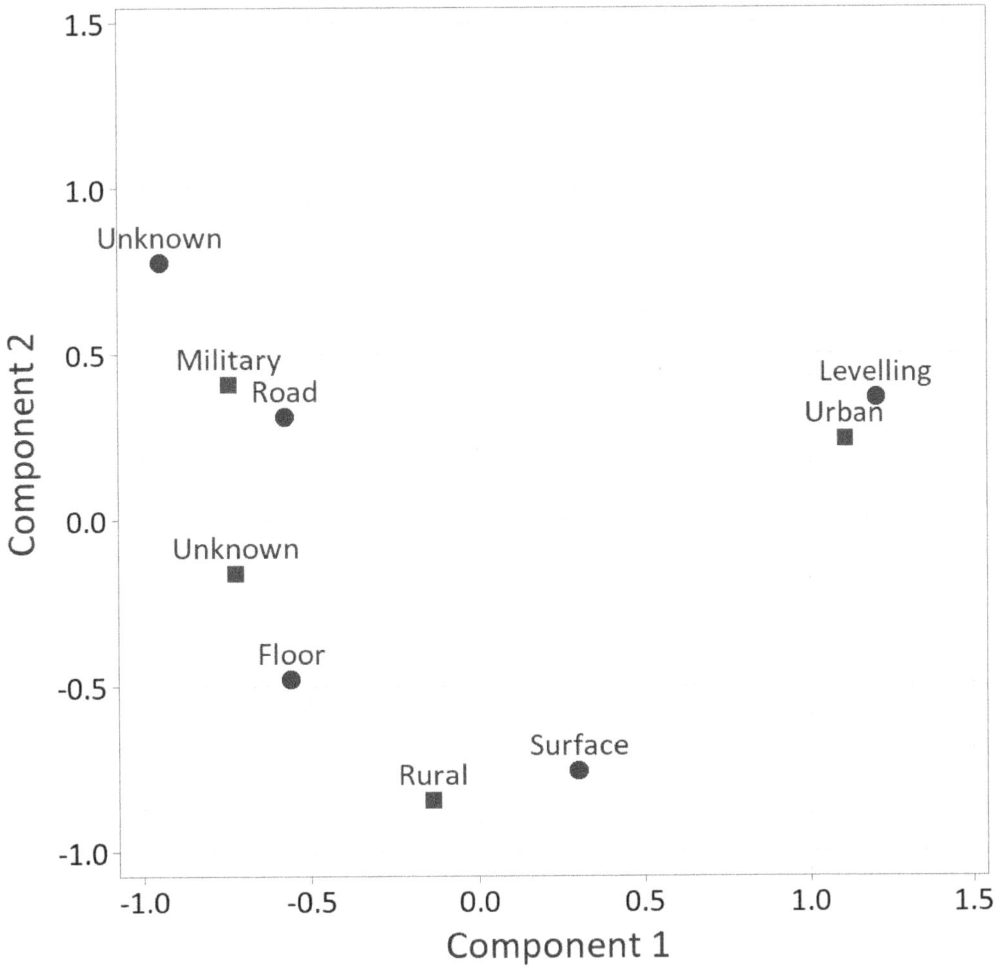

Figure 5.24. Simple correspondence analysis completed on the frequency of occurrence for deposition of lava in specific contexts at different site types. Total inertia for component 1 is 67% and for 2 is 31%.

which is less likely to survive in urban environments that have seen multiple post-Roman phases of occupation. The fact that detailed stratigraphic information was available for Silchester is also a factor for consideration. Similarly, the greater occurrence of deposition in road surfaces at military sites has, to some extent, been related to the fact that Roman period infrastructure is more likely to become the focus of excavations at military sites.

Finally, data was examined to explore whether regional variation in deposition practices existed. Figure 5.25 shows the frequency at which deposition occurred in specific contexts across the different regions of Britain. Initial observations suggest that pit and ditch deposition occurred most frequently in the south-east and eastern regions, with pit deposition being significant in Wales also. However, the data is difficult to compare as each region yielded different volumes of data, creating problems with visibility and interpretation. Correspondence analysis was, therefore, used to compare the datasets (figure 5.26).

The results show that deposition in floors, roads and for levelling, alongside demolition/rubble was more common in Scotland, the north-east and north-west. This is an interesting distinction, as it suggests that this form of deposition occurred more often in the north, and is, therefore, likely to be strongly associated with military sites, in relation with various fort abandonment and reconstruction phases. The south-east correlates most closely with deposition in pits, postholes, and wells, while the east is associated with enclosures, ditches, corn dryers/ovens/hearths, as was the south-west. The East Midland sites clustered most closely with burials, wells, and pits, as did Wales, though to a lesser extent. The West Midland sites do not appear to have any specific strongly associated features. Again, these patterns can be explained by the higher incidence of certain features at particular site types and regions. For example, the east saw much more frequent use of lava milling tools at rural sites (see Chapter 6), and so higher occurrence of deposition in ditches and enclosures is expected. However, the fact deposition for the northern and the eastern regions correlate does suggest that some pattern in practices could exist. The analysis does suggest that military users had a different relationship with the lava milling tools than populations in the south, and that this was more functional and less personal. Further data and analysis would clarify this in future.

Looking at deposition by site type (figure 5.27), pits are the most common feature for lava deposits across all site types. Deposition in ditches and enclosures occurred most frequently at rural sites, and floors/roads/levelling and ditches were most common at military sites, again, the result of there being more of these types of features at these respective sites. This was, again, examined using correspondence analysis to establish the extent to which site type was affecting the overall interpretations of deposition behaviour associated with lava (figure 5.28). These results confirm some of the observations made of the regional variation in deposition; rural sites are closely correlated with enclosures, ditches, and ovens/corn dryers/hearths; urban sites are associated with wells, within pits and wells; military sites are clustered with roads/floors/levelling, rubble, and demolition. These closely

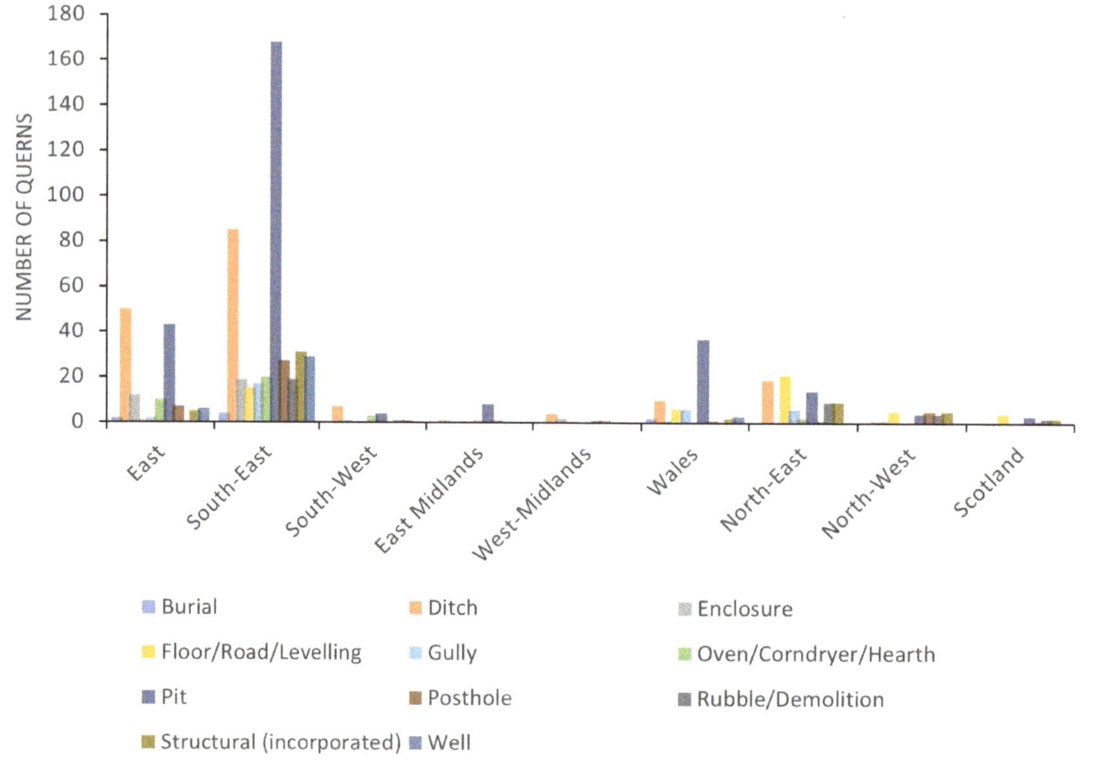

Figure 5.25. Chart showing the frequency at which lava milling tools are deposited in different archaeological features by region (T= 794).

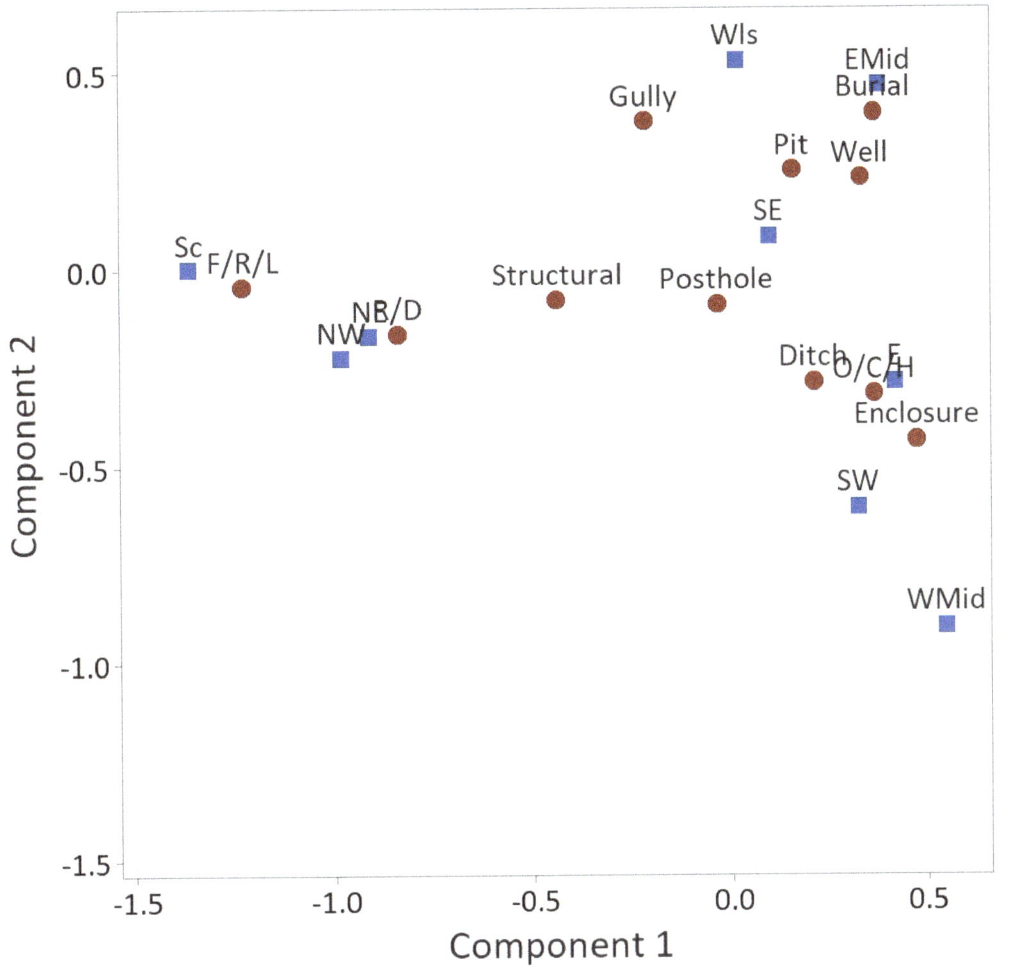

Figure 5.26. Simple correspondence analysis completed on the frequency of occurrence for deposition of lava in specific contexts at different site types. Total inertia for component 1 is 67% and for 2 is 31%.

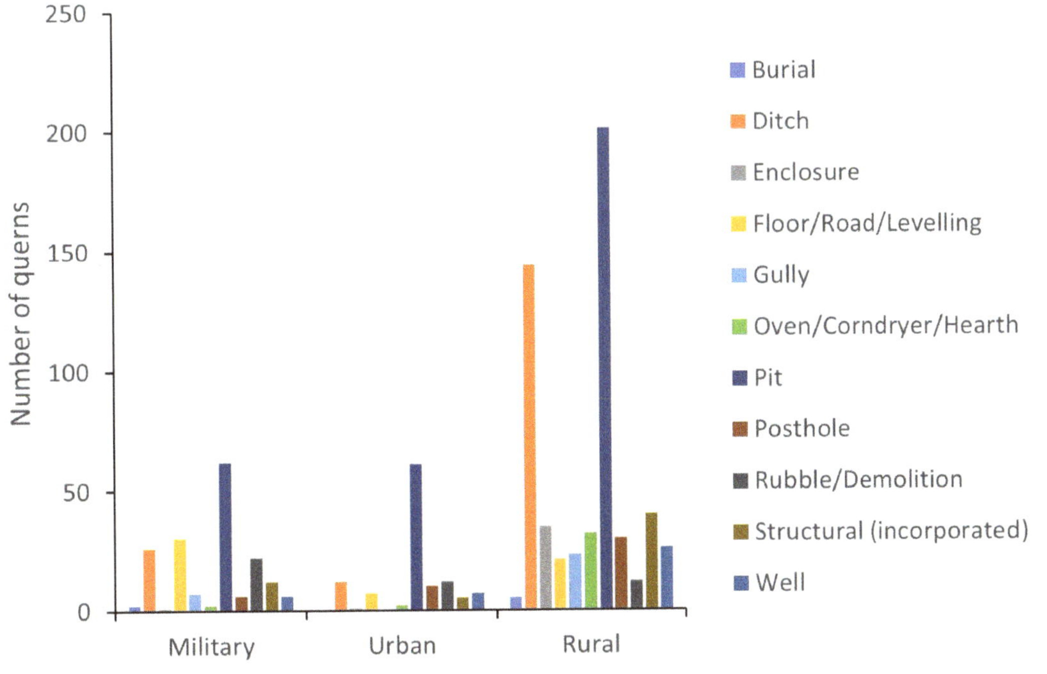

Figure 5.27. Chart showing the frequency of deposition of lava in specific archaeological features by site type.

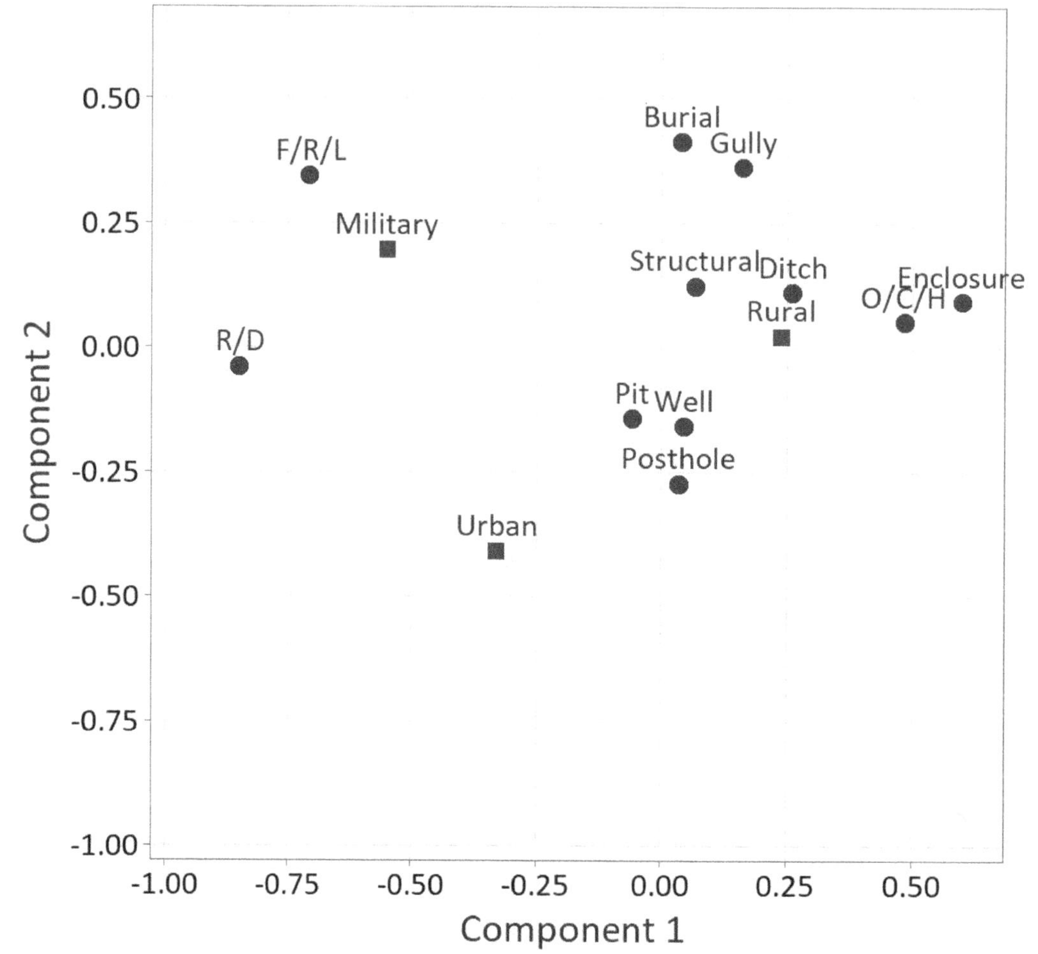

Figure 5.28. Results of correspondence analysis showing the relationship between deposition context type and site type. Inertia for component 1 is 78.73% and 21.27% for component 2.

correspond to what was observed for the regional variation, and site type has the biggest influence on the type of deposition experienced by lava objects, and often relates to the types of features that are most common in those contexts. However, this does not explain all the patterns seen. For example, deposition of lava into possible 'ritual' pits in urban contexts is a behaviour that may have been a specific practice associated with lava querns (Fulford, et al., 2006), and the clustering of urban sites with pits on the graph does suggest that this is being highlighted in the data. However, the fact that pitting is a common urban practice of rubbish disposal (Fulford, pers. comm.) may also explain the strong association seen. This has been explored further in Chapter 7.

5.7. Conclusion

Overall, the full dataset analysis has identified patterns and trends to the data, some of which has been examined in more detail within the individual case studies that follow. In relation to the manufacture stage of lava milling tool biography, a diameter measurement of 500mm has been identified as the division between lava querns and lava millstones. This differentiation needed to be established at an early stage in the data analysis to ensure consistency, but in-person data collection has confirmed that the functional features of lava milling tools agree with where this separation has been defined. This provides a useful benchmark for all future lava milling tool recording for Romano-British assemblages. When considered separately from the probable lava millstones, stone diameter measurements have demonstrated the elevated level of standardisation that exists for lava querns in the province, 50% of which fall into the size range of 400–420mm. This further supports the idea that these objects were centrally produced and may have been made to set specifications and standards.

This standardisation is also visible in the functional fittings of lava querns, which predominantly had elbow handle fittings, overdrift rynd fittings that were fixed in a consistent way and raised annular kerbs. Kerb width did not appear to display a consistent pattern but there did appear to be some correlation between possible 'iron band' handle fittings and indistinct, faint kerbs. This may relate to the changing role of the kerb from a functional to a decorative element for querns with iron band fittings, but further work would need to be undertaken to verify this theory. Iron band handle fittings were identified in two cases, with a further four possible examples. The lack of other types of handle fitting on these querns suggests that the iron band may have been intended at the point of manufacture, making this a previously unidentified typological variation from those listed by Röder (1953) (see Chapter 2).

The extent of use has been explored by measuring the thickness of upper and lower stones and comparing these by region and site type. It was found that there was probable inequality in the supply of lava milling tools that effected populations differently depending on whether they were living at a rural, urban or military site. Military sites may have had priviledged access to lava products, as these sites presented the thickest querns at the point of deposition. Urban lava querns were slightly thinner, while rural sites produced very thin and worn quern stones. Rural populations must have experienced greater difficulty with regards to replacing their milling tools, as these were worn to extreme levels in some cases.

Spatial distribution of lava milling tools has shown a clear east-west and north-south divide. Lava milling tools were more likely to be used at sites in the east and south-east of Britain. This is especially the case around London and Colchester where lava volumes are the most dense, and these are the ports that were probably accepting cargoes of lava querns for further distribution. In the east of the province, rural sites of any status were consuming lava products, probably due to the lack of suitable local stone sources for producing milling tools. Elsewhere in the south, lava consumption is centralised on urban and military sites, with some 'seepage' into the wider hinterlands. In the north and at Hadrian's Wall, only military sites were found to have lava, with very little being distributed amongst the rural populations. This may show the scale to which there was integration, connectivity and exchange between different communities in different regions of the province.

In terms of chronology, lava was introduced to Britain in the first century, where it was first used at military sites and urban sites that had early military foundations. Lava became much more widespread and peaked in distribution in the second and third centuries, after which a contraction is seen with a long trail of continued deposition into the fourth and fifth centuries. The continuation into the later Roman period is likely to relate to the scale of reuse and redeposition of lava milling tools, shown by high levels of depositon in floors, roads and surfaces and reuse as building material, especially at military sites. Other patterns relating to deposition context are harder to identify and have been examined for individual case studies.

6

Foreign Stone in the Hands of Farmers: Lava Querns and Millstones in Rural Roman Britain

6.1. Introduction

Rural life was the lived reality for most of the population of Roman Britain, though early research focused much more on military and urban sites. In recent years, recognition of the significance of the rural past has transformed research objectives and perspectives within the discipline (Hingley, 1989; Rippon, et al., 2015; Taylor, 2007). One of the most important and influential projects to harness Roman rural research objectives, the 'Rural settlement of Roman Britain' Project (RSRB), involved large-scale collection of data relating to site characterisation, finds evidence, burials, and environmental evidence from c. 2500 Roman rural sites in England and Wales. The outcome of the project was an open access database containing data collated during a comprehensive literature sweep of excavation reports and county journals of any date, and 'grey literature' reports produced as part of developer funded excavations post 1990.

This chapter addresses research questions relating to lava querns and millstones by utilising the RSRB data to gain greater insight into the role of these objects within the wider context of rural life, while also considering the inter-relations between rural, urban, and military sites. As strong indicators of agriculture, food production and processing activities, querns and millstones are significant objects that are likely to have held important roles in rural life. This is recognised in the RSRB project (Allen & Lodwick, 2017, pp. 142–177; Brindle, 2017, pp. 71–72; Brindle & Smith, 2017, pp. 198–200), though distinction between querns and millstones of lava from those made from more local sources was not made within the original study. This chapter will therefore utilise the RSRB dataset (relating to presence, absence, and volumes of querns at a range of different rural sites) but add the further dimension of a comparison between imported and local querns. The structure of this chapter does not follow the object biography approach, as potential individual case study rural sites could not be explored due to Covid restrictions. However, a lack of access to suitable primary data for rural sites has created an opportunity to explore the data in other ways.

6.2. The Dataset

A total of 920 rural sites in the Rural Settlement of Roman Britain database yielded information on querns, out of *c.* 2500 sites recorded by the project. These were identified by a database search followed by thorough examination of the original site reports and publications to extract data not captured by the RSRB database. This included petrographic identifications or descriptions, more detailed information about individual examples of lava querns, context of deposition, chronology and physical features relating to typology. Sites that had lava querns or millstones present numbered 363, with a total of 1467 objects recorded from those sites, and those with querns or millstones present but none of lava numbered 557, with a total of 3210 objects[1] (figure 6.1). This dataset was used in combination with the site and settlement data produced as part of the RSRB project to develop a greater understanding of the role of lava querns in Romano British rural socio-economic life. The data analysis was structured to first explore the details pertaining to lava querns and millstones as objects: sizes, types, frequency, context of disposal etc. Following this, wider patterns of distribution and use were explored.

Figure 6.1. Percentage of sites in the dataset with lava milling tools present compared to those with milling tools, but none of lava (T=920).

It should be noted that although the site designations used as part of the RSRB project have been utilised as much as possible, it has not been practical to mirror these precisely in every stage of the data analysis. For some areas of this study, the number of site designations was reduced significantly to accommodate the needs of this specific project. Sites were classified as either a small town/settlement, a vicus, a villa, or other. It was felt that this provided the means to explore different social groups within a rural context; civilian, elite, and military. With respect to the socio-economic approach applied as part of this research, it was determined that this differentiation was more important in terms of understanding lava quern distribution than whether a settlement could be described as 'complex' or 'enclosed.' The 'other' group

[1] Although 1286 sites from the RSRB database had querns present, it was not possible to include every site; excavation reports did not always provide geological determinations, and some reports could not be accessed due to Covid restrictions.

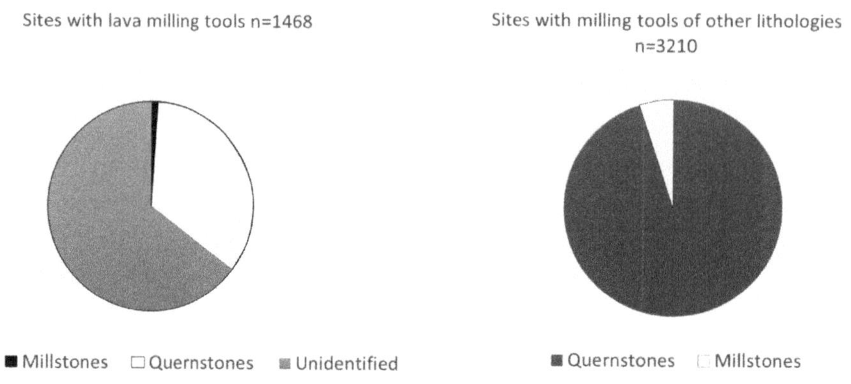

Figure 6.2. Percentage proportions of millstones to quern stones for sites with lava and non-lava milling tools.

describes sites that were not associated with a settlement at all, such as a road, field system, or single building of unknown purpose, while the small town/settlement group incorporates a wider variety of minor and major settlement types, including farms, farmsteads, defended towns, roadside settlements, and nucleated settlements.

For the distribution analysis in this chapter, the minor and major site types were utilised to a greater extent in a more nuanced examination of lava distribution. This analysis differentiated between roadside settlements, religious/ritual sites, defended small towns, funerary sites, villae, vici, small towns/settlements, farmsteads and 'other.' These were chosen as groupings, again, in an attempt to present how lava quern distribution may have been influenced by social differences and levels of connectivity with other parts of the province. For future researchers, the full breakdown of minor and major site designations as provided by the RSRB project is included for each site recorded within the project database.

6.2.1. Millstones vs Quern stones

For the purposes of this research, lava millstones have been defined as being over 500mm, though features typical of millstone operation should also be sought on individual stones as larger querns may have existed (Chapter 5). For other stone types, there is less clarity in terms of where the division between querns and millstones exists (Chapter 1). The RSRB project used 600mm to establish this threshold. However, it was found within the site reports that millstones were often identified using other means; occasionally resulting in a millstone identification when the diameter was less than 600mm, and sometimes when diameter measurements could not be taken. For example, signs of fixtures or fittings to suggest mechanical rotation can also be used to categorise a stone (see Chapter 2). Therefore, the dataset presented here contains 'millstones' that may not fall into this category based on size alone. Designations have been determined based on individual object features, sizes, and descriptions.

From the 363 sites with lava present, 1467 querns or millstones were recorded; 17 identified as millstones, 502 as quern stones, 949 unidentified (figure 6.2). The fragmentary nature of lava creates problems when attempting to identify diameter, as seen from the large number of unidentified examples. As diameter is a fundamental measurement that determines the separation between a quern stone and a millstone, those identified constitute a sample of what may have existed in the Roman period. From the chart it is possible to see the scale of the issue relating to identification. Once fragmented, millstones are much harder to identify and it is difficult to know whether the proportion of lava millstones present in the dataset reflects preservation, or if lava millstones truly were as rare as the data suggest. From the non-lava data, 3048 querns and 162 millstones are represented. Other stone types, such as sandstone and conglomerate (figure 6.3), are much less affected by preservation and identification is less of an issue, which is reflected clearly in the data. This may have resulted in better identification of millstones for those of non-lava types, though the numbers of millstones are still heavily outweighed by the volumes of querns. In terms of proportions, as a ratio of lava millstones to lava querns there are approximately 1:33, while non-lava examples show a ratio of 1:20.

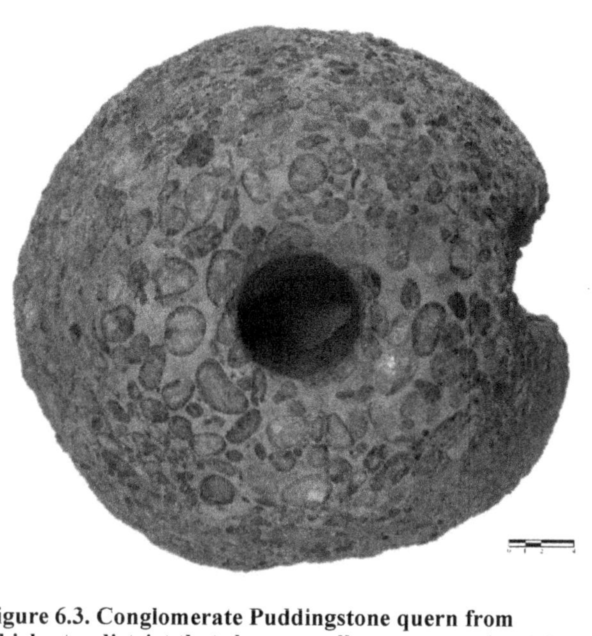

Figure 6.3. Conglomerate Puddingstone quern from Chichester district that shows excellent preservation (photo by author, ©Novium Museum).

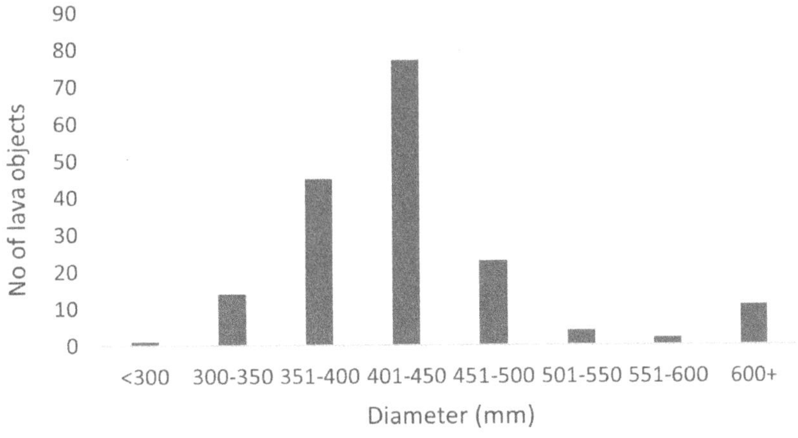

Figure 6.4. Frequency of different diameter groups for lava quern stones/millstones (T=177).

The issue regarding sizes of millstones should be investigated further, however. The physical limitations and abilities of the human body have been used to determine the threshold of what constitutes a quern and a millstone; anything too large cannot be operated by hand as a person would not be able to create the movements necessary. Although this provides us with a suitable measure with which interpretations can be guided, it is not drawn from archaeological data. As has been shown for lava milling tools, it may be possible to better distinguish querns from mills by systematically comparing the diameters of milling tools of specific types to see where measurements diverge. There is no requirement to separate upper and lower stone measurements, as diameters for paired stones are known to have been almost identical.

The high level of standardisation present in lava querns highlights those objects that do not fall within the 'normal' specifications, making it possible to determine a difference in diameter between standardised querns and less common specialist millstones (see Chapter 5). From figure 6.4 this standardisation of size is clearly visible for rural sites also, with the size range of 401–450mm being the most common amongst the 177 specimens complete enough for measurements to be taken. This standard size of a lava quern does not appear to exceed 551–600mm, which is where the classification of millstone can begin to be applied. Looking in more detail at this data using only smaller size categories up to 580mm and discounting the 'obvious' millstones (figure 6.5), it is possible to refine this further. From this, it appears that most lava querns are between 381–460mm in diameter, with some larger sized querns/smaller sized millstones occurring in 461–500mm range. The clear separation from a standardised diameter occurs after 500mm, where sizes are both inconsistent and infrequent. As millstones must be specially manufactured to fit the mechanisms already in place in mill structures, millstones are less likely to be subject to the same degree of standardisation as their hand operated counterparts. The lack of clear conformity between diameters after 500mm appears to match what could be expected of millstones. These results agree with what has been identified within the wider dataset, showing that the source of lava milling tools for the rural parts of the province were probably the same as that of the whole of Britain.

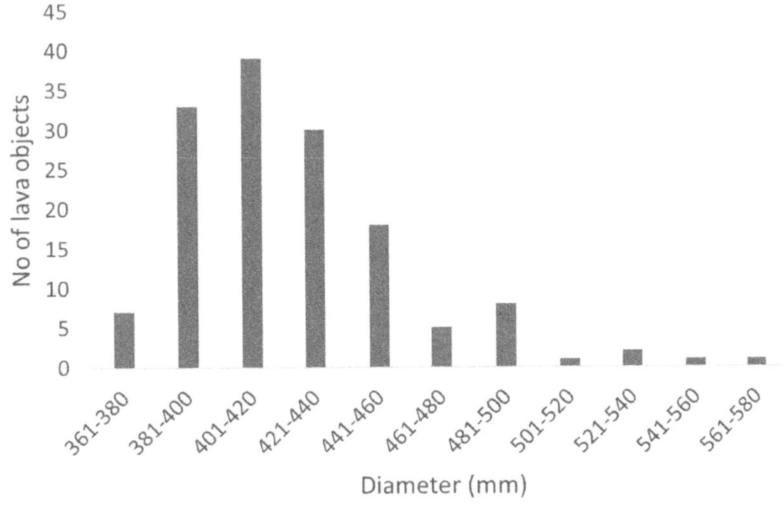

Figure 6.5. Frequency of different diameter groups for lava quern/millstones up to 580mm (T=165).

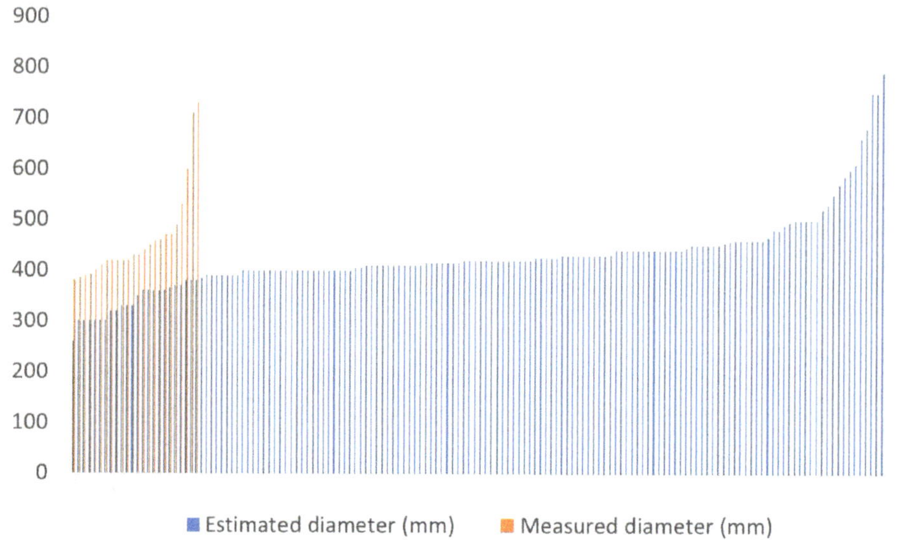

Figure 6.6. All measured and estimated diameters of lava querns and millstones (T=176).

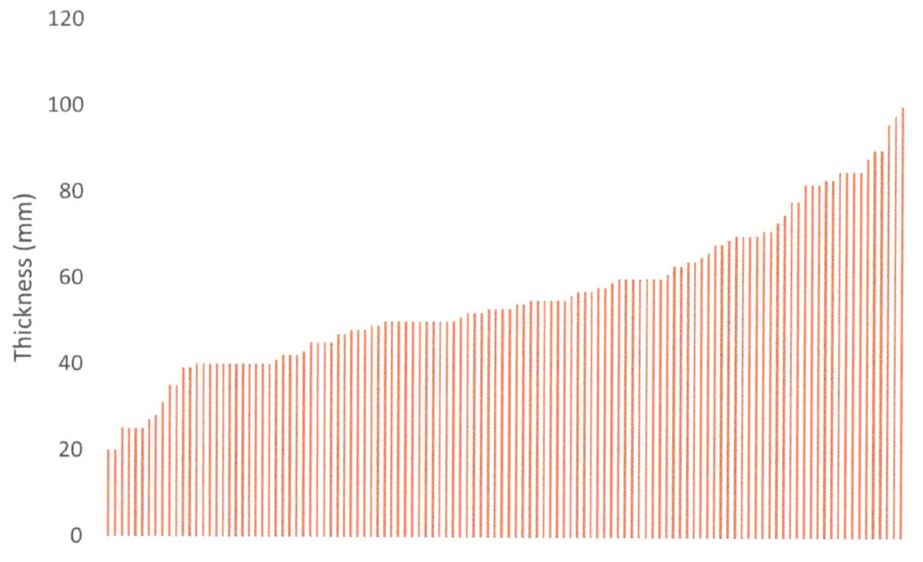

Figure 6.7. Max Thickness at Edge for Upper stones (mm) (T= 114).

In completing this analysis, it is also important to recognise the degree of error that could be present in the data. Diameters could only be precisely measured for stones that were at least 50% complete, and greatest accuracy could be achieved only for the most complete fragments. A considerable proportion of the data for diameters was estimated using the curvature of the outside edge, or by assuming a centre point and extrapolating the possible size. Therefore, a degree of inaccuracy must be assumed. Figure 6.6 displays the diameters of each individual quern or millstone in the dataset as a histogram, identifying the measured and estimated examples. Although there is a great deal of agreement with regards to the standardised sizes of querns between the measured and estimated stones, it is quite clear that measured diameters do not measure less than 380mm. There is a strong indication that the smaller estimated querns in the dataset may have been underestimated, as these do not fit within the normal measured range. This range lies between 380mm to approximately 500mm, and there is close agreement across the entire dataset. With both estimated and measured stones, there is a clear deviation after 500mm, and this correlates with what has already been surmised about the separation between quern stones and millstones.

6.2.2. Rural Quern Stone Thickness

The thickness of a quern stone is indicative of the extent of its use prior to disposal or deposition. As discussed in Chapter 4, this could demonstrate perceived value of the object and relate to the economy of supply with regards to the replacement of old and worn querns. The thickness at the edge of quern stones was recorded in the dataset for a total of 62 lower stones and 114 upper stones. These values were separated according to upper and lower stone, as these will wear differently. For the upper stones, the thickness was plotted onto a histogram for each individual stone (figure 6.7) to show the range and frequency of thickness within the dataset. They ranged from 20mm in the smallest instance, to 100mm in the largest. As unused upper lava quern stones are known to have had a thickness in the range of 100–120mm, some of these clearly represent little used or unused examples. The largest proportion, 60% (n=71), fall in the range of 39–60mm thick and demonstrate the thickness at which upper stones were more likely to be disposed of. As the upper stones were more vulnerable to wear around the central hole that housed component parts for enabling the rotary movement, these were more likely to show a consistent pattern of disposal due to wear. The data support this, though the thicker querns may not have been completely worn and are worth investigating further to see if there could be other reasons for their deposition.

Quern stones with a thickness of over 60mm represent 36% of the total number of upper stones (n=42). Of these thicker quern stones, 50% (n=21) were recovered from vicus sites, 38% (n=16) from small town/settlement sites and 12% (n=5) from villa sites. For clarification, the RSRB project refers to vicus sites as settlements close to military forts, and these would have been both inhabited and used by civilians, military personnel and people connected to military life. Within each site type, thicker querns comprised 23% of the total number of querns recovered from small town/settlement sites, 32% of the total at villa sites and 75% of all querns from vicus sites (table 6.1). This shows that the querns recovered from vicus sites were more likely to be thicker than 60mm than those recovered from villa or small town/settlement sites. This can also be seen in figure 6.8, which plots the thickness of each individual quern in the dataset as identified by the site type. Although the range for the thickness measurements are similar for each site type, there is a greater density of those over 60mm for the vicus sites. The close connections that must have existed between the military community and those inhabiting the vicus sites provides one possible explanation for the results seen, and this correlates closely with the data for quern thickness at military sites presented in Chapter 5.

Table 6.1. Comparison of quern thickness by site type

Site type	Total querns recovered.	No of querns over 60mm thick.	% of thicker querns from each site type.	% of thicker querns within each site type
Small town/ settlement	71	16	38%	23%
Villa	15	5	12%	33%
Military Vicus	28	21	50%	75%
Total	114	42	100%	-

A similar process was undertaken for the lower quern stones (n=62) (figure 6.9), which showed a 'normal' disposal thickness at around 20–50mm. 79% of lower querns (n= 49) fell within this range, with 20% (n=13) in the upper range over 50mm. However, unlike with the upper stones, the lower stones do not demonstrate such an obvious difference between those within the 'normal' range of thickness, and those in the upper range. The variation is less pronounced and a shallower gradient on the graph. This is marginally steeper after a thickness of approximately 55mm, with a much lower frequency of querns with these thicknesses.

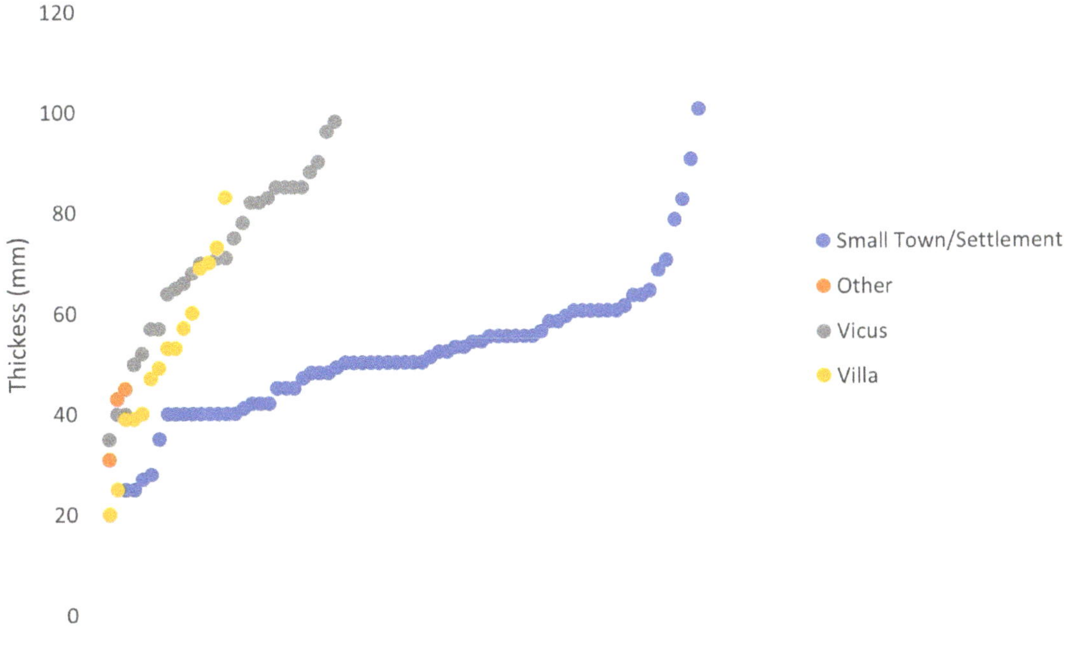

Figure 6.8. Individual thickness for each upper quern stone in the dataset, identified according to the site type from which it was recovered (T=114).

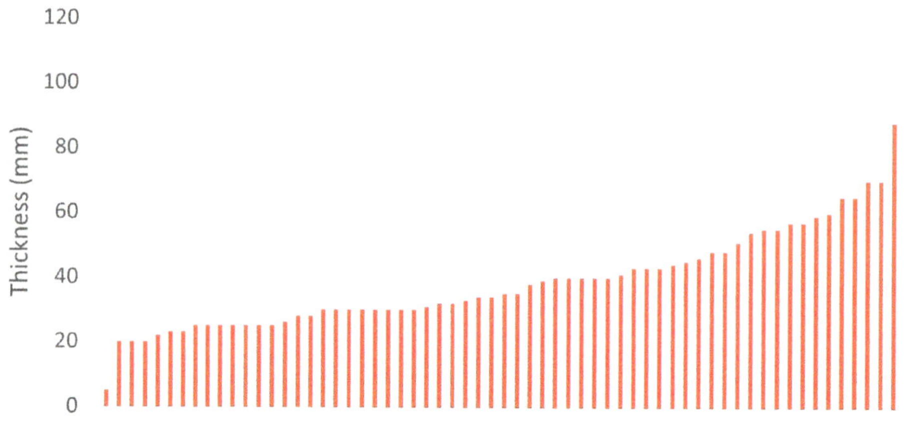

Figure 6.9. Max thickness of lower quern stones at edge (mm) (T=62).

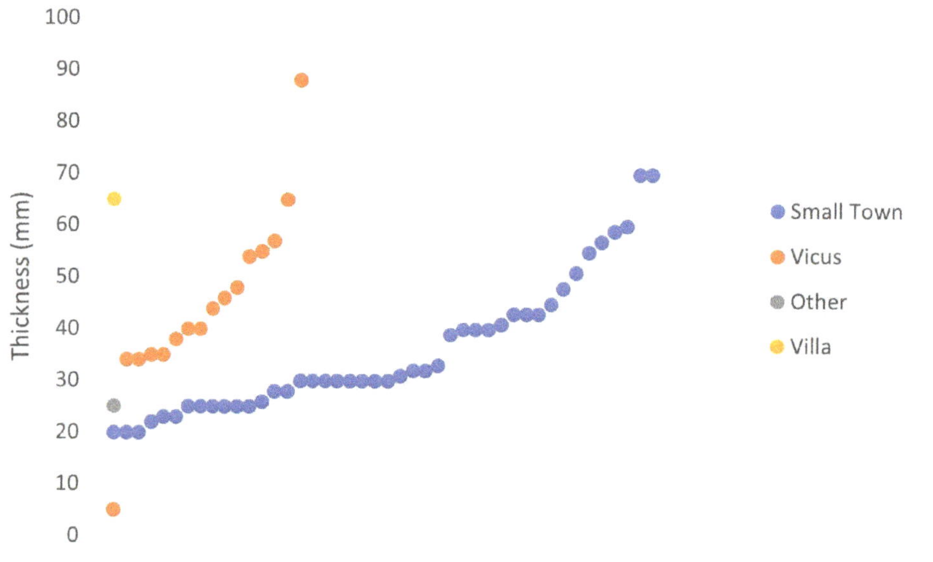

Figure 6.10. Individual thickness for each upper quern stone in the dataset, identified according to the site type from which it was recovered (T= 62).

Although the trend is less clear for the lower stones, it was still worthwhile to examine the thicknesses in relation to the site types from which they were recovered. These were also added to a scatter plot to demonstrate the individual thickness measurements for each lower quern in the dataset, identified according to the site type from which it was recovered (figure 6.10). Again, although the ranges of thickness for small town/settlement sites is like that of the vicus sites, the vicus sites have a slightly higher threshold of thickness when they were deposited. The thinnest querns at vicus sites were approximately 35mm thick before they were thrown away, whereas 61% (n=27) of all querns recovered from small town settlement sites (n=44) were less than 35mm when they were disposed of, a clear difference.

These results present some interesting possibilities with regards to the perceived value of a lava quern between different social groups, and the extent to which a quern could be used before it was deemed acceptable to dispose of/deposit. The pattern seen relates to the supply of replacement querns and differing accessibility to exchange networks. Other considerations might include the ease or difficulty posed in replacing a quern, and the question of expense/relative wealth of the consumer. It is important, therefore, to understand how different site types experienced connectivity, and how this may have had consequences on the value attributed to distinct types of material culture. Though it is well-known that vicus, villa, and roadside settlement sites had good access to a range of goods, as shown by the volume and types of material culture present, how this relates to the perceived value of an object is less well-researched. As objects with a quantifiable way of measuring use-wear, querns may allow us to explore this concept. Though the results shown here are merely suggesting a correlation between access to exchange networks and object value, more data would provide much greater resolution and would enable other trends, such as that of regionality, to be explored.

6.3. Distribution Analysis

A record was made of both the locations of lava querns/millstone sites and sites where querns/millstones of other lithologies were found, but none of lava. It should be noted here that the sites with lava present usually also had querns of indigenous stone types within the same site. These are not presented on this distribution, though a more detailed analysis of this variable will be undertaken as part of the examination of the province-wide dataset (Chapter 5). To explore the economic aspect of quern/millstone transport, trade and exchange, the location (assumed or confirmed) of Roman period roads were also plotted (road data derived from Talbert & Bagnall (2000), as used by Ancient World Mapping Centre (2021), alongside the location of the ports or harbours that are thought to have been in use during this period (data compiled by Cleere (1978)). It must be noted that the chronology of the roads and ports/harbours is not identified on this map, and they will not all have been present or in use throughout all phases of Roman occupation. However, due to the scale of the data being examined and to enable better visualisation, it is necessary to simplify some aspects on the maps and to instead deal with nuance, including chronology, within the discussion and interpretation.

Similarly, location data for the presence of military forts/fortresses was added to the map using a variety of source data, though this is only visible for the close-view maps. These are all confirmed military sites, though not all have been excavated and none are identified here with regards to chronology or phases of use. Those military sites identified solely as 'Saxon shore forts' have not been included, as these are less relevant to current investigations concerning early to mid-Roman period lava quern trade and exchange. The reason for the addition of military sites is due to the long-standing hypothesis that lava querns and millstones were initially introduced to Britain with the arrival of the Roman military (Gluhak & Hofmeister, 2011, p. 1604; Green , 2017, pp. 171–173). These objects are thought to have been imported as part of military contracts and used primarily within military contexts (Peacock, 1980, p. 50). Therefore, any relationship between military and rural sites, in terms of distance and interconnectivity, is a key factor to consider when examining rural distribution.

6.3.1. General Distribution: Examining Presence/Absence of Lava Milling Tools

The initial means of analysis was to use 'point' data with indications of volume for the number of lava and non-lava milling tools from all rural sites. The main reason for this was to provide a way of displaying both presence and absence data on one map, something that cannot be done easily using a heatmap. This, therefore, allows for an assessment to be made with regards to 'missing data' and regions of Britain where we are unable to make interpretations relating to quern or mill use. Querns and millstones were vital tools for daily food processing activities wherever grain was consumed, and should be present, in some form, in most regions where domestic occupation is evident. Centralised methods of grain processing may have negated the need for querns to be present at multiple sites in a region, but this does not account for some of the large gaps in data seen within this dataset. The gaps in distribution mark a much wider trend in regional deficits of archaeological data and highlight some of the issues faced when attempting to identify provincial-wide patterns in distribution and consumption. Missing data in central East Sussex, the north-west, central-west and Wales is not a problem that is confined only to this project (Fulford & Brindle, 2016, p. 9) and more work must be done to understand the role these regions played in rural Roman Britain and to better contextualise the data that we currently have.

From figure 6.11, there is an east-west and a north-south divide in the data, which follows expected trends according to the distributions of other finds types, for example coins (see Allen, et al. (2018) to map other finds groups from the RSRB data). Other expected patterns specific to lava querns and millstones are also visible (see Chapter 5, distribution section). It has long been speculated that one of the places that lava milling tools entered the province was at Colchester, and that this occurred in the early post-conquest period as demonstrated by pre-Boudican examples in the city (Buckley & Major, 1983, p. 73–75). This has been interpreted as the result of Colchester's early associations with the military. Meanwhile, previously explored widespread distribution of lava querns at nearby settlement sites have been taken as indicative of 'civil/commercial' trade involving these objects (*ibid*). Although the distribution shown here does not absolutely confirm these findings, it does also present a high concentration of lava around Colchester and other parts of East Anglia and supports this theory. Similarly, several researchers have suggested that London was one of the primary locations that received lava querns and millstones via the Rhine using North Sea trading routes (Morris, 2010, p. 79). Therefore, it is not surprising that there is a dense cluster of lava present at settlement sites around the Thames estuary and the Greater London area. The urban material will be discussed in Chapter 7, but it is interesting to note here that rural finds are clustered around the Thames estuary region suggesting the impact of water transport rather than London acting as a centralised distribution area for these objects. However, the elevated level of modern development in this area could be over-emphasising the concentration of lava distribution.

In areas with high numbers of lava querns (especially the south-east), there are few sites that have querns but no lava present. It would be possible to surmise that trade and exchange mechanisms involving lava milling tools had consequences for the distribution of non-lava and/or vice versa. It is likely that there are complex inter-relationships at play between the use of different stone types and the choice between imported and locally sourced querns. These are unlikely to have been uniform across space and time. Aspects of this distribution that need to be explored in more detail include chronology, local geology, preservation of artefacts and issues related to excavation and recording. Nonetheless, the data provide an opportunity to understand how and why this may have occurred. In the

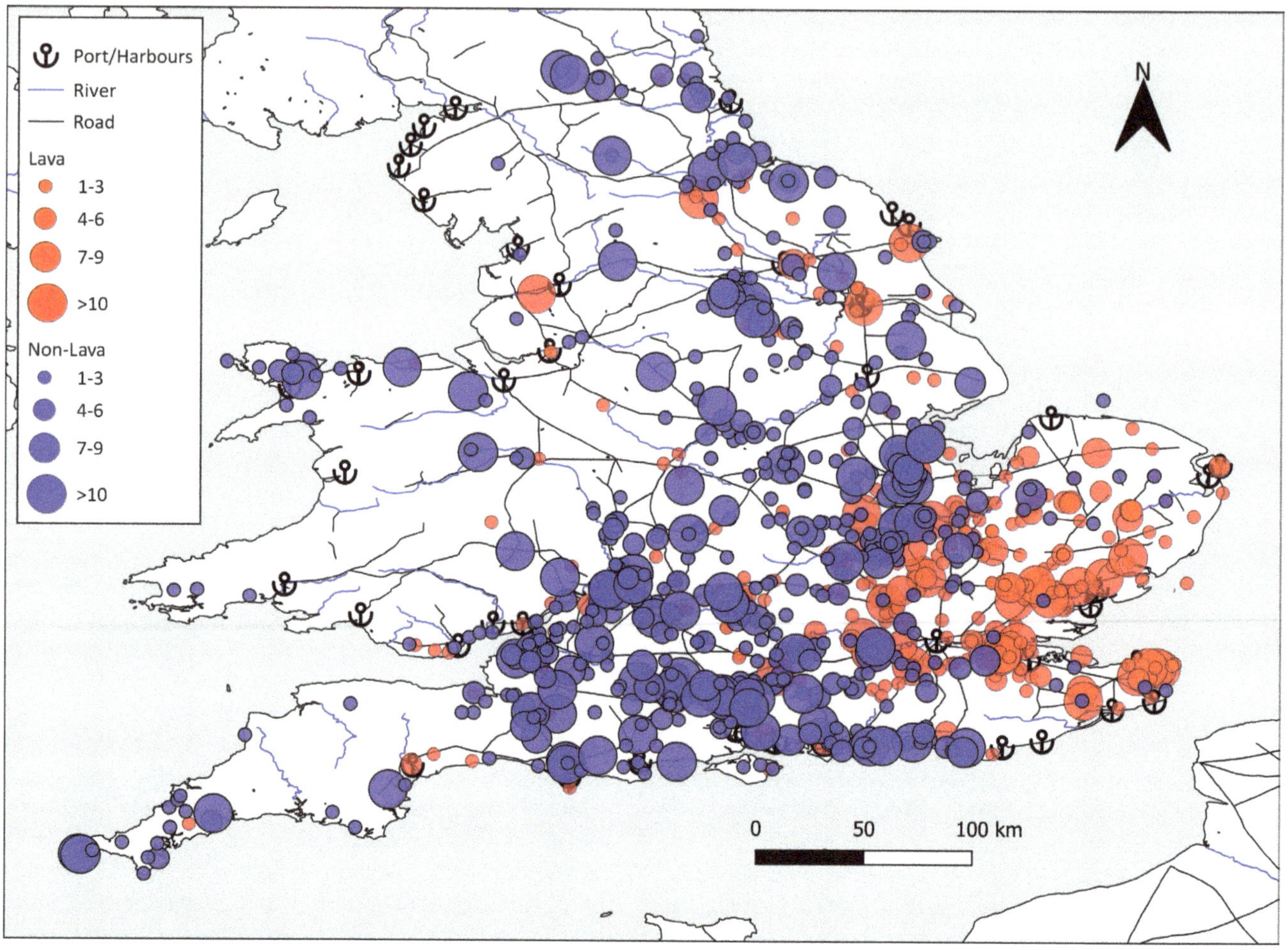

Figure 6.11. Rural distribution map to show the presence of lava milling tools in Roman Britain and their relative volumes, alongside locations of sites with milling tool, but none of lava (Background GIS data courtesy of the Ancient World Mapping Center).

South-East, sites located away from Roman roads appear to have no access to lava querns but have produced small numbers of querns of other lithologies. Examples include Nonington in Kent, where two sandstone quern fragments were recovered, and north-east Horley in Surrey, where four greensand querns were found. Both these sites are dated from the Late Iron Age to early Roman periods, indicating that the chronology is the variable affecting the absence of lava at these sites, and that the road systems may have not been as relevant in these specific cases. The clear absence of lava along the south coast is another trend that is worth considering further and will be addressed in later discussions.

Lava querns occur in a wide band from the dense distributions around London and Colchester, moving westwards towards Wales and the Bristol Channel where there is a presence of lava at multiple sites. The route of the Thames has some correlation with this data though there are sites to the north and south of the river that follow the Roman road network west out of London where the presence of lava is also clear. The distribution makes it difficult to distinguish between the role of road vs river transport along this route, as either or both may have been involved. Unlike the eastern region of Britain and the south-eastern area around London and the Thames estuary where lava sites are dominant, the presence of lava is more frequently interspersed with non-lava sites further towards the west. If we assume that London and Colchester were main distribution centres for lava milling tools, it is expected that volumes and concentrations would decrease away from these centres. Another way of examining and visualising this is through a heatmap to display the quantities of lava milling tools by location at the recorded sites within the database (figure 6.12).

As seen in the previous distribution, the heatmap shows a visible difference in distribution between the north of Britain and the south, as well as between east and west; the south has greater volumes than the north, while there is also a marked greater volume in the east compared to the west. There are clear focal areas of distribution, which cluster around the urban centres of Colchester, London, and the Thames Estuary region. There is also a heavy concentration around St Albans and Chelmsford, which needs to be investigated more fully. These areas may be within the catchment areas of London and Colchester respectively and so, it is worthwhile exploring this further with the addition of the urban and military site data (see Chapters 7 and 8). At present, it appears that urban centres

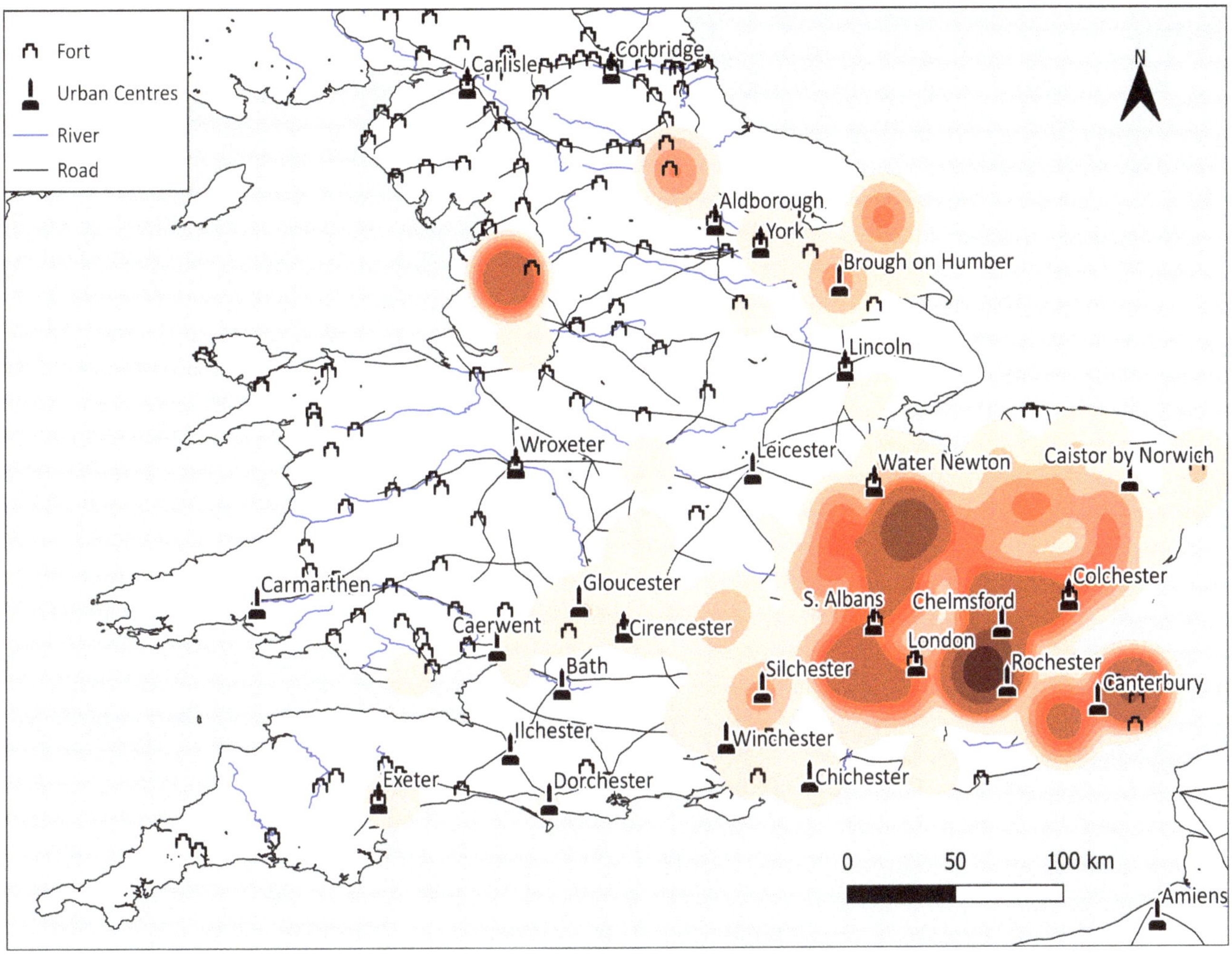

Figure 6.12. Heatmap showing quantities of lava querns at rural sites (Background GIS data courtesy of the Ancient World Mapping Center).

in the south-east and east of Britain were foci for rural site lava distribution, while in the north, the clusters mostly centre on military sites. The southern distributions centred on urban areas is of interest. Shaffrey (2021) concluded that at Silchester, querns of other stone types were not distributed to the hinterland via the urban area. Similarly, rural distributions of pottery in the hinterlands of Silchester and Wroxeter have supported the idea that the urban areas were not market centres of distribution (Evans, 2007; Fulford, pers. comm.). However, as the Wroxeter hinterland did not yield much in the way of imported pottery at all, it is unclear whether this signifies a cultural and social separation between the consumption of imported goods in the urban vs the rural populations in these regions. Nonetheless, the distribution of lava in urban hinterlands of eastern parts of Britain suggests that lava querns saw quite different distribution methods and mechanisms than those of indigenous quern types. However, these mechanisms were not unilaterally replicated across all regions of the province, especially in the west. The distribution seen may also be related to the fact that many of the urban areas in the south are also ports. However, it is possible to see that landlocked Silchester (*Calleva Atrebatum*) also displays a similar pattern of lava quern distribution, as does St Albans (*Verulamium*). Volumes of lava decrease further west and north, and away from these centres. It seems clear that the ports at Colchester, Rochester and London had important roles to play in terms of civilian lava consumption, while other ports on the southern and western coast see much lower volumes of lava spreading out to rural sites. The ports at York and Brough-on-Humber show high concentrations of lava reaching surrounding rural settlements.

Urban centres further west also show some clustered lava distributions, such as that seen at Silchester and Gloucester, though these are significantly less concentrated than around the more easterly urban areas. It is interesting to note that in the south of England, many of the rural concentrations of lava tend to be focused round urban centres, while in the north, excluding Hadrian's Wall, this occurs more often in association with military sites. This may simply be due to a militarised north and an urbanised south, though it will be worth exploring further to assess whether different parts of the country were supplied with lava through different mechanisms of exchange. The addition of the urban and military data will help to distinguish this more clearly. In the area of Hadrian's Wall, there are virtually no lava milling tools at rural sites within the dataset, other than those around the urban area of Carlisle.

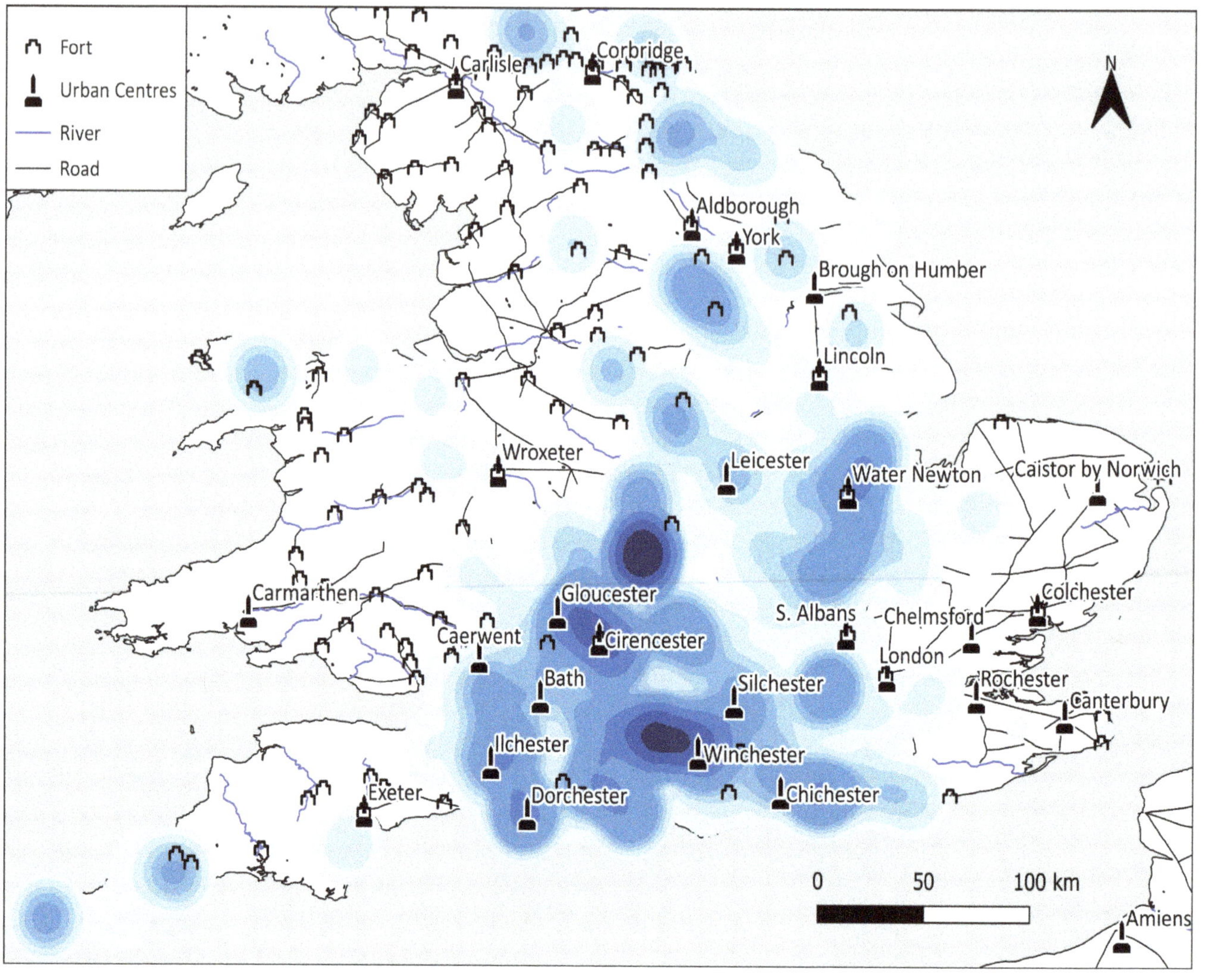

Figure 6.13. Heatmap distribution of non-lava querns and millstone volumes at sites with milling tools present, but none of lava (Background GIS data courtesy of the Ancient World Mapping Center).

The distribution of sites with only non-lava milling tools (figure 6.13) shows us that finds have been recorded from the rural hinterland areas, especially on the eastern part of the frontier, but that lava specifically was absent. The lack of lava in these areas suggest that quite different distributive mechanisms were in place, and that these appear to have excluded the civilian populations. However, there are relatively few sites from the hinterland regions, and as a largely understudied area, it is difficult to know if this is a true representation of past activities (Breeze, 2018, p. 2–4). A shared geology for neighbouring rural and military sites at Hadrian's Wall would result in similar soil conditions, and preservation bias during comparison is not an immediate concern on an inter-site level. Though the data are sparse for rural settlements, it should still provide a comparative picture of what was present or absent, though data resolution is an issue, especially towards the western hinterland regions of Hadrian's Wall.

The heatmap showing the locations and concentrations of non-lava milling tools at sites where lava was not present shows very different locations of clustering compared with lava presence, being most concentrated in a wide band stretching from the central southern coast northwards, through the midlands, heading eastwards past Lincoln and following the trajectory of the north-eastern coast before petering out slightly north of Hadrian's Wall. Unlike the lava milling tool heatmap, any centralised distribution areas are much harder to identify, and distribution occurs as a spread, with a much larger range of smaller clusters across a wider area. This is to be expected due to numerous quern quarries and production areas within the province. The known or assumed locations for some of these quarries is shown in figure 6.14. This allows us to view how, for example, the concentration of non-lava querns around the Chichester district probably relates to the local production of Lodsworth Greensand querns at Pits Copse (Peacock, 1987), while the clusters seen around the Bristol Channel region may relate to Old Red Sandstone, thought to have been quarried from several locations within this region as observed by detailed examination of distribution and other evidence (Shaffrey, 2006).

Although not related to this current research on lava querns, knowledge of other quern production areas is fundamental to our understandings of how any trade

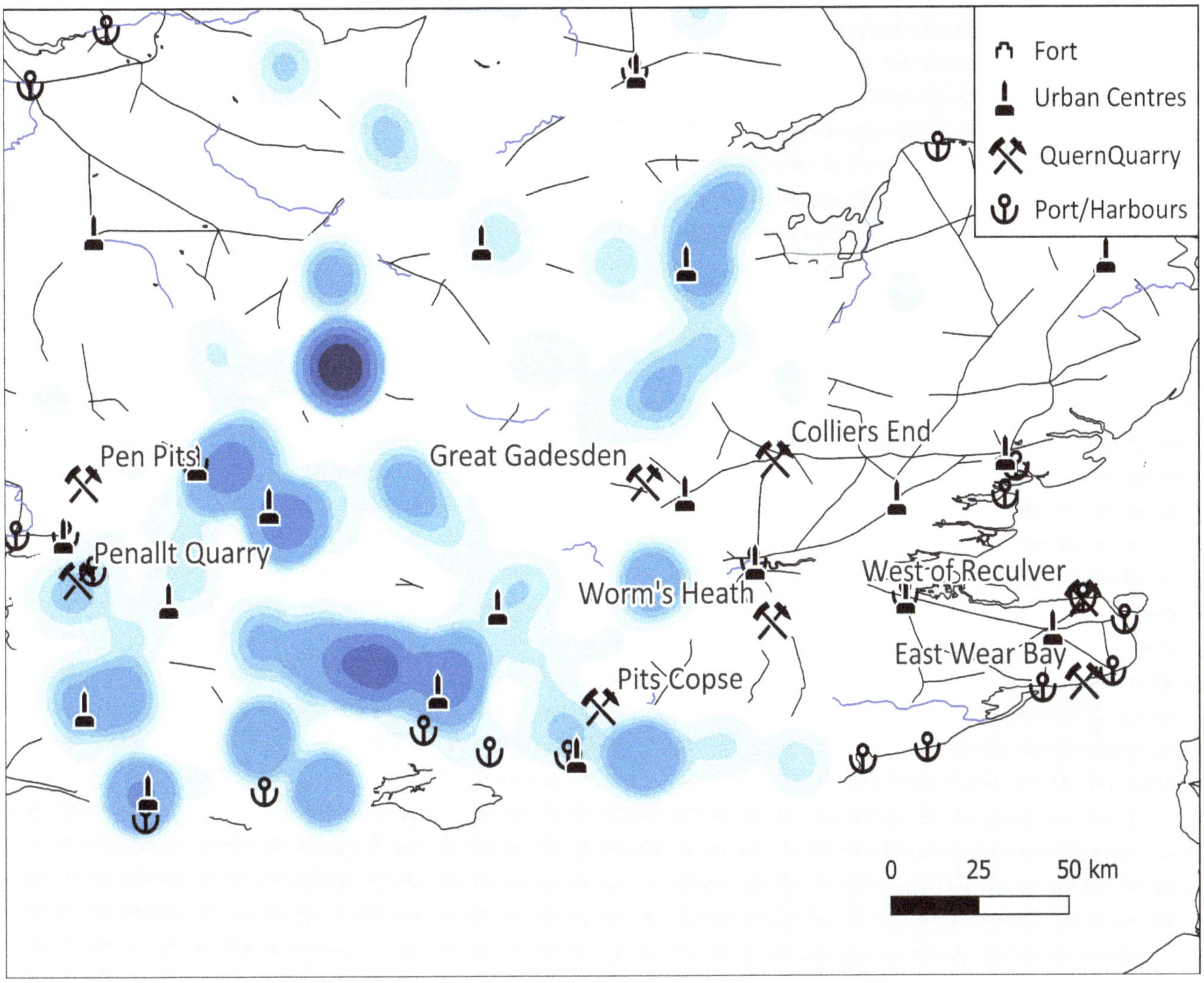

Figure 6.14. Heatmap distribution and volumes of non-lava querns at sites with no lava present and the location of known or assumed quern quarry sites of indigenous stone types. 1-Pen Pits- Lower Greensand, 2-Pits Copse- Lodsworth Lower Greensand, 3-Colliers End- Hertfordshire Puddingstone, 4-Worm's Heath- Puddingstone, 5-Great Gadesden- Hertfordshire Puddingstone, 6-Penallt Quarry- ORS, 7-West of Reculver- Calcareous Sandstone, 8-East Wear Bay and Copt Point, Folkstone- Lower Greensand (Peacock, 2013, pp. 136–137) (Background GIS data courtesy of the Ancient World Mapping Center).

involving lava operated within the wider Romano-British economy. It should be expected that interrelationships existed with regards to the use of lava or non-lava milling tools from dissimilar sources, and that any trade or exchange of these goods would have been affected by proximity to stone source/distribution location, costs of import/transport, availability, and desirability. It would be desirable, therefore, to consider these carefully and to include them within a wider discussion of the stone trade in Britain.

6.3.2. Close-up Analysis of Distributions in East Anglia and the South-East

The main area of interest, in terms of lava querns on rural settlements, is in the east and south-east of England where there are large clusters of lava milling tools at a high number of sites. As previously discussed, it is unsurprising that these appear to be focused round Colchester, Kent, and London, as these have been long recognised as possible ports where lava was imported into the province, and locations for redistribution. However, using a closer view of the east and south-east region (figure 6.15), it is possible to see that in the direct vicinity of London there are fewer lava milling tools present. This may be the result of site misclassification in the wider London area, or the possibility that the area was not as intensively occupied or farmed during the Roman era. In direct comparison, there are some large clusters of lava in the immediate areas surrounding Colchester.

The main concentration of lava in the Thames region is further east, approximately 10km to the west of the 'civilian port' of Rochester (Cleere, 1978; Fryer, 1978, pp. 150–153; Rippon, 2008, p. 89), and this cluster is visible on either side of the Thames. This gives an indication that civilian trade and exchange of lava objects in certain regions may have occurred via different ports to that of

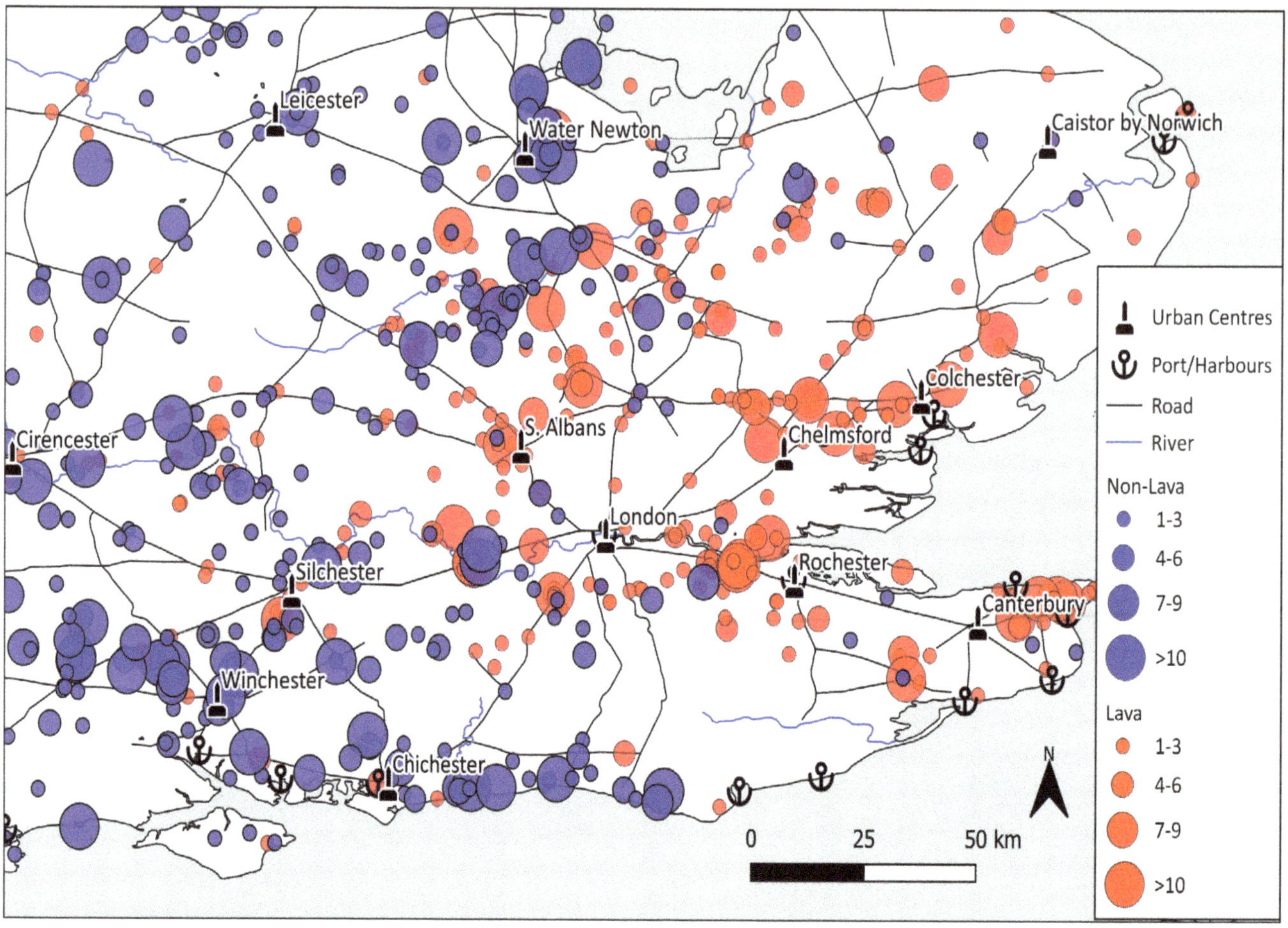

Figure 6.15. Distribution map to show the rural presence of lava milling tools in the eastern and south-eastern region of Britain, and rural sites with milling tools, but none made of lava (Background GIS data courtesy of the Ancient World Mapping Center).

military and urban supply. However, intensive modern construction in the Thames Estuary region as part of the Thames Gateway Project and HS1, alongside high levels of gravel extraction may have skewed the visibility of finds at large-scale excavations. Examples include Mucking and Springhead, which is where these clusters are focused. Nonetheless, the large volumes of lava querns and millstones at these locations, both rural and civilian in their context, may warrant further investigation. That the Thames was a means by which lava milling tools were conveyed into Britain seems to be indisputable. What is less clear are the mechanisms by which these found their way to rural settlements and whether the density of finds are indicative of a 'commercial' trade in these objects.

It is also interesting to note the small cluster of lava around Richborough and Reculver, both of which were ports and early military sites (Historic England, 1999; Millett & Wilmott, 2003, pp. 184–185). This distribution fits well with traditional interpretations relating to early military use of lava for milling (Gluhak & Hofmeister, 2011, p. 1064) and serves to demonstrate how commodities brought to Britain for military use could 'spill' out to nearby settlements that were geographically or strategically connected to military supply, and especially to ports. Again, however, the area has recently seen some intensive construction and development that could be providing a false impression of concentrations in comparison with other regions.

Colchester, London, and St Albans also were urban centres, though each had vastly different characteristics. Colchester was a presumed port with an early military presence. The port was likely to have been in the area now known as The Hythe, in which case seagoing vessels will have needed to offload cargo to lighter craft downstream of the town (Crummy, 1997, p. 72). London was also a port, with a speculated early military character, while St Albans was civilian. This makes it difficult to determine which aspects of these sites that produced the driving force behind their role in rural distribution, or whether it was simply one factor of many. The fact that Colchester and London were some of the earliest occupied sites in Roman Britain may have some bearing on the early introduction of lava to the surrounding rural areas. If lava became less popular and imported in lesser volumes in later periods, this would have a clear impact on the number of lava milling tools that turn up on rural sites elsewhere in the province.

Another feature of this south-eastern distribution is the vital role that roads and rivers must have played in the movement of lava milling tools. It is apparent that both modes of transportation were utilised and, despite the bulky nature of the goods, transport via road appears to have been just as vital as movement over water. This is not a new concept with regards to the movement of querns, as similar patterns of distribution via the road network were observed by Shaffrey (2006) for querns of Old Red Sandstone. The distribution does appear to suggest that the probability of having a lava quern in the south-eastern region of Britain is strongly associated with proximity to major Roman period transport infrastructure. What is most significant is that many of these routes via road or river connect clearly with the locations of the main lava clusters, in many cases, along a direct path. For example, the route from Colchester to St Albans presents multiple sites where lava is present. This appears to confirm that there were centralised distribution areas, such as Colchester, where lava milling tools were brought into the province and that it was from these locations that redistribution occurred via road and/or river to rural sites. This differs from the distribution of querns of other stone types, which were not dependent on urban sites in the same way that imported lava was (Shaffrey, 2021).

6.3.3. Distribution by Site Type

The map in figure 6.16 shows the distribution of lava milling tools with identification of the type of rural site from which they came, using the RSRB project site categories. In the east and south-eastern areas of Britain, there is less discrimination in terms of the types of sites where lava was found. Virtually every site type is represented, and none is more dominant than another. Further west and away from the Colchester and London catchment areas, where the site clusters become less dense, there are more roadside settlement sites, defended settlements and villas with lava present, and the small town/settlement and farmstead sites become much less likely to have one of these objects present. This is the pattern seen as far west as Gloucester, where farmsteads are the least likely site type to have a lava milling tool. In the south-west, data are very sparse and there is only enough to say that two farmsteads close to a vicus site had lava present. Wales and the north-west of Britain show similar distributions. Lava is predominantly found at roadside settlement sites and vici. This is also true for the area around Hadrian's Wall, though defended small towns have also shown a lava presence. In the north-east and sites north of Lincoln, there is a scattering of different site types where lava was found. There are several farmsteads and three small town/settlement sites, though these mostly are near to either a vicus, villa, or roadside settlement site.

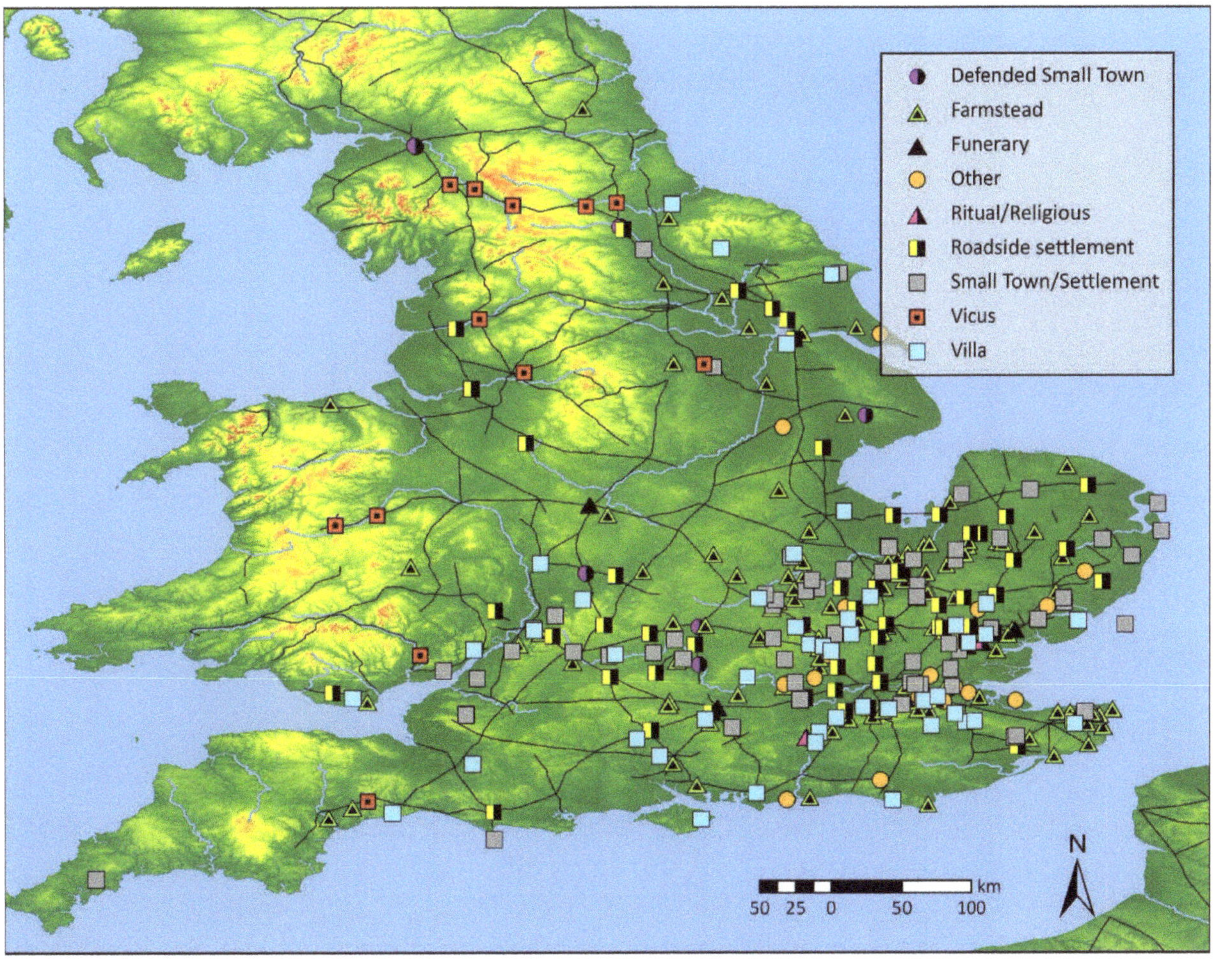

Figure 6.16. Distribution of lava milling tools by rural site type (Background GIS data courtesy of the Ancient World Mapping Center).

For comparison, Figure 6.17 shows the distribution for non-lava sites, again by site type. It is interesting to note that the sites in northern East Anglia without lava present are predominantly roadside settlements. Examples include Woodcock Hall and Crownthorpe, both in Norfolk. This is unexpected, as this site type tends to be more likely to have more finds in general and more likely to have objects that have been imported or manufactured elsewhere (Smith, 2016a, p. 235–236). This will be worth some further investigation to determine why this may have occurred. Other than this, there are very few villae or roadside settlements within the London and Colchester catchment areas that have milling tools, but none of lava. West of London, it is mainly small town/settlement and farmstead sites that do not have lava querns or millstones, though south and west of Winchester, there is less discrimination based on site type, and more villa and roadside settlements become included. In the south-west, it is mostly farmstead and small settlement sites that use non-lava querns only, though this should be expected due to a decrease in the range of settlement types in this region more generally. Again, Wales and the north-west of Britain show similar patterns of distribution; several farmsteads, small town/settlement sites and some vici do not have lava. In Wales, these focused in the coastal areas or on the main river routes, though this may be due to the availability of data and site information. In the north-east of Britain, farmsteads and small town/settlement sites without lava are the most prolific, showing that alternative means of grinding and milling were used instead of lava. Lava was clearly not a popular choice in this area, either due to restricted access or choice.

A close-up view of the east and south-east region is shown in figure 6.18, to provide clearer detail of the area with the highest density of sites. There are many small town/settlements, farmsteads, and roadside settlements, especially in the northern part of East Anglia, with lava present. There is a clear widespread distribution of lava, probably from Colchester where high volumes must have been imported for commercial exchange, and this reached the rural areas of eastern Britain to the extent that all types of rural settlement had access to the product. This is also true for the area around London, though there are more roadside settlements and villas that show a lava presence than smaller settlement types. This may just be due to the different settlements that existed within this region. The small cluster of sites around Richborough and Reculver with lava present are, also, of a range of site types. This is like the pattern seen in East Anglia and could be indicative of large numbers of lava objects being available at this location for wider commercial exchange with other nearby rural settlements. The overall larger numbers recorded from Kent are, however, also a reflection of the large-scale road development and construction in the region in recent years, which has resulted in better documented assemblages from rural sites.

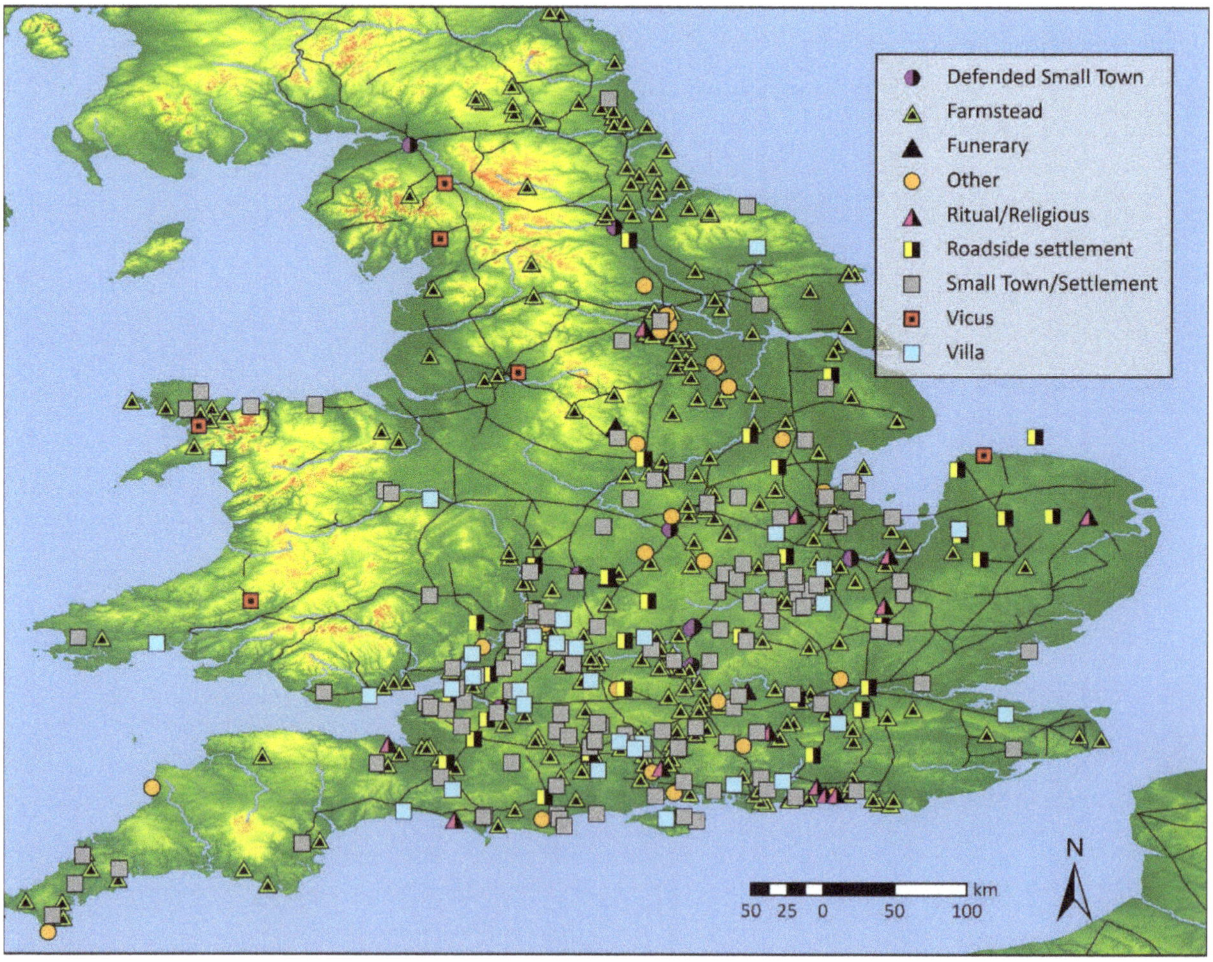

Figure 6.17. Distribution map showing rural sites with milling tools present, but none of lava by rural site type (Background GIS data courtesy of the Ancient World Mapping Center).

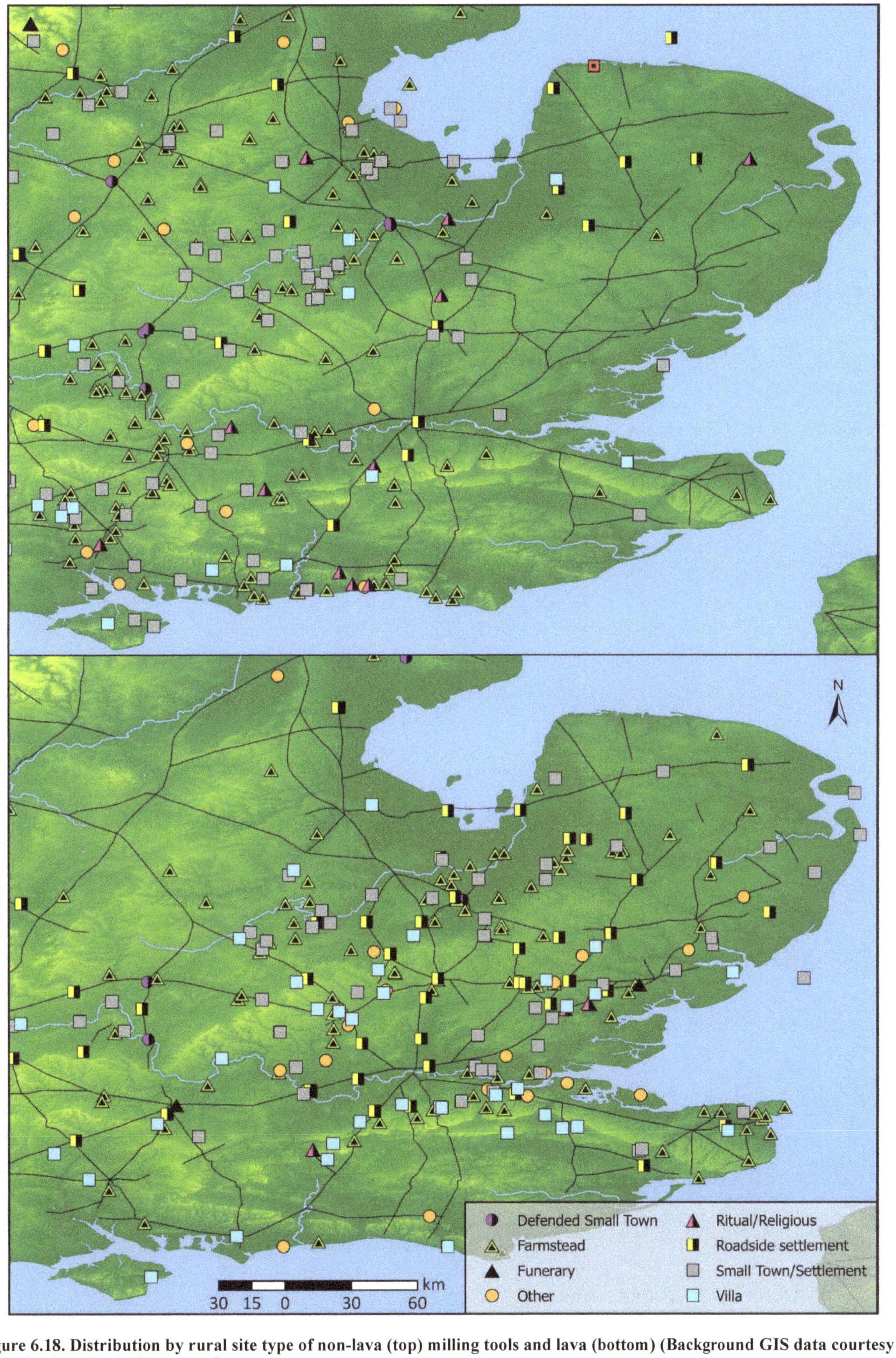

Figure 6.18. Distribution by rural site type of non-lava (top) milling tools and lava (bottom) (Background GIS data courtesy of the Ancient World Mapping Center).

Further west from London and Colchester, though fewer sites in general have lava milling tools present, the mix of site type also changes. As shown in figure 6.18, the site types with lava present west of Silchester comprise the following: 10 farmsteads, 6 villas, 3 defended small towns, 7 roadside settlements, 5 small town/settlements and 1 ritual/religious site. Proportionately, it is expected that there would be a much higher number of farmsteads and small town or settlement sites in existence and so this shows that there is a greater emphasis on villa and roadside settlements and that there is a higher chance of finding a lava quern or millstone at these sites than on a smaller rural settlement site type. In comparison to the data for this same area that shows sites with milling tools present but with none of lava, we can see that farmsteads and small town/settlement sites were more heavily reliant on other stone sources for their milling needs. However, a few kilometres west of Winchester, the situation changes again, and villas also become more likely to have a non-lava milling implement than to have been using lava.

In this region, it is also possible to see that non-lava milling tools are more likely to found than lava at sites that are characterised as primarily ritual/religious in nature (figure 6.19). This seems to occur in a localised area to include Chanctonbury Ring, Lancing Down, Slonk Hill, Reigate Road, Wanborough Temple, Silkstead Sandpit, Weycock Hill. A smaller cluster of cases also exists between Water Newton and St Albans, with the possible shrine at Aldwick and the ritual/religious sites of Haddenham Shrine and Colleyweston. Lava occurs at four religious/ritual sites, the Romano-Celtic temple at Farley Heath, Great Chesterford Temple, the shrine and pottery production site at Ivy Chimneys and at Boreham Principia, the latter of which is only possibly a shrine and more popularly recognised as a *principia*. Lava is very rarely present at ritual and religious sites, with indigenous stone querns being much more commonly associated with these contexts, though mainly in a very restricted area south of Silchester, with a smaller group north of St Albans. This could simply be the product of proximity to sources of these indigenous stone types. For example, the cluster of non-lava querns close to Chichester could be the result of its location near to the Lodsworth quarries that were used for querns and mills in the local areas (Peacock, 1987). However, it would also be possible to argue that there was a deeper social and cultural association with these objects at these locations due to their production or distribution in these areas. For lava, these connections may not have existed due to it being an imported product.

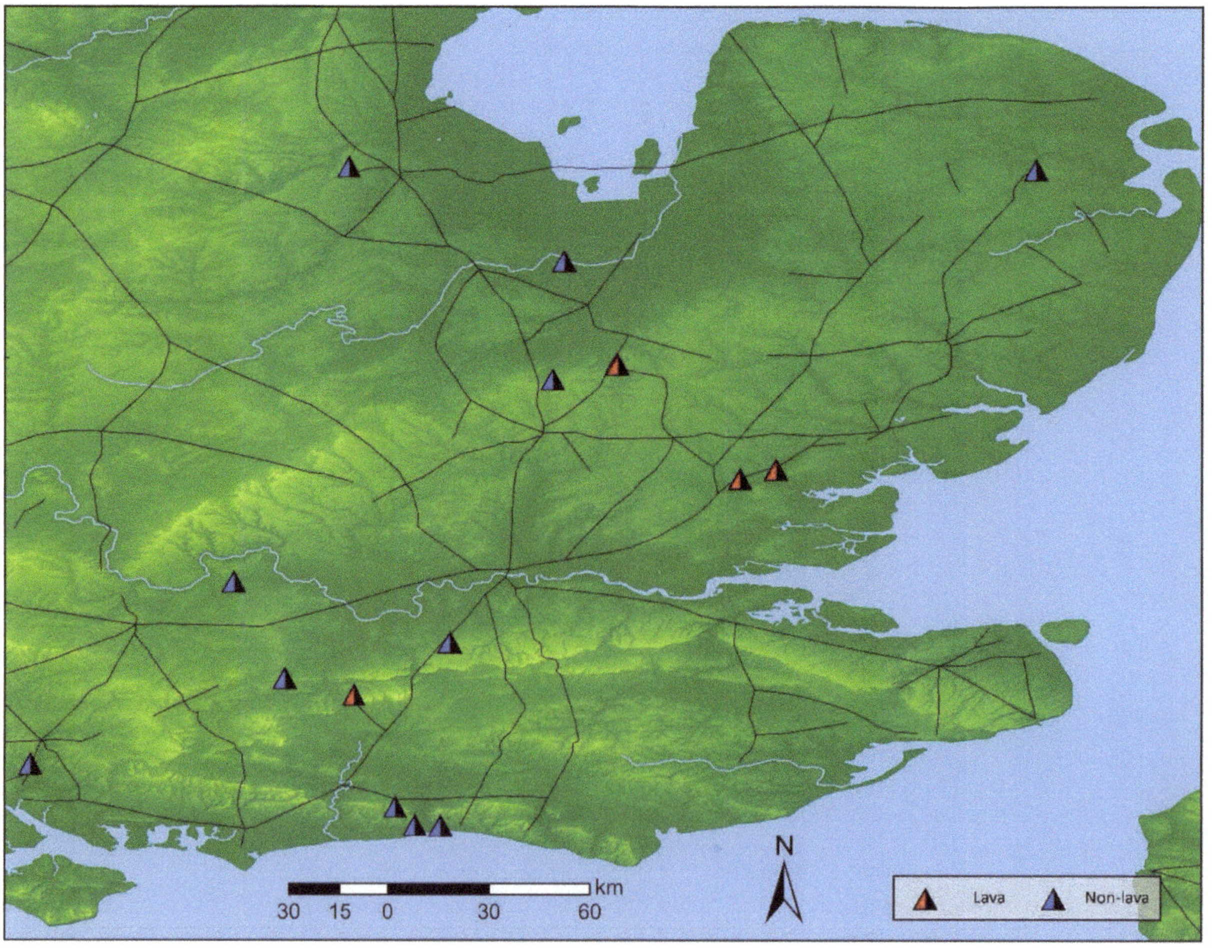

Figure 6.19. Location of rural religious/ritual sites with lava and non-lava milling tools in the east and south-east of Britain (Background GIS data courtesy of the Ancient World Mapping Center).

6.3.4. Chronological Distribution

Due to the often-residual nature of lava finds, with querns often reused or recycled, chronological definition is difficult to establish. However, as there is a large volume of data within the rural database, this investigation hopes to provide a broad overview of chronological distribution. Lava querns or millstones that are likely to be residual or recycled have been omitted, and attempts have been made to only include those with well-dated, secure contexts with a narrow date range. Initially, date categories were split by century; first, second, third, fourth, and fifth (figure 6.20). This was to get a general impression of the pattern of chronological change and to highlight periods where better definition should be sought. Due to some dating ranges spanning more than one century, some objects have been included on more than one distribution map. Unfortunately, phasing is different for each site, and it is not possible to correlate all dating information within the same scheme.

As expected, the rural areas around Colchester, London, Richborough and St Albans see the earliest concentrations of lava. As the first three are early Roman ports, this fits well with what is currently known about the modes by which lava entered the province. The role that St Albans may have played will be subject to further investigation as part of the urban dataset in Chapter 7. Other sites that have lava from this earliest period include rural settlements in the north and Wales that are close to forts or fortresses, though none is present at the sites around Hadrian's Wall. The concentration of lava seen in the south-east of England continues into the second and third centuries, which follows the trend expected when considering the importance of North Sea trade during this period, and the movement of goods between the Rhineland and Britain as seen by other material culture types. Other than the south-eastern region of Wales, there is extraordinarily little lava to be found in Wales after the second century, while it is more prevalent in the area around Hadrian's Wall in the second and third centuries, which fits well with current historical narratives relating to the movement of troops during these periods. Large volumes of lava were reaching rural settlement sites in specific areas of the province in the first century, while in the second century, distribution was more widespread but not at the concentrations seen in the earliest period.

There is a pronounced contraction in the number of sites showing a presence of lava from the third century onwards and it is likely that much of the lava still visible on the heatmap for the fourth century represents querns or millstones that are residual. However, there is no clear way of defining the extent to which residuality has affected the distributions seen. As explained in Chapter 5, a time lag probably existed from when a lava quern was last used till the point at which it reached its final deposition location. It has been speculated that the import of lava milling tools largely ceased after the third century and the distribution maps support this idea, though they do not prove it completely. One way of exploring this further is to look at sites where occupation commenced after the mid-third century and the probability that lava was present. This reduces the chance that any lava present could be residual, though it does not eliminate the possibility that lava was brought from elsewhere as recycled or reused stone. Overall, there were 23 sites within the dataset that were dated to AD 250 or later, 16 of which produced only non-lava milling tools, while 7 produced lava (figure 6.21).

Of the sites that produced lava, most cannot be confidently shown to have produced securely dated lava of a late Roman date. Two, Old Shifford Farm in Oxfordshire and Totternhoe in Bedfordshire probably had earlier phases of occupation, raising the possibility of residual material being present (Matthews, et al., 1992; Hey, 1995). The site at Bletsoe in Bedfordshire is only tentatively dated to the mid-third century through sparse dating evidence (Dawson, 1994), while the small fragment of lava recovered from the surface of a ditch at Beech House Hospital in Suffolk was not from a secure context (Muldowney, 2008). The excavation report for Hartismere School in Suffolk suggests that the lava from the site was old before it was deposited, as shown by the condition of the fragments (Craven, 2012). Nonetheless, two of the four fragments from Wickham Barn in East Sussex appear to be late Roman in date, and all must post-date the earliest occupation phase that commenced after AD 250 (Butler & Lyne, 2001). The site at Summerton Way in Greater London raises some interesting questions in relation to the distribution of lava at rural sites due to the presence of Mayenware and Eifelkermik pottery. Although the site did yield some second century pottery, suggesting that the lava fragments recovered are residual, evidence strongly suggests that occupation at the site began after AD 250. It is possible that broken lava querns were brought to the site for reuse, and the initial date of their use predates the occupation of the site. Nonetheless, Lakin (1997) has suggested that directed trade with the Eifel may have occurred at the location, and this might explain the presence of lava querns in such a late Roman context. If previous exchange mechanisms that moved lava querns and millstones into Roman Britain had slowed or stopped, it is possible that lava continued to enter the province via other means, such as alongside the movement of late Roman pottery from the same manufacturing region. It is likely that this did not occur on the same scale as previously, as Mayenware pottery is not abundant in Britain (Fulford & Bird, 1975), and is mainly distributed in the south-eastern region of England, with 90% of sherds recovered from Canterbury, Colchester, London and Richborough (Tyers, 1996c).

However, the exchange of lava querns may have continued under different conditions, as opposed to the redistribution to rural areas via urban and military sites as seen in the earlier periods. The finds and other evidence at Summerton Way suggest a high-status site, which may have offered the occupants more options in terms of trade and exchange with regions further afield. This is a situation that is not likely to be replicated for most Roman rural sites and could explain the relative deficit of rural sites with securely dated late Roman lava. It is possible to examine late Roman lava distributions with that of late Roman Rhenish and East Gaulish pottery, and this may help to establish whether such a correlation may have existed.

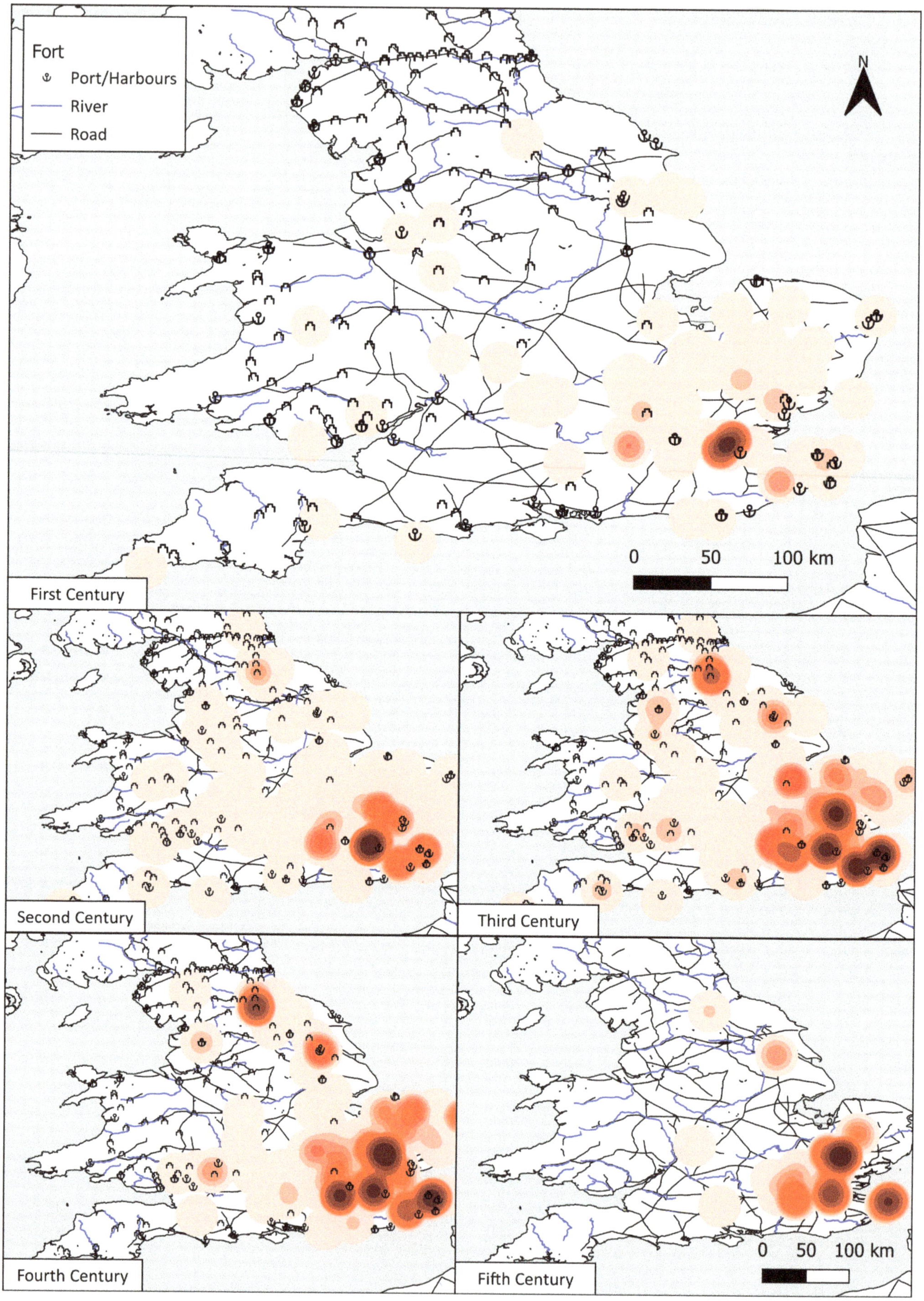

Figure 6.20. Chronological distribution of lava at rural sites, top first century, middle left second century, middle right third century, bottom left fourth century, bottom right fifth century (Background GIS data courtesy of the Ancient World Mapping Center).

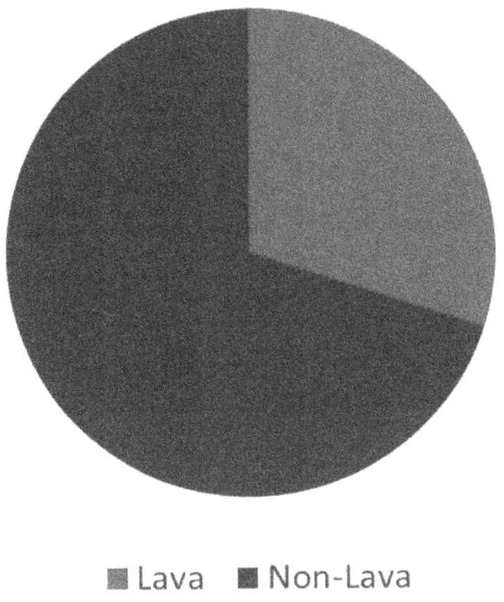

Figure 6.21. Sites with occupation dating after AD 250 showing percentage with lava (T=7) or only non-lava (T=16) milling tools present.

6.4. Conclusion

The analysis of the rural data has produced some interesting results, some of which will require further investigation to fully determine their potential significance. As objects that have long been associated with military life, this is the first time that lava querns from rural contexts have been subject to detailed examination, and it is apparent that they were just as significant within the rural sphere as the military. Large volumes of Roman lava milling tools have been identified on a range of rural sites in Britain and their use was common, especially in areas of south-east England and East Anglia. It is probable that in the east and south-east, lava querns entered civilian circulation via ports on the east coast, predominantly London and Colchester, though Richborough, Reculver and Rochester may also have been involved. Lava was redistributed from the ports and other urban areas, reaching rural settlements by river and road networks. In the north, west and Wales, redistribution appears to have occurred via military sites, and mechanisms of exchange seem different to those experienced in the south of the province. Rural sites around Hadrian's Wall provide an exception to this. Though data is scarce for the region, rural sites in the north-east show that lava was not reaching rural settlements in the areas immediately surrounding the frontier, implying that the movement of imported goods beyond military sites is subject to different processes than in the south. The distribution of lava fits closely with the model of North Sea trade, as described by Morris (2010), in that lava milling tools were transported from the Rhineland to the east coast of Britain. However, the south coast shows a deficit of lava, where the production of querns of indigenous stone types may have impacted the exchange of imported milling tools.

Lava distributions by settlement type broadly follow imported and 'Roman' type objects, such as writing equipment (Eckardt, 2018; 2015), though their prevalence at farmstead sites in the south-east and East Anglia is unexpected if considering this pattern. However, the natural geology of some regions does not yield the stone varieties necessary for grinding and milling, and suitable stone would have been brought from other areas to fulfil this requirement. East Anglia and London are such regions, and the concentrations of lava in these areas conforms to the general theory of supply and demand. This is supported by the fact that significantly fewer farmsteads away from these regions appear to have used lava milling tools. From this we can explore the theme of connectivity, and what this may have meant to people living in different social groups. Roadside settlements and villa sites had greater access to imported lava, regardless of geographical location, while farmsteads and small town/settlement sites experienced connectivity and access relative to their proximity to urban and military sites. Data relating to the thickness of a quern prior to deposition also supports this idea, as lava querns were deposited at vicus and villa sites when they were thicker and less used than at small town/settlement sites, where wear was much greater prior to deposition. Better access to replacement querns is likely to have been a factor that determined the intrinsic value attributed to a lava quern, and relative connectivity to exchange systems would have had a significant role to play.

The chronological data has also confirmed some suppositions with regards to the use and distribution of lava milling tools in Britain and shown that it fits the patterns seen of other imported finds categories. The initial influx of lava into Britain in the first century is confirmed, and the concentrations around London and Colchester show where it was entering the province for redistribution. Exchange of lava is widespread by the second century and begins to contract in the third and fourth centuries, though high volumes in the south-east and East Anglia continue to be deposited during this period. The site at Summerton Way, however, has inspired some new thought with regards to late Roman period exchange of lava milling tools and the possibility for continued movement of lava into the province, albeit on a smaller scale. The import of pottery from the Mayen region of Germany is a late Roman venture and has not previously been related to the Mayen lava quern trade. The evidence at Summerton Way suggests that exchange of these products may have been related, though this is likely to have occurred through different mechanisms of exchange to the earlier redistribution via urban and military sites. The occurrence on the same sites of Mayen lava, Mayenware or Eifelkeramik pottery should, therefore, be the focus of future research. Similarly, other future work should include a detailed study of the interplay between urban, military, and rural sites with a thorough exploration of the chronological shift in distributions between different site types within similar exchange and trade catchment areas.

7

Day to Day Grind of Urban Life: Lava Quern and Millstone Use in Romano-British Towns

The distribution of imported lava milling tools at urban sites in southern Britain has been well-attested and is likely to have been a result of connectivity to ports and centres of manufacture or distribution via state-imposed infrastructure (see previous chapter, where this issue has been discussed from a rural perspective). This may have allowed for access to a range of goods that did not always exist for more remote rural or northern locations (Laurence & Trifilò, 2015, p. 104). Major towns, such as London and Colchester, show us that the Roman period urban populations were consuming an array of imported and exotic goods from many parts of the empire (for example see Laurence & Trifilò, 2015; Eckardt, 2015). This chapter aims to examine the use, distribution and deposition of lava milling tools from urban contexts to more fully understand the social and economic significance that they may have held in Roman towns. It investigates themes relating to the urban economy, social and ritual identities, changing technologies and scales of food processing activities, and changes in the consumption and distribution of imported goods. The chapter presents data that sits alongside that of the rural and military sites to provide comparison for the different processes and patterns that are identifiable across the datasets.

Before examining the role of urban sites in the distribution and consumption of lava milling tools, it is first necessary to identify what is meant by an urban site. Urban life in the Roman world is often associated with a large population size, industrial activity, commerce, administration, and government. However, a degree of ambiguity is generated when the term 'town' is employed to describe small towns, *vici*, *municipia*, *civitas* capitals and *coloniae*, as the social and economic conditions experienced and created by populations within each would have been different. The extent to which these were controlled or managed by the state would also have differed enormously; the status of the town and the period of governance would have been significant factors (Wacher, 1995, pp. 17–23; Creighton, 2006, pp. 70–78; Mattingly, 2006, pp. 255–291). Determining the given status of a town is not always possible and can be further complicated by chronological changes in status, of which we rarely have any knowledge. Verulamium provides one example in that it was referred to as a *municipium* by Tacitus but when this promotion was given is still a matter for debate, as is whether Tacitus fully understood the constitutional meaning attributed to the word (Bogaers, 1967; Frere, 1964; Niblett, 2001, p. 66). Similar debates exist for London, Colchester, and many other towns; mostly due to the complexities of dealing with incomplete and anachronistic historical accounts (Mattingly, 2006, p. 260–286). This study does not aim to scrutinise the legal status of towns, nor to determine which should be applied to the case studies in question. For the purposes of the analysis here, the most important archaeological distinction to note is that the two selected case study sites, Verulamium and Silchester, are major towns; identifiable as such from the presence of major public buildings, the implementation of a grid system, defensive structures, and their roles as administrative centres (Wacher, 1995; Jones, 2004, pp. 162–165).

7.1. Verulamium: Introduction and Background

The Roman town of Verulamium is situated on the south-eastern edge of the Chiltern dip slope on a chalk plateau (figure 7.1). Evidence suggests that the area was settled prior to the development of the town during the Late Iron Age (Wacher, 1995, p. 214–215) and the surrounding countryside was dominated by arable fields and grassland (Dimbleby, 1978). During the pre-Roman Iron Age, other sites in the vicinity, including Braughing, Welwyn and Baldock, held good connections with continental Europe as shown by the presence of imported pottery, coinage, metalwork, and wine, though the settled regions immediately around Verulamium did not see such an influx of imported goods or traditions (Niblett, 2001, p. 32–33).

Verulamium is often described as an *oppidum* at the time of the Claudian conquest (Hunn, 1992, p. 48; Woolf, 2007; Moore, 2017) and is thought to have been within the territory of the *Catuvellauni* (Frere, 1983, p. 1–28). It is possible that a faction of the tribe located in Verulamium was more sympathetic or apathetic to Roman rule, and the town may have been treated favourably after the invasion. This is suggested as an apparent continuation of wealth in the hands of a local elite, who may have maintained their power post-conquest (Niblett, 1999, p. 412–413). A current lack of structural evidence for a fort at Verulamium reinforces the narrative that Roman military intervention was not required. However, a more recent investigation to examine and compare site finds profiles completed by Pitts (2014) has suggested that the Roman period foundation of Verulamium had a military association. This is supported by the recovery of other military material culture from Verulamium, such as weaponry and armour. How this evidence relates to military occupation of the town, domination, or merely a military presence is yet to be determined fully (Frere, 1983, pp. 3–5; Niblett, 2001, pp. 56–58).

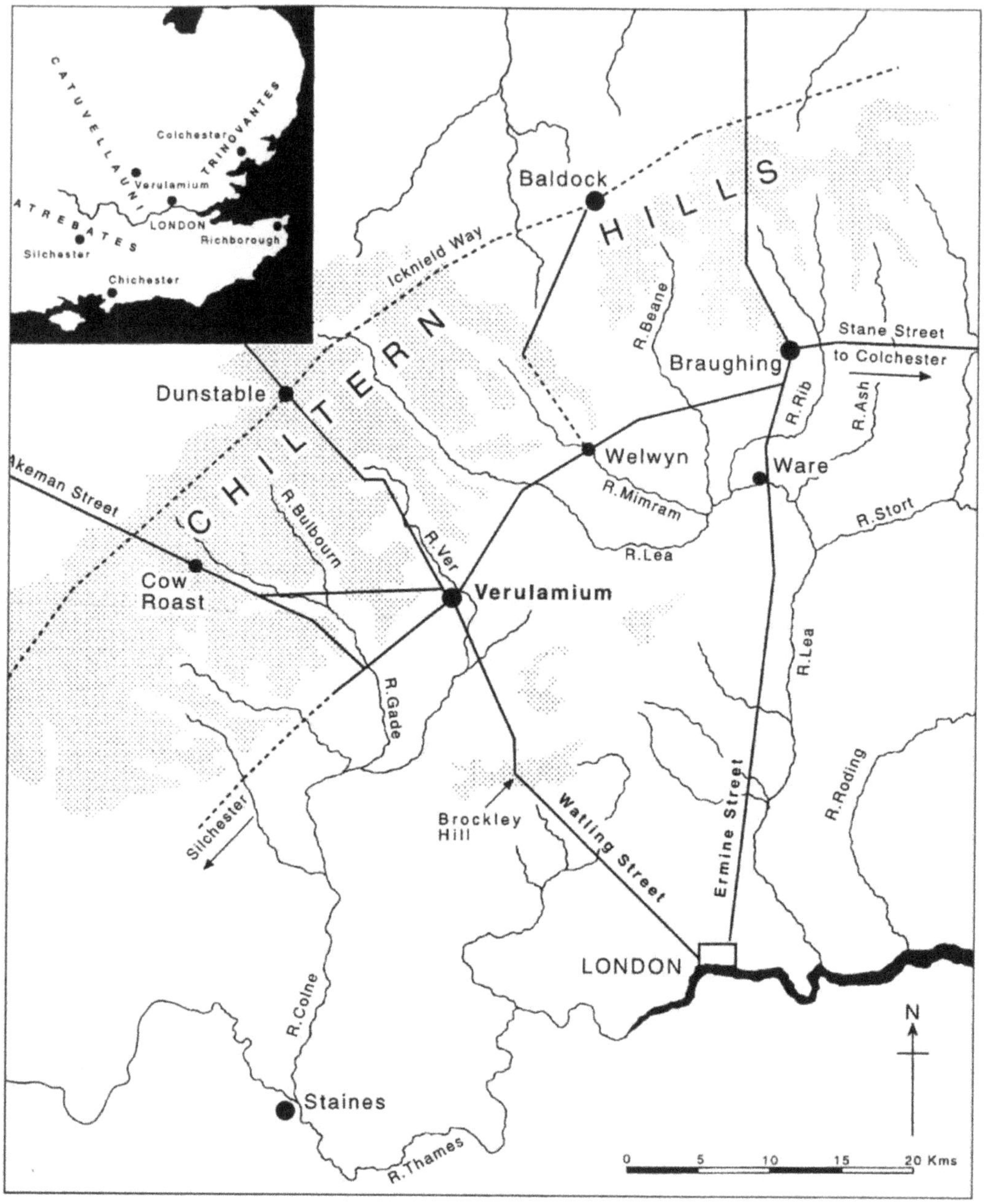

Figure 7.1. Location of Verulamium and other nearby important sites (Niblett, 2001, p. 30).

Historical accounts tell us that Verulamium was sacked, and presumably burnt, during the Boudiccan revolt of AD 61, alongside Colchester and London (Webster, 2004, p. 93–95). However, unlike Colchester or London, the 'burn horizon' that often defines the Boudiccan fire in these cities does not exist at Verulamium. After the Boudiccan revolt, probably around the Neronian and early Flavian periods, expansion of the town appears to have occurred rapidly and suggests an increase in population, continuing into the early second century (Niblett, 2001, p. 67–81). During this time, craftworking, industrial and commercial activity increased in intensity (Frere, 1983, pp. 10–12; Niblett, 2001, pp. 89–96). Aside from being a market town, Verulamium was also an important administrative and religious centre and would have served multiple roles to the inhabitants and the neighbouring populations (Niblett, 2001, p. 108–112).

Meanwhile, the products of agriculture from the wider area are also likely to have been brought to the town for exchange and taxation, all of which contributed to the wealth of the town. There must have been large-scale arable agriculture undertaken in the rural areas around Verulamium, as shown by the large granary at Gorhambury villa (Neal, et al., 1990, p. 40–48). Three smaller granaries have been found within Verulamium, but these were destroyed by the Antonine fire, while a larger tower granary appears likely to have existed from the third century (Niblett, et al., 2006, p. 173–180). Multiple corn-drying ovens have also been found in the town and its surrounds. Environmental evidence shows that grain was probably threshed and dried prior to storage and that these processes had probably occurred by the time the grain had arrived at the town (Niblett, 2001, p. 106–107). It has been speculated that a mill existed on the Ver from foundations recovered that may have belonged to such a structure, adding to this evidence that indicates centralisation of food processing was occurring (Rogers, 2013, p. 139).

A large fire occurring between 155–160 AD destroyed much of the lower part of the town, though recovery and continued occupation is evident (Frere, 1983, p. 13; Niblett, et al., 2006). The town walls were built in the early third century (Frere, 1983, p. 16–17), during which the Ver was canalised. This may have provided a sufficient head of water for the operation of mills further downstream (Rogers, 2013, p. 89–132). Fortunes appear to have changed slightly by the end of the third century, and parts of the town and neighbouring villas show evidence of decline. It is possible that the collapse of the nearby iron industry may have impacted the local economy. Despite this, decline was not uniform across Verulamium. Activity, rebuilding and other signs of occupation around the forum area continues well into the fourth century, continuing in some cases into the fifth century (Niblett, 2001, p. 125–136).

7.2. Verulamium: The Dataset

Lava from querns or millstones was recorded from various locations within the town and three sites from the extra-mural areas: Verulam Hill Field, King Harry Lane, and Folly Lane (see figure 7.12 for locations). Most of the data for the town and all the data for Verulam Hill Field was recorded by in-person examination of the lava milling tools. Publication information was used to add a further thirteen examples to the town dataset. Data from the latter two sites was compiled using only published information. In total, 126 groups of lava querns were recorded from in and around the town: 71 from Verulamium itself, 34 from Folly Lane, 18 from King Harry Lane and 3 from Verulam Hill Field (see Chapter 3 for description of what constitutes a 'lava group'). Though this appears to be a substantial assemblage of lava querns and millstones, it should be noted that 31 of these lava groups do not have site or context information, making it impossible to establish the likelihood of them belonging to the same object. This missing context information also caused some issues in terms of using supplementary quern data from publications, as there was the potential to duplicate the data for undiagnostic fragments recorded in person. For example, the publication by Niblett, et al. (2006) lists eight lava quern fragments recovered from *insulae* III and XIII. As many of the fragments recorded at the museum were missing context information, it was impossible to know if these were the same fragments in the publication. Therefore, the published examples may have been recorded twice, though those without findspots do not feature heavily in the discussion relating to distribution and undiagnostic fragments were not used to examine typological differences. Due to the age of the excavations within the town itself, context information was generally patchy, and it is entirely possible that fewer objects are represented than is currently shown here. The issue of redeposition must also be considered as many of the lava groups are heavily fragmented and one quern or millstone may easily have been deposited within multiple contexts.

Conversely, it is highly probable that the collections stored by Verulamium Museum are only a sample of what was excavated, as querns and millstones tend not to be fully recorded and have often been relegated to spoil heaps, especially in older excavations. The excavations at Gorhambury Villa, a site close to Verulamium, can be taken as an example (Neal, et al., 1990). The excavation report details that Millstone Grit and lava were the most common materials used for querns at the site. Only two samples of lava were retained, but it must be assumed that much more was excavated at the time. Unfortunately, no quantification of the lava appears to have been carried out, making it impossible to determine the extent to which it was present and the minimum number of querns or millstones that may have been there. The numbers within the dataset may, therefore, significantly under-represent the true picture of lava quern use in and around Verulamium, which also has implications for comparisons with other sites. Considering the proximity of Verulamium to London and Colchester and the long chronology of excavation at the site, large volumes of lava should be expected.

7.3. Verulamium: Manufacture

Of the 126 groups recovered from in and around Verulamium, 38 could be identified as either a quern stone or a millstone (see Chapter 2 for definitions), with the percentage of the assemblage breakdown shown in figure 7.2. As has been

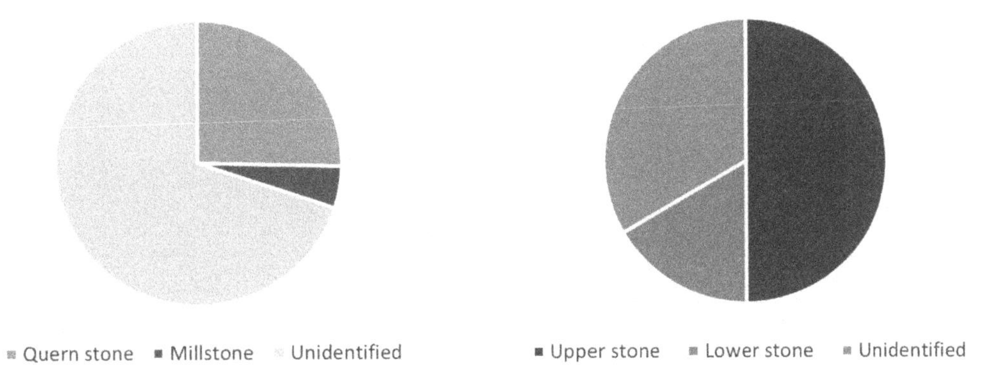

Figure 7.2. Left, percentage of assemblage identified as quern or millstone T= 126. Right, percentage of millstones identified as upper or lower stones T=6.

seen in other assemblages, the number of unidentified lava fragments is high; the largest identifiable group is hand operated querns (25% of the total), with a smaller number of millstones (5% of the total) identified.

7.3.1. The Millstones

Though the number of lava millstones for Verulamium seems very low (6 examples), it is relatively high when compared to sites such as Chichester, where only one lava millstone was recovered from within the town walls. In fact, it is a high number of millstones of any material type from an urban context, as can be seen from sites such as Cirencester and Gloucester; both towns produced only a single individual indigenous stone millstone despite extensive excavation (Shaffrey, 2021, p. 23–25). From the six millstones, 3 (50%) were upper stones, one (16%) was a lower, and two (33%) were unidentified.

The types of millstone from Verulamium vary and though the sample size is small, it is possible to make comparisons and observations. The range of diameters is relatively broad but generally within the 'normal' range of sizes when compared to the diameters of lava millstones from all Roman Britain, though allowances must be made for the estimation of diameters from fragments (figure 7.3). The variety indicates that different types of millstones were entering the town, and this may have occurred as a development of millstone type over time from the same manufacturing region, or because of differing supplies of millstones due to it being a specialist product. In the Eifel region where Mayen lava is sourced, millstones were probably produced near the harbour of Andernach and multiple workshops may have been involved to create a greater range of types (Mangartz, 2008, p. 90). Two of the millstones are just at the upper quartile, or the top 25%, of sizes in Britain indicating that they are slightly more unusual. The lava millstones in the upper quartile of size are distributed throughout southern Britain and do not show any specific pattern of use in terms of site type or location (see Chapter 3). These would have been manufactured to fit the specifications of the mill they were to operate within and are likely to have been expensive and high-status tools. They show that grain must have been processed at scale in Verulamium.

Another indication that large-scale grain processing took place is the presence of a highly unusual millstone fragment recovered during excavation work in 1938 (figure 7.4) (CAT 0995). The findspot was within *insula* XXVIII, just southeast of the museum (see figure 7.12 in distribution). During this work, part of a metalled Roman road was cut through, implying that material closely associated is likely to be Roman in date, including the millstone fragment illustrated. The projected diameter is large at 717mm and falls within the dimensions attributed to a millstone. It has a thickness of 70mm and there is no indication that there was ever a kerb present on the upper surface. The central hole is likely to have been circular in shape, with an estimated diameter of 165mm. The edges are straight and vertical with no signs of linear tooling and the grinding surface has radial dressing, a characteristic that is typical of the Roman period. The decoration has no known parallel in Roman Britain and does not appear to be related to the functioning of the object as a milling tool. It consists of a six-pointed star pattern with spear-shaped leaves radiating outwards from the points in low relief. Only one section of the pattern is visible on the fragment, and the rest is reconstructed below, which assumes a regular and repeating pattern.

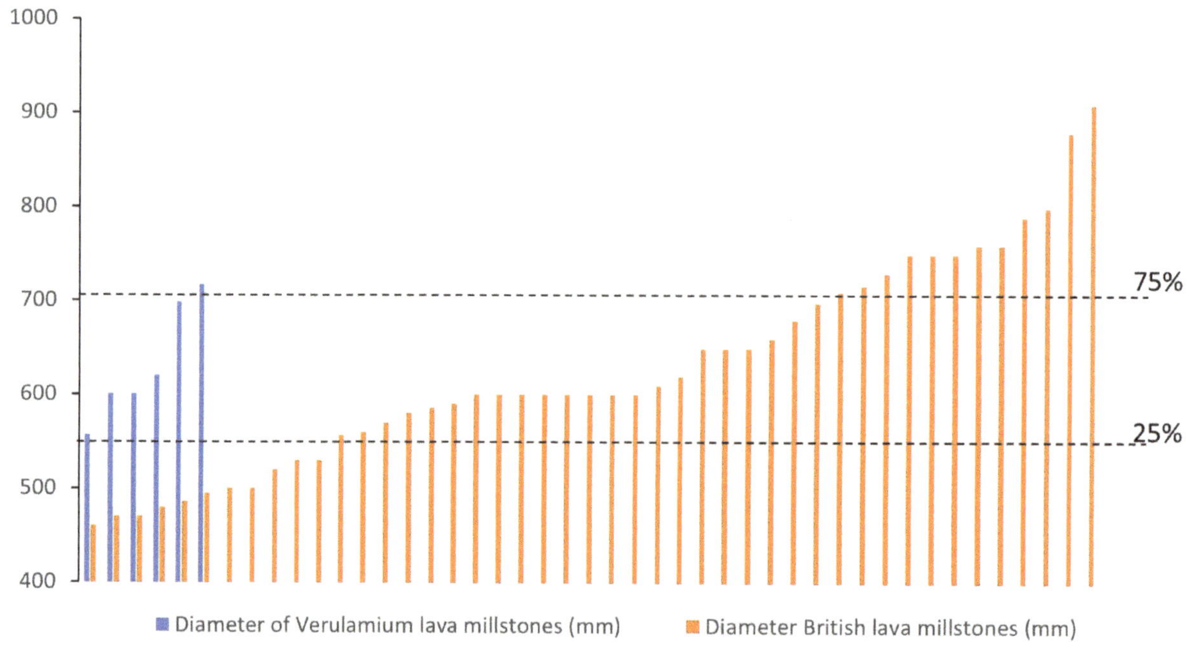

Figure 7.3. Histogram to show the diameters of the lava millstones recovered from Verulamium (T= 6) compared with those of Roman Britain (T= 45). The lines indicate the upper and lower quartile of diameters from the British lava millstone data.

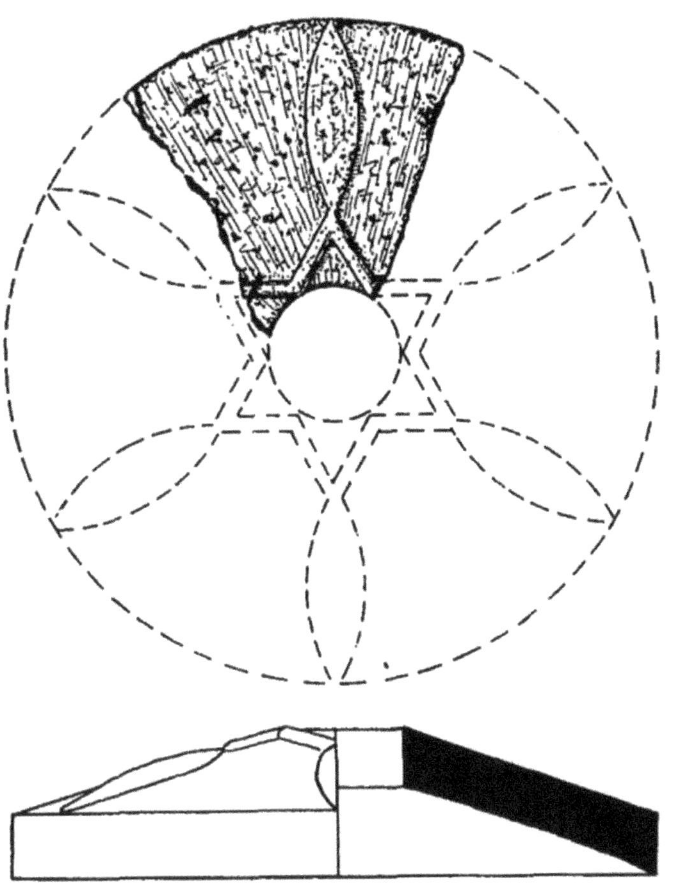

Figure 7.4. Fragment of decorated lava millstone (left) (© St Albans Museum Service), reconstructed view of the decoration on the millstone if complete (right) (Corder, 1943, fig 1) (CAT 0995).

The most interesting aspect of this fragment is the way that it does not at all conform to the standardised lava querns or millstones found on other Roman sites and, other than the concave grinding surface and surface dressing, it has none of the typical characteristics. However, millstones of lava appear to have been less standardised, probably due to their unique functioning and operation as specialised equipment. Raised kerbs are not necessary if there is no handle to be fitted, which could be the reason why it is absent on this stone. Unfortunately, there is nothing visible on the fragment to indicate how the stone may have been rotated, and its identification as a millstone is based on its projected size from the curvature of the outer edge. The upper surface design is unique. Querns and larger sized querns/small millstones are sometimes inscribed with text, though these tend to appear predominantly on military sites and may have been used to identify individual milling tools by *contubernium* or unit (Carroll, 2005). Although this presents a different form of personalisation, a degree of functionality can be attributed to inscribed forms of decoration using writing, which is different to that seen on this example.

Though the precise provenance of this lava used to make the millstone is not known, there is a high probability of it being Mayen in origin. Geochemical analysis would need to be completed on the stone to confirm this. It is possible that the millstones were finished closer to the site of the mill, in order that precise measurements to accommodate the mechanism could be met more easily. The decoration could, therefore, have been the addition of a local stonemason who was employed to carry out the specialist work required to finish production of the millstone, and the design may have been unique to the stone, or to the craftsperson who carried it out. It is likely that the design was added during the original stages of shaping/finishing the stone, as it is of a moulded form (Shaffrey, pers. comm.). This means that it was probably associated with its use as a millstone and is unlikely to represent adaptation of original form or be a later addition to the stone.

In terms of the mechanical operation of the millstones from the assemblage, there is, unfortunately, very little evidence. Few features relating to the mechanical parts were present on the surviving fragments and millstones were predominantly identified based on their size alone. Two examples provided further evidence in the form of possible millstone fittings. The first does qualify as a millstone with a diameter of 557mm (Shaffrey, 2015), but could be identified as a large quern if the fittings indicated that it was operated by hand (CAT 0997). However, the description for this fragment in Frere (1983, pp. 80–81) suggests that it may have had a hole into which a lifting mechanism would have been fitted (figure 7.5), resulting in a more confident millstone classification. The other larger millstone fragment from the same publication that measures 698mm in diameter is described as having the same characteristic (CAT 0998). Though described in the

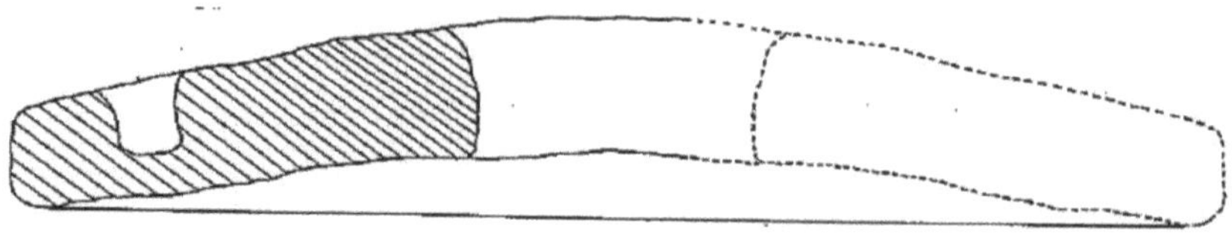

Figure 7.5. Millstone of lava from insula XXVIII showing a hole in the upper surface that may have allowed for a lifting mechanism to be fitted (Goodburn & Grew, 1984, pp. 80–81) (CAT 0997).

finds report as being a possible handle socket, this feature is more likely to be related to its use as a millstone as lava querns from Mayen were not fitted in such a fashion. Nonetheless, lava querns were often mended or adjusted to work in different ways to the original standardised designs and the holes on the upper surfaces of these two upper stones may have been added later to allow for continued or modified use. An example of this is seen in the Housesteads assemblage, where lead 'plugs' were added on either side of an iron bar fitted over the central hole, possibly to act as a rynd or to allow for an adjustment of the angle of the grinding surface (see figure 8.20 in Chapter 8). The holes of these millstones could have existed to accommodate a similar type of fixture, but interpretation becomes difficult without viewing the completed object. It does also raise the possibility for a different type of lava quern or millstone to be present here, as those manufactured away from Mayen may have not been standardised to the same criteria as those produced in the Eifel region. However, the most likely interpretation is that the presence of this feature corresponds to use in mechanically rotated mills, especially when considered alongside the size of the stones.

7.3.2. The Miniature Milling Tool

In stark contrast to the larger millstones in the assemblage is one example of a complete miniature lava quern upper stone, with a diameter of 235mm (figure 7.6) (CAT 2679). This is a highly unusual object that is sadly without stratigraphic information. As the artefacts stored at the museum are predominantly from within the Roman town, it is probable that it is of Roman origin, but another period cannot be ruled out. The condition of the quern stone is excellent, which could imply that it is of a later date, but the smaller size of the object makes it more robust and less vulnerable to damage meaning it may have preserved better. The dressing on the grinding surface of well-defined harps is characteristic of the Roman period, though little else of the quern resembles that of the standardised lava stones manufactured in the Mayen district.

Figure 7.6. Miniature lava millstone upper stone (CAT 2679) viewed from the top and bottom surfaces (© St Albans Museum Service).

Many miniature querns of indigenous stone types have been found in Iron Age and Roman period contexts in Britain. Miniature querns are often misidentified or misreported, as has been shown by McLaren & Hunter (2008) for Iron Age and Roman period examples from Scotland, meaning that there is likely to be a much larger number of these artefacts in museums and archives than currently known. Roman period miniature querns of indigenous stone types in England have been tentatively identified at Ramsgate, Kent (Fry, 2009) and in the South-East Cheviots, Northumberland (Topping, 2008, p. 353). In mainland Europe, miniature querns of sandstone have been found in the Lower Danube region (Watts, 2019, p. 391–408) and the Paris Basin of France (Lepareux-Couturier, 2014, p. 155–156) and are contemporary with British examples.

Viking era versions are known from Norway, and these have been correlated with those recovered from Shetland to show parallel periods of use and similar functions (Hansen & Larsen, 2000, p. 115–119). Other examples are known from Lagore, Ireland (Co. Meath) (McAlister, 2013, p. 4). The Roman and Iron Age miniature querns from Scotland, and those from Viking and Medieval Shetland tend to mirror their larger counterparts with some precision, reflecting the design, function, and mechanical operation of full-sized versions. How these objects were used and their meaning, however, has seen interpretative differences. Traces of wear on the Roman and Iron Age miniature quern stones shows that they were used, possibly for grinding seeds or similar small volume substances. Conversely, those from Shetland do not present wear traces and are often made from different materials to those of locally produced full-sized querns or millstones and have, therefore, been interpreted as toys to help socialise children into future adult roles (Hansen & Larsen, 2000, p. 114–115).

The miniature milling tool from Verulamium would ordinarily be identified as a millstone due to the underdrift method of fixing the rynd, but the method of rynd fixing may not necessarily define its method of rotation (Shaffrey, 2015, p. 73–78). It does not resemble or reflect the highly standardised design of the hand operated lava querns, all of which have an overdrift rynd fitting, and cannot be identified as a miniature version of these. Unfortunately, the object was not available for examination and was not recorded in-person, meaning that it is not possible to confirm that signs of wear are present on the grinding surface. From the image provided by the online collections, wear does appear to be visible, but

Table 7.1. Summary of other miniature lava milling tools recovered in Britain

Collection/ Excavation	Findspot	Context Date	Diameter (mm)	Upper/ lower stone	Under/ overdrift	Dressing
St Albans Museum, Hertfordshire	Not known	Unknown	235	Upper	Underdrift	Harping
A14 road scheme, Cambridgeshire	Fenstanton Gravels	Modern	210	Upper	Underdrift	Harping
Norris Museum, Huntingdonshire	Local find	Unknown	210	Upper	Underdrift	Possibly pecked
Norris Museum, Huntingdonshire	Local find	Unknown	225	Lower	N/A	Not visible but signs of vertical tooling at sides
	Southampton French Quarter	Post-medieval	250	Upper	Unknown	Harping with vertical tooling at sides
Thameslink, London	Not known	Post-medieval	Unknown	Unknown	Unknown	Not visible but signs of vertical tooling at sides

this would need further investigation. The holes on the upper surface may have been used to accommodate other fittings and are not dissimilar to those found on runner (upper) millstones, possibly used to lift stones or to fix the stone to the rotational device. Due to the small size of the stone, such lifting mechanisms would not be necessary, indicating that this is possibly a model, replica, or toy millstone.

From the full dataset of lava milling tools from Roman Britain, there are several parallels (table 7.1). Although similarities exist between these objects, none could be described as identical, and they all have variations in design. The example from Verulamium is probably the best preserved and well-manufactured and the closest parallel is the upper stone fragment from the A14 road scheme excavations (figure 7.7). It was recovered from a modern context but may be residual Roman; no milling tools were recovered from post-Roman contexts other than the miniature quern, suggesting an earlier date. This was also produced in the style of a miniature millstone, with an underdrift rynd and a harped grinding surface dressing. It had clearly held an iron rynd, discernible from the residue left behind (Banfield, in prep.). Unlike the Verulamium miniature, the upper surface had been roughly worked and there was no further evidence for millstone fittings. The rough upper surface is like an example from the Norris Museum in Huntingdonshire (Shaffrey, pers. comm.), which is near to the site of Fenstanton Gravels where the A14 miniature millstone was found. This also presents an underdrift mechanism and holes for a possible lifting mechanism in the upper surface (figure 7.8). Together, these three objects should be classified as millstones and though they cannot be dated with any confidence, a Roman date should be considered as a strong possibility.

Figure 7.7. Miniature upper lava millstone recovered as part of the A14 road development scheme in Cambridgeshire (CAT 1483). Provided here as a parallel to CAT 2679 (photo by author).

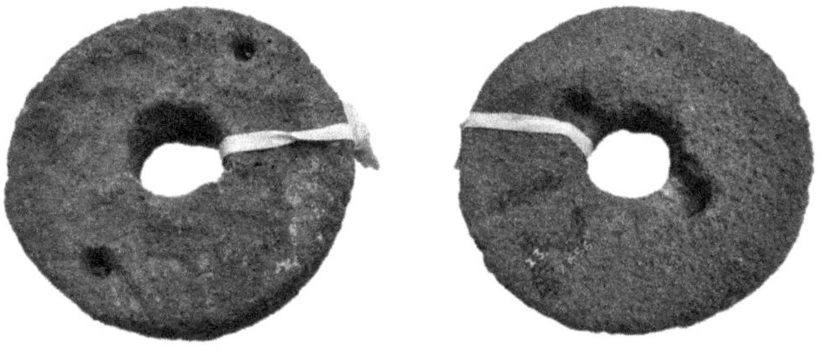

Figure 7.8. Miniature upper millstone recovered from the local area of Huntingdonshire and stored at the Norris Museum (photo by Shaffrey, n.d).

The other three miniatures are less comparable with the Verulamium stone, though they all have vertical tooling on their sides. This is a feature that is typical of a Roman period Mayen lava quern, and it is possible that these represent replica versions of the hand operated lava querns. For the Viking era miniatures recovered from Shetland, it was also found that both miniature querns and millstones were present, and these were interpreted as gendered toys: mills for male children and querns for female (Hansen & Larsen, 2000, p. 115). How well this theory fits with the evidence provided here would require further investigation. As Roman lava is not usually found in burials, this cannot currently be established. XRF analysis would help provide a method by which the date of the milling tools could be explored, as Mayen lava was not exported to Britain after the Early Medieval period and was superseded by lava from the quarries of Niedermendig (Kling, 2008, p. 141). Where and why these objects were produced would also require further thought. The lack of standardisation present implies that there were multiple locations where these objects were produced and it seems likely that they were made specially, in small numbers and probably local to the place of use. The level of skill applied in making the miniatures appears to vary from one example to the next, and there is nothing to suggest centralised production. Use wear analysis should also be conducted to provide a clearer idea of how these items may have been used.

The biography of the miniature millstone from Verulamium is one that is worth exploring in more detail. Although the earlier stages of its lifecycle are unknown to us, gaps in our knowledge can be filled using known trajectories of similar objects. In this case, a considerable amount is known about the Mayen lava quern and millstone trade (Chapter 2). The provenance of the miniature millstone as a lava from continental Europe is highly likely and its journey to Britain as an import is assured. Regardless of whether it arrived at Verulamium in its miniature form, or as a standardised quernstone, an extensive web of interactions must have taken place to make its journey possible. If we assume that the miniature millstone was formed through the reworking of a standardised quernstone, its initial form was originally conceived and imagined by quarry workers and stone workers, whose relationship with the lava was habitual and related to subsistence (see Chapter 2.5). Exchange relationships between people resulted in the movement of the lava quern to Britain, during which time its potential as an item for commercial or strategic military gain caused it to be reinterpreted and revalued, imbuing it with different meaning (Chapter 2.10–2.12). This renewed concept of value motivated people to transport the lava quern many miles across the sea and land until it could be exchanged at the location of its primary use as a milling tool.

The primary use life of the lava quern may have been several decades long, though no traces of this part of the lifecycle can be identified on the object. However, once the lava quern ceased to be of use as a milling tool due to wear or breakage, its meaning and relationship with humans changed significantly. An object that no longer functions is often reinterpreted as being the sum of its component parts and is no longer viewed as a whole entity. It is possible that the quern was disassembled and metal or wooden parts recycled. The lavastone could then also be reinterpreted as a raw material for reuse. However, the raw material did not provide a blank canvas in terms of its reuse potential, as the material properties of the vesicular lava and memories of its life as a milling tool continued to act as a bridge between its previous life and its reincarnation as a miniature millstone. A psychological attachment to the memory of the original milling tool may have existed, shaping the decision to rework the lava into a representation of the former object. The new miniature millstone, therefore, was created to reference its past relationships and interactions with people, whilst simultaneously utilising its new form and reimagined materiality to communicate the mechanical aspects of grain processing practices to a different audience, possibly a child.

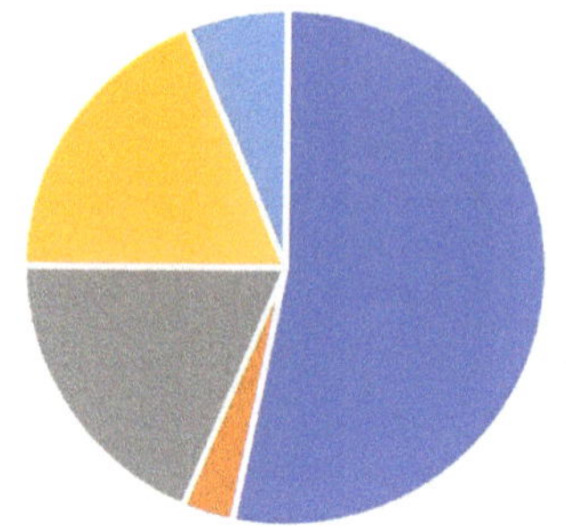

Figure 7.9. Verulamium quern assemblage showing percentage of upper and lower stones present (T= 32).

7.3.3. The Querns

A total of 32 quern stones were identified, all of which were recovered from within the town except one example from King Harry Lane. Of these, 30 were diagnostic and could be identified in terms of whether they comprised fragments of upper or lower stone. A large percentage of these (56%) were upper, or probably upper stones, with a smaller number of lower or probable lower stones (38%) (figure 7.9); it should be noted that lower stones tend to be less identifiable due to their lack of features. Meanwhile upper stones are both more identifiable and are more likely to be fragmented due to the way they wear and their mechanical fixture points, meaning they are more likely to be counted multiple times when recording. The percentages shown here are, therefore, representative of other sites in Roman Britain where fragmentation is common, as shown in the comparative discussion later in this chapter. It is probable that most of the undiagnostic fragments of lava once belonged to quern stones, as opposed to millstones, as these are more common in the archaeological record and that querns are highly underrepresented here.

There were 21 querns that had estimated or measured diameters, the majority of which were estimated due to the fragmentary nature of the querns from the town. These ranged from 360 to 460mm, which is within a relatively normal range for a Romano-British lava quern assemblage. As Mayen lava querns are highly standardised objects, only 25% of all British examples fall below a diameter measurement of 400mm, and 25% measure over 425mm. This means that half of all measured lava querns in Britain have diameters that fit within a very narrow range of 400–425mm (figure 7.10). A similar pattern of measurements is present in the Verulamium assemblage, with 50% of the querns measuring between 400–420mm. These are clearly as standardised as those found elsewhere in Britain and are very likely to have been sourced from the same location, Mayen in Germany. The location of Verulamium within the catchment area of the North Sea trade routes (see Chapter 2) is the most probable explanation.

A similar picture is presented by the other features of the quern fragments, and most display the attributes of the standardised

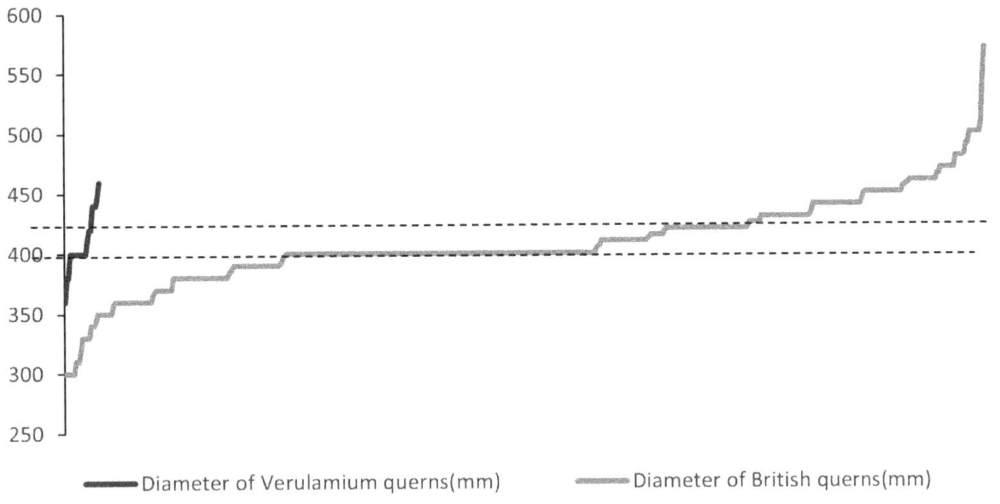

Figure 7.10. Graph showing the different diameters for lava querns from Verulamium compared with Britain. The 75th and 25th percentile for the British assemblage is show as lines.

design of a Mayen lava quern. Of the 18 upper stones, 11 had a kerb and the same number had the typical quartered upper surface dressing, while 25 from the entire assemblage showed signs of vertical tooling on the outside edge. There were a few exceptions observed, including one upper stone fragment that had an unusually shaped rynd chase for a lava quern (figure 7.11). This may represent a different style of fitting, be a way that the object was mended to extend its use life or been altered to allow for it to be reused for a different purpose. The other features of the object are very standard, and so it probably shows a refitted rynd, in a similar way that some of the querns at Vindolanda were found to have been repeatedly mended, especially in relation to the rynd, which was vulnerable to damage by wear (Chapter 8).

7.4. Verulamium: Distribution

From within the town walls, lava was retrieved from ten of the *insulae* (figure 7.12). Due to the lack of context data for some of the lava groups, it is not possible to discount the possibility that lava was also found at other *insulae* at Verulamium. Similarly, as there is no excavation or sampling strategy outlined for any of the sites, the dataset should be treated with caution as it likely presents incomplete

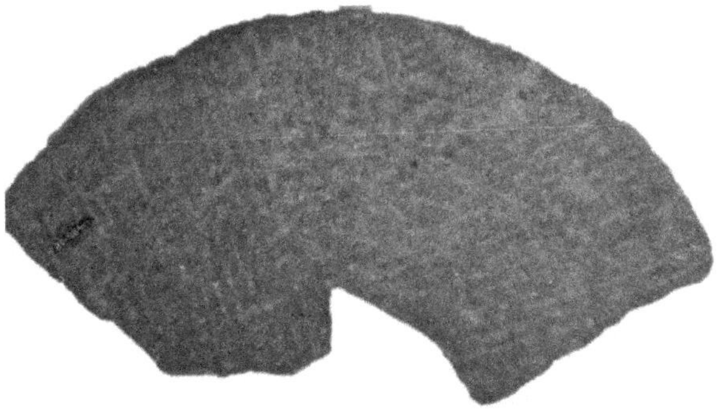

Figure 7.11. Fragment of lava quern (CAT 2594) showing an irregularly shaped rynd chase that may have been the result of mending or reuse (photo by author © St Albans Museum Service)).

information. It should be noted that no lava was originally recorded for the earliest excavations, those completed in *insulae* I-III by the Wheelers (Wheeler, 1932; 1936; 1938). The one fragment that was catalogued for *insula* III was only published in the later report by Niblett, et al., (2006). This indicates that it was not standard procedure to record, conserve or curate querns during these early investigations. However, these *insulae* were also excavated in subsequent projects during the 1950s and 1980s, and so it becomes challenging to examine the possible reasons for an absence of data in certain areas of the town.

Folly Lane, which like Verulamium, had a large area of excavation, was similarly zoned (figure 7.13) and finds from the site were catalogued according to their findspot within these areas. It should be noted that although this site is best known as a ceremonial and burial site, it saw changing uses at different stages of the Roman era and can be considered alongside the urban data as being related to urban activity. Distribution of deposition on an intra-site basis can be useful to examine as it can help us to understand the relationship between material culture and specific activity areas, structures, economic zones, or ritual places. King Harry Lane and Verulam Hill Field were smaller scale projects, meaning that zoning was not used as part of the excavation strategy. Finds could be correlated with their findspots at King Harry Lane and analysed in terms of context of deposition, which was also carried out for the larger sites where context information is available. Unfortunately, the information from Verulam Hill Field was not sufficient for distribution analysis to be undertaken. The number of lava groups from each area of the town and extra-mural areas is shown in table 7.2. Although Verulam Hill Field, Folly Lane and King Harry Lane are described here as extra-mural, the town walls were constructed in the third century, and its presence is not applicable to all periods of the town's history. The term extra-mural is used for ease of description.

Across the three extra-mural sites at Verulamium, a similar number of lava groups were recovered as were recorded from the excavations within the town. Details from the

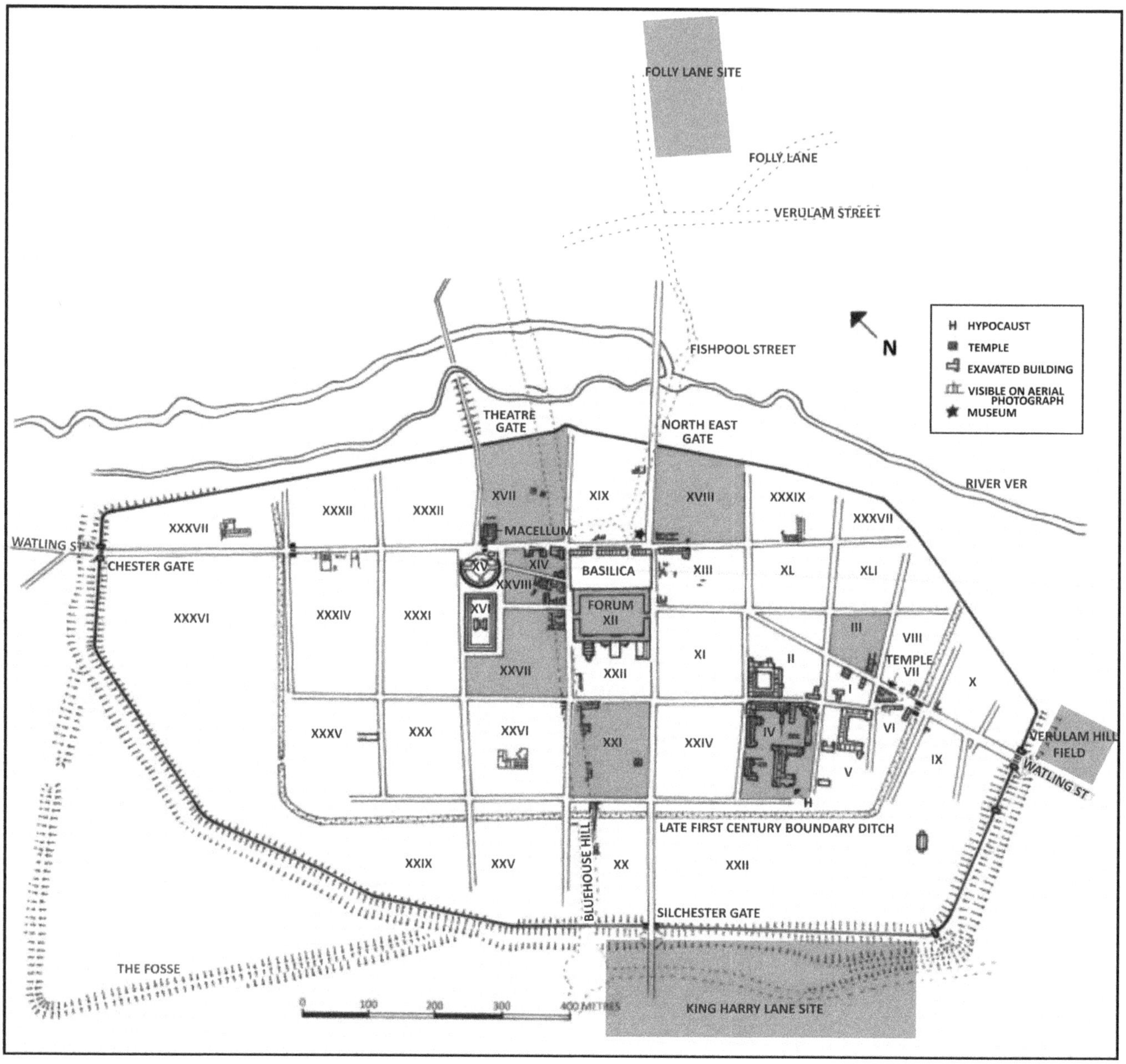

Figure 7.12. Map of Verulamium showing the different insulae of the town and areas of excavation covered in the data analysis. Locations where lava was recovered is shown shaded (after Niblett, 1999; St Albans and Hertfordshire Architectural and Archaeological Society, 2019).

publications at Folly Lane and King Harry Lane show that lava comprised approximately half of the overall quern assemblages at these sites. Unfortunately, the Frere publications for Verulamium do not provide details for all querns and millstones excavated, but the sample of stone types that have been catalogued present a similar picture in terms of lithologies and volumes when compared with the data from the more recent excavations at Folly Lane and King Harry Lane. Overall, the data suggest that lava milling tools were abundant at Verulamium, and that lava and non-lava types occur within the town and in the extra-mural areas, possibly at similar ratios. Though Folly Lane, Verulam Hill Field and King Harry Lane have generally been classed as funerary and ritual sites, these areas of the town have also provided evidence for industrial activity, the dumping of domestic refuse (Niblett, 1999, p. 416–417) and domestic occupation (Stead & Rigby, 1989, p. 4–11) and probably served different functions at different times. These sites and their surrounding locales may have seen occupation 'overspill' from the main part of the town at times when the population was expanding beyond the previously designated town limits, such as during the late first and early second centuries. The presence of querns in these areas is, therefore, not surprising, and indicates that grinding activities occurred both within and outside of the town, where different parts of the population resided or worked. As stone is often recycled and reused for construction or for resurfacing, this could alternatively simply indicate the locations where stone objects were deposited as part of building projects, or refuse areas where

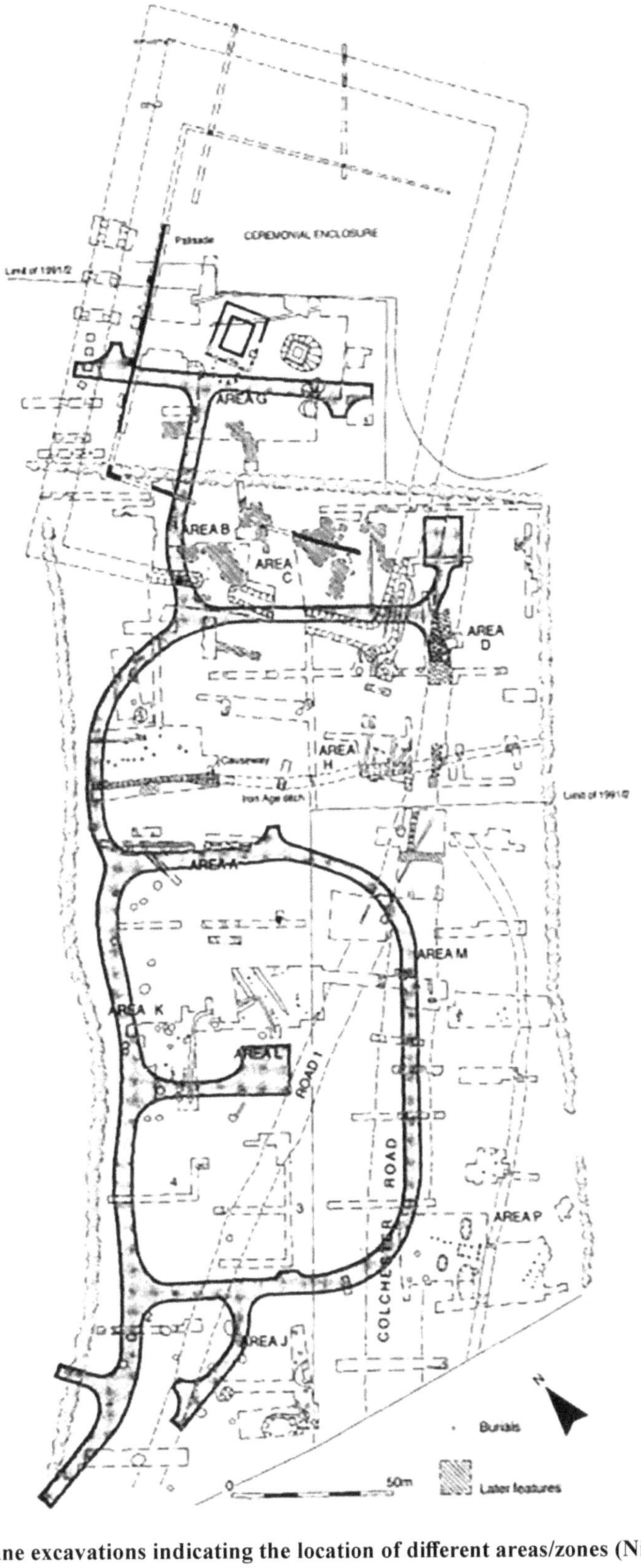

Figure 7.13. Site plan of Folly Lane excavations indicating the location of different areas/zones (Niblett, 1999, p. 3).

rubbish was dumped. More detailed analysis of context of recovery could help to highlight if this was the case, and this has been completed as part of the investigation into deposition.

At Folly Lane, zones P and J show the greatest number of lava groups, and these are located towards the south-eastern corner of the site. Zones B, C and G also show relatively high volumes, and these zones are clustered in

Table 7.2. Number of lava groups recovered from sites in and around Verulamium, quantified by site zone where information is available

Site Location	Site Zone	Number of lava groups	
Folly Lane	A	1	
	B	4	
	C	4	
	G	4	
	H	1	
	J	5	
	K	1	
	M	3	
	P	7	
	2	2	
	Not specified	2	**TOTAL=34**
Verulamium	Insula III	−1	Possibly recorded twice
	Insula IV	1	
	Insula VII	5	
	Insula XIII	1+ (7)	7 possibly recorded twice
	Insula XIV	14	
	Insula XVII	2	
	Insula XVIII	1	
	Insula XXI	1	
	Insula XXVII	1	
	Insula XXVIII	3	
	Site 9	1	
	SW of forum	2	
	Not specified	31	**TOTAL=71**
Verulam Hill Field	Not zoned	3	
King Harry Lane	Not zoned	18	

the northern-most end of the site. The data imply that there are small concentrations of lava in these parts of the site, and that this may relate to the use or deposition of lava milling tools within these zones. Zones B, C and G are within the ceremonial enclosure, making a total of twelve lava groups recovered from this part of the site. Tempting as it is to derive a ritual association with lava querns from this distribution, it is necessary to emphasise that most examples were dated to the mid-to-late third century or fourth century when the ceremonial role of the site was in decline (Niblett, 1999, p. 417). This is more likely to represent the dumping of rubbish during phases when the infilling of ritual shafts and levelling of the site occurred, with parts of it turned over to agricultural use by the early fourth century (*ibid*).

Overall, 25 out of the 34 lava groups for Folly Lane belonged to these later phases, indicating that querns may have been broken up to reuse for construction or as hard-core for resurfacing and levelling. In zones P and J, two groups of *in-situ* lava were found incorporated into floor surfaces dating to the fourth century, while the excavation report comments on the use of broken up querns, predominantly of lava and Millstone Grit, as post-packing in the late Roman agricultural buildings that were built after ceremonial use of the land ceased (Niblett, 1999, p. 100). Another example of reuse of lava during this period is found in the inclusion of lava quern fragments in the flue of a corn drying oven. Querns are sometimes reused to line ovens and hearths, and to backfill corn dryers, an observation that is not always commented on in excavation reports. A similar pattern of lava accumulation is seen in zones P and J, where ten of the twelve lava groups were dated to the later phases of the site and are likely to have been used in similar ways. Niblett (1999) discusses that the number of querns available for reuse is indicative of the scale of grain processing in the area, yet the condition of the querns suggests that this occurred in an earlier phase. Though this is likely to be true in terms of chronology, it is difficult to establish where the grain processing may have taken place. The movement of refuse over large distances for recycled building material should not be underestimated. Huge volumes of ironworking waste in the form of slag were moved from Southwark to areas north of the Thames in Roman London (Starley, 2003). The broken querns at Folly Lane may have come from any area in or around Verulamium and may be more relevant to interpretations of the wider economy of the town than that of a specific area.

The data for the lava milling tools at King Harry Lane is more precise in terms of where the lava was recovered, but little detail is available on the fragments themselves and the chronology of deposition. All the querns were retrieved from phases of occupation that occurred after the site had ceased functioning as an Iron Age cemetery and had become a settlement (Stead & Rigby, 1989, p. 4–9). This occupation phase existed between AD 60–250, giving a very broad date range for the deposition of the querns, which could not have been deposited prior to the Roman period due to the date they started to be imported into Britain (see Chapter 5). Four lava groups can be more closely dated to the second century and one to the late first century. However, this does not provide the precision necessary to explore chronological changes in quern distribution at the site. Nonetheless, spatial distribution does provide a glimpse into possible patterns of consumption and deposition. This has been mapped onto the site plan in figure 7.14, alongside the locations for querns of other stone types (i.e., not lava). The locations for lava are marked in red, while those of non-lava are marked in blue. Exact locations are not provided in the excavation report, and so entire features or approximate findspots have been highlighted.

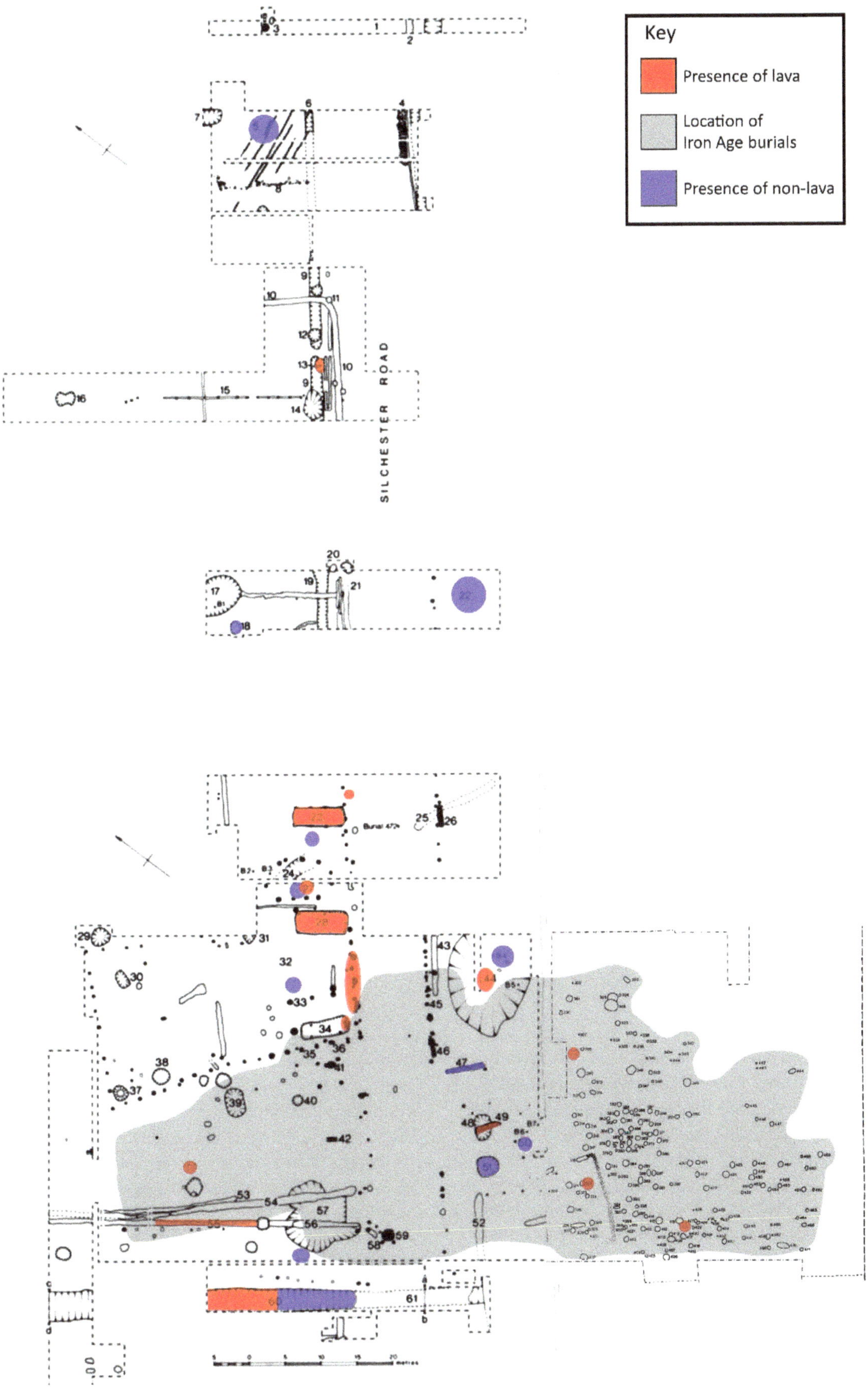

Figure 7.14. Location of lava and non-lava milling tools at King Harry Lane site (after Stead & Rigby, 1989, fig. 3, 4 and 182).

Figure 7.15. Processing of rice at the side of the road in the central highlands' region of Dalat in Vietnam, specifically drying, de-husking and bagging of rice (photos by the author).

The placement of the Late Iron Age and early Roman cemetery that predated the Roman settlement phase of the site has been shaded. The plan in figure 7.14 only allows for the location of earlier graves to be illustrated where there are fewer later Roman features, which is on the right side of the plan. On the left, the extent of the cemetery is demonstrated by the shaded area, but the density and location of the burials cannot be shown due to the high number of later Roman features that obscure the ability to visualise the interaction between the earlier and later phases. The cemetery appears to have had clear limits on the north, east and west sides, while on the southern edge, it was bounded by a ditch (Stead & Rigby, 1989, p. 80). The location of the cemetery is significant despite the repurposing of the area for habitation, as it is highly probable that its previous use would have been known to the new site occupants. It is interesting to note that the evidence for occupation in the Roman period is more intensive on the western side of the road, where burials were sparser. Though it is likely that the later phases destroyed some of the burials making the density appear less, it is also possible that the most significant area of the cemetery, towards the east, was respected by the subsequent occupants.

This also correlates with what is seen of lava quern distribution at the site, with twice as many recovered from the western side of the road than the eastern. It is not known whether this is also the case towards the northern part of the site where Roman period occupation may have occurred more intensively on either side of the road, as only the western side was extensively excavated in this area. It would be interesting to see if quern deposition becomes more evenly distributed away from the location of the burials. It is likely that quern deposition coincides with the location of domestic and industrial activity areas, as opposed to the distribution necessarily reflecting selective deposition patterns that avoid the cemetery. However, there is a possibility that the use or disposal of domestic or commercial food processing tools in mortuary contexts was not deemed appropriate.

The distribution of querns close to the road also provides some interesting data for discussion. Though it cannot be assumed that these demonstrate activity areas where food processing occurred, it could be used to argue that grinding and milling may have taken place there. Unfortunately, the excavation did not cover a wide enough area to determine whether the concentration of querns finds decreases away from the road, but the current picture does imply this was the case, for both lava and non-lava milling tools. Roads and roadsides are often utilised in traditional societies to dry and process crops after harvesting (figure 7.15), and it is possible that similar behaviours occurred in the past. They provide the hardened surfaces required and the ease of access to centralise the process of gathering the crop from various fields in the vicinity before transporting the finished product elsewhere. This theory would need to be supported with environmental evidence but provides an interesting research question worth exploring. Currently, the trend seen from the distribution can only be taken as indicating possible deposition of querns close to the road, which could suggest accumulation of rubbish in these areas or a particular deposition practice or preference.

Within the town itself, the distribution of lava milling tools is difficult to establish due to incomplete data, especially findspot information. Furthermore, there is a high probability that different excavation strategies were used in different *insulae*, and that sampling and curation of material culture was not uniform over the long duration of archaeological excavation by different teams of archaeologists. This makes the data analysis problematic, as it becomes challenging to compensate for these differing strategies without knowledge of what they may have been. However, as most of the dataset comprises lava querns retrieved during the Frere excavations, it can be assumed some uniformity of approach was applied to the collection of the lava. The sometimes very small fragments that were retrieved and kept in the museum collections indicates that we are examining a significant proportion of what was present at the time of excavation. What cannot be assumed is that lava was absent in the *insulae* where excavations took place but where lava was not recorded, especially in the earlier excavations. We may only examine the presence of lava and its possible concentrations in specific locations.

Overall, this does not furnish us with a great deal of information, though it is worthwhile to examine the distribution of lava in correlation with other evidence for food and grain processing, and grain storage within the town to attempt to corroborate the possible interpretations. From the volumes of lava recovered from the town, the most significant concentration can be seen in *insula* XIV where 14 lava groups were recovered. This is of importance as *insula* XIV is located adjacent to Watling Street, which would have served as a major thoroughfare and was located opposite the *macellum*. The *macellum* was constructed around the Neronian to early Flavian era and was a major public building, with maintained use through the whole Roman period of the town (Niblett, 2001, p. 105). It saw several phases of rebuilding and existed as a stone structure by the second century. Volumes of butchery waste indicate that this was an area where meat was processed at scale, possibly prior to its sale at the *macellum*, where other types of food were probably also exchanged or sold, and this is most prevalent in the third century. A commercial level of enterprise relating to the processing of meat was probably being undertaken at Verulamium (Niblett, 2001, pp. 105–106), and it is possible that this was the centre for other similar large scale food processing activities. The volume of querns recovered nearby could be indicative of grain processing at a similar scale occurring close to the markets, possibly for sale or redistribution via the *macellum*. This should be compared with the evidence at Chichester, where large numbers of lava and other querns were found at the Cattlemarket site outside of the town. Environmental evidence and butchery suggested large scale food processing occurring within a designated area at the site and were indicative of a thriving market area where livestock and grain were brought from surrounding rural areas for processing and probable sale or exchange (Down, 1989, pp. 82–83). Whether similar activities took place at the Verulamium *macellum* should be investigated further.

The surrounding rural areas around Verulamium are likely to have been used extensively for arable agriculture, and a large granary at Gorhambury Villa can be used to argue for this idea (Neal, et al., 1990). Three smaller granaries existed in the town that were destroyed by the Antonine fire, all within *insula* XIII. It is not known whether replacement structures were built afterwards, and if they were, these did not occur in the same location. However, a large tower granary is known to have existed by the third century in *insula* II, and this would have held a significant volume of grain brought from the rural hinterland. Environmental evidence indicates that the grain had been dried and threshed prior to storage, and that these processes were undertaken before the grain was brought into the town (Niblett, 2001, p. 106); possibly at the roadsides as suggested by the quern distributions at King Harry Lane where some grain may also have been ground to fulfil more immediate needs. The cluster of lava querns near the *macellum* and the large-scale storage of grain in the town is indicative of centralised food processing, a pattern that is typical of what would be expected of an urban environment with a population engaged in crafts or industry that preclude the ability to engage in self-sustained food production or processing. Structural remains on the River Ver have been speculatively interpreted as a mill (Rogers, 2013, p. 139) and the recovery of millstones from the town also supports the idea of mechanically rotated mills in the vicinity. The querns from near the *macellum* suggest that these were used at scale for processing, and whether these were largely replaced by mechanical means of grinding is difficult to ascertain without a better-defined chronology of quern and millstone use in the town.

Of the fourteen lava querns recovered from near the *macellum*, eight could be assigned dates that were more specific than simply classed as Roman (table 7.3). These predominantly fall within a late first to mid-second century date range, which matches our current understandings of the peak of the lava millstone trade in Roman Britain. This means that most of the dated lava is from contexts that predate the fire that destroyed large parts of the town, including the *macellum* around AD 155/160 (Frere, 1983, p. 13; Niblett, et al., 2006). Though it seems likely that artefacts that predate this period are much easier to date, it does raise the question of whether the event of the fire altered the use of the space around the *macellum*. The previous structure was replaced with a masonry building and it probably took some time for this to be completed. The new *macellum* was a grander design but was smaller when compared to its predecessor, indicating possible alterations to its previous uses. However, the possible reduction in lava does not necessarily denote a reduction in the number of querns and more work would need to be done to see what was occurring in relation to the chronological change in milling tools of other stone types in the vicinity.

Related to the possible large-scale grain processing occurring at the site near to the *macellum* is the grain storage area in *insula* XIII, which was subject to detailed analysis by Niblett, et al., (2006) in an article that aimed to produce a more cohesive

Table 7.3. All fourteen querns recovered from insula XIV with their allocated date period

Lava quern catalogue No	Museum accession No	Allocated date range
999		fifth century
2557	1980.932	Roman
2588	1980.979	175-275
2594	1980.984	Roman
2595	1980.981	Roman
2596	1982.163	Roman
2597	1982.16	150–155/160
2598	1980.98	Roman
2609	1980.978	85–105
2611	1982.162	85–105
2612	1980.982	150–160
2614	1980.991	Roman
2616		155/160
2617		155/160

and interpretative account of earlier excavations that were reported in a more piecemeal way at the time of the original separate investigations. Examination of environmental and structural evidence concluded that both wheat and barley had been stored at the site prior to the second century, with a change to only barley for malting by the third century once rebuilding had replaced the storage area after the Antonine fire (Niblett, et al., 2006, p. 180). It appears that the event of the fire may have resulted in some changes in the way that commercial activities were organised, including that of the storage, processing and possible sale of grain products. It is interesting to note that at the grain storage site in *insula* XIII most lava was recovered from reuse contexts; specifically those associated with surface make up and levelling. These were predominatly dated to later phases; from the fourth century onwards. This contrasts with those found near to the *macellum*, which were mostly dated to before the fire.

However, it would seem that the groups of lava querns from *insulae* XIII and XIV may have been in use at similar times, but with those in *insula* XIII being reused. It is worthwhile to note that the change in the use of the site to that of a malthouse and brewery would probably have negated the requirement for lava milling tools, as these are likely to have been unsuitable for the job. Lava was a popular tool for grinding grain as the natural vesicles in the material could cut the grain more easily to produce flour. In beer production, the barley needs to be dehusked, an outcome that is best genrated by crushing the grain. Sandstone, grit or conglomerate querns would have been more suitable for this processing, and it is possible that lava was rejected in favour of these. This could explain the absence of lava from post-fire contexts when barley became the dominant grain type and wheat ceased to be stored at the site. However, there were no querns of indigenous stone types recovered that may have replaced those of lava, and it may be the case that all processing took place elsewhere, making the absence of lava in secure later contexts less relevant.

Of the six millstones that were recorded at Verulamium, all were recovered from within the town itself, which could provide further evidence that centralised processing of grain occurred there. Only three of these could be allocated a location and two of these are within close proximity to the granary at *insula* XIII, both being found in *insula* XXVIII, not far from the possible grain processing area near to the *macellum* (figure 7.16). Again, care must be taken as the distribution only identifies the locations where lava millstones were

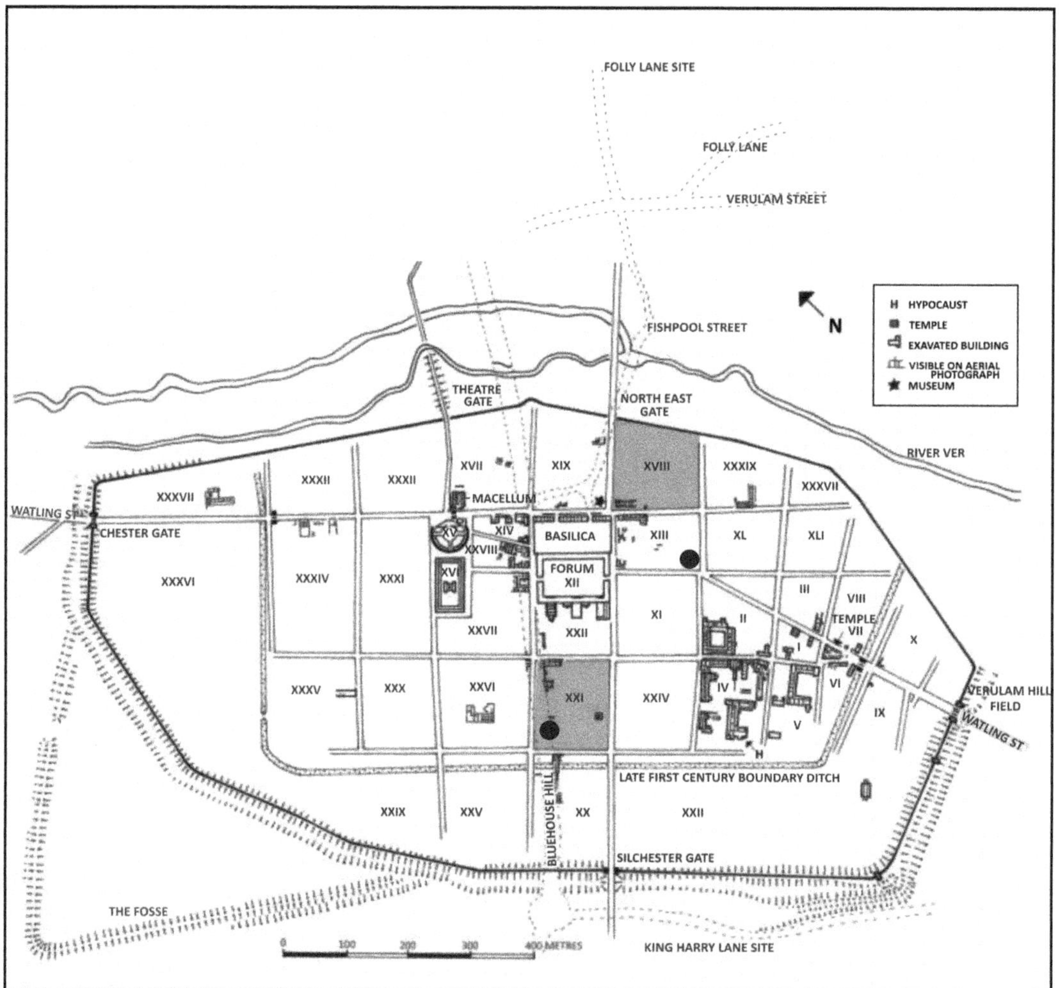

Figure 7.16. Plan of Verulamium showing the locations where lava millstones were recovered in red. Approximate locations of the granaries are indicated by dots. (Plan after Niblett 1999; St Albans and Hertfordshire Architectural and Archaeological Society, 2019).

deposited and may not relate to activity areas where grain was processed. However, it is curious to note the general clustering of lava milling tools in the northern part of the town. The fact that millstones are present at Verulamium should indicate that grain was processed at scale, as it is unlikely that these objects would have travelled to the site as recycled building material unless they did so from the area directly around or inside of the town. Unfortunately, only two of the examples are dated: One from *insula* XXVIII to the period of the Antonine fire and within fire debris (CAT 0997), the other from *insula* XXI to the mid-fourth century and recovered from rubble indicating a possible earlier date for original use (CAT 0998). Again, it is interesting to note that in the northern part of the town in *insula* XXVIII, the lava millstone must have been in use at the time of the Antonine fire, or the period preceeding it. The consequences of the fire appear to have caused a major change in operations and the use of lava, for both querns and millstones, possibly ceased or decreased significantly in this part of the town at around the same time.

7.5. Verulamium: Primary Use

For the Verulamium dataset, there were 13 upper stones and 9 lower stones that provided suitable measurements for thickness (figures 7.17 and 7.18). These have been

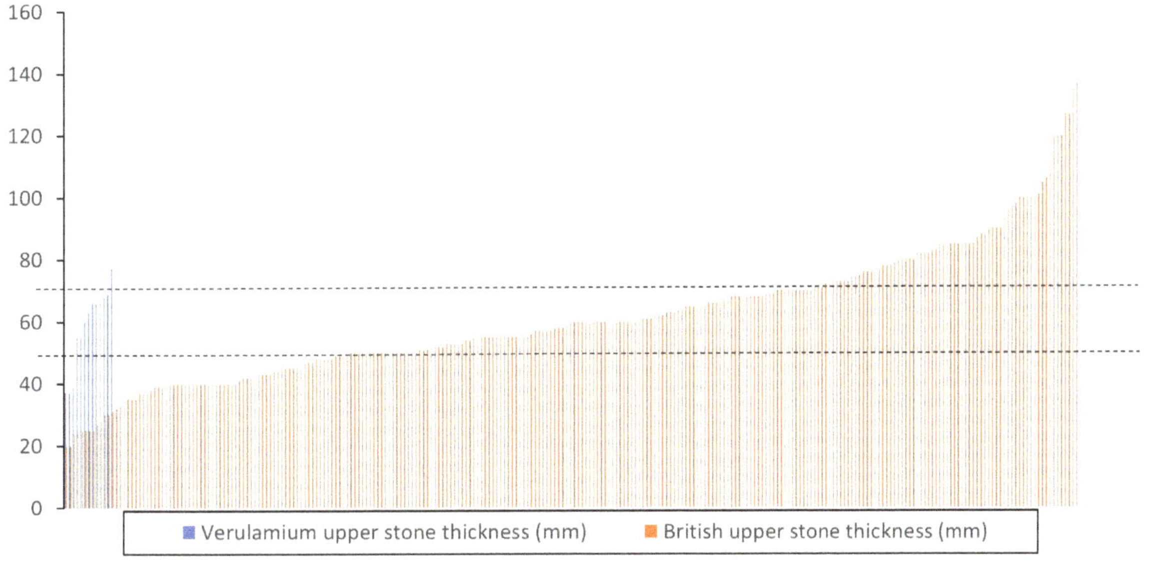

Figure 7.17. Graph showing the edge thickness of upper stones from Verulamium (T= 13) compared with those from all Roman Britain (T= 248) excluding those from 1 Poultry. The 25th and 75th percentiles are indicated with lines.

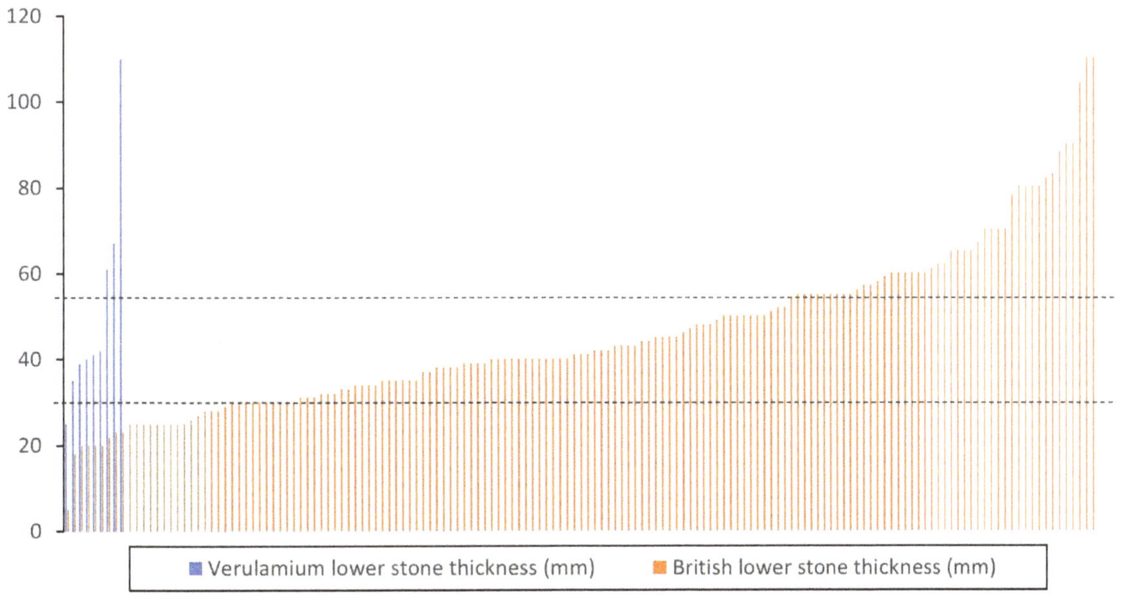

Figure 7.18. Graph showing the edge thickness of lower stones from Verulamium (T= 9) compared with those from all Roman Britain (T= 138) excluding those from 1 Poultry. The 25th and 75th percentiles are indicated with lines.

plotted alongside the data from the entire lava quern dataset from Roman Britain, with a total of 422 upper stones included but excluding the probably unused querns from the excavations at 1 Poultry that would skew the data considerably. For Verulamium, the sample size is not very large making it a challenge to interpret the data with confidence, though for both upper and lower quern stone thicknesses suggest that these are largely within the 'normal' range when compared with the data from all lava querns in Britain, which lies between 48mm and 72mm. This may seem relatively thick, but as it is a measurement taken at the edge, this is the part of the upper stone that is least effected by wear and so would remain thicker than the centre that tends to wear away completely prior to the quern becoming defunct. Three examples from Verulamium fall within the lowest 25th percentile for upper quern thickness, CAT 2612, 2603 and 2599 and these were all excavated from within the town. One upper stone was within the highest 25% percentile for thickness and was also from within the town walls. However, these examples are not very different from the normal range, suggesting that there was a standard thickness that a stone would reach before it was no longer used as a quern and deposited.

There are some differences seen in the lower quern stone data, though again, the small sample size provides some interpretative issues. Most of the lower stones fall within the 'normal' range for thickness, which lies between 32mm and 55mm, with one falling slightly below. There are three stones that are within the top 25% of lower stone thicknesses, two of which are not greatly different from the normal range. One, however, is very thick (CAT 2559) and is an anomaly within the Verulamium dataset, having few parallels within the British dataset with a thickness at the edge of 110mm. This is a possible unused, or very little used quern stone and has a similar thickness to the 'new' querns from 1 Poultry. The reasons why this exists within the assemblage has been investigated in the section on reuse, as evidence on the stone implies that it may have been repurposed. Whether this example was broken on arrival is a possibility. The chance that this may represent a quern roughout may also be considered, though this is unlikely. The high level of standardisation for Mayen lava querns implies that they were centrally produced and the evidence at Mayen fully supports this theory (Chapter 2).

7.6. Verulamium: Reuse/Modification

Though there is little to be identified from use-wear analysis due to the fragmented state of most of the lava groups, there are two examples that show some possible adjustments that may have been related to their primary function as milling tools. These are both lower stones that appear to have had part of the lower part of their sides chipped and removed, with clear tool marks present on one (figure 7.19). Quern stones with chipped edges have been identified on quern stones of indigenous stone types, such as Lodsworth Greensand. The intention behind this behaviour could relate to deliberate breakage of the stone as part of a ritualised act, as this is a phenomenon that

Figure 7.19. Lower quern stones(left) that have been chipped at the bottom edge. Top image CAT 2605, bottom CAT 2592 (photo by author © St Albans Museum Service).

can be observed for querns from the Iron Age and earlier (Watts, 2014, p. 9). However, this chipping does not appear to have been completed with the general aim of breaking the entire quern stone and looks carefully and deliberately performed. The resizing of upper or lower quern stones is another possibility, to 'marry' them to other stones once their original pair stone had worn out. As upper stones are more likely to wear faster, it is likely that lower stones were reused and reworked to allow continued use. However, as lava querns are highly standardised, the extent to which reworking was necessary should be questioned. As imported products, it may also be worth considering the likelihood that a single unpaired stone could be procured to replace a worn upper for a reusable lower. Lava lower stones were possibly paired with uppers of other stone types, though the practicalities of doing this and the effect it had on the processing of grain should be explored.

Observations of the Verulamium chipped lava lower stones do not suggest that the aim was to resize the stones, as the focal area of chipping is near the underside of the quern, leaving part of the original edge vertical and intact. The result is a 'bowl' type profile as the lower surface gradually cuts away underneath the stone, a slightly less extreme version of the 'Napoleon hat' type lava querns that predate the rotary versions manufactured at Mayen (see figure 2.12, drawing 3). The interpretation for the unusual shape and style of this type of quern is that it was bedded into the ground to provide greater stability for when the stone was in use. A similar purpose may have existed for the chipping of the sides of the rotary querns, and these may have been adjusted to fix them to the floor or other surface to provide greater stability during the grinding process.

In terms of object biography, modification of an existing quernstone is an interesting stage in its use life. The standardised form of a lava quern is something that was generated by those involved in the manufacturing stages of the quern lifecycle and reflects a centralised and probably

Table 7.4. Summary of lava querns showing signs of reuse in Verulamium

Catalogue number	% of stone remaining	Description	Evidence for reuse
2557	2	Very small fragment with thickness of 11mm	Both top and bottom surfaces are worn very smooth and not identifiable as original surfaces.
2559	100	Possible unused lower stone that has been broken into multiple pieces.	Original harped dressing is visible on the grinding surface, but this is altered with a heavily inscribed swirled pattern. Central hole appears to have been widened and there are visible signs of redrilling that may have fractured the stone (figure 134a).
2566	1	Small fragment that has been reshaped and 'squared off'.	The stone has been reshaped for reuse and wear indicates that processing other than grinding may have taken place (figure 134b).
2594	25	Large piece of upper stone with unusually shaped rynd chase.	The shape of the rynd chase could indicate that it has been adjusted to accommodate a different use (figure 124 in 'manufacture').
2616	Unknown	'fragment…reused since it has 2 adjacent planes which lie at an acute angle. Upper surface is fairly flat, but others are rough' (Frere, 1972, p. 158).	Object not seen in person, but publication describes surfaces that may have been reused for other processing activities.
2684	3	Small fragment with a smoothed and curved shape.	The shape and texture of the surfaces must have been produced from other processing activities as these are would not have occurred from grinding (figure 134c).

controlled process of production (Chapter 2.5–2.7). Those involved in the stone working and quern finishing stages of the object life cycle may not have had the freedom to innovate or to express their own ideas on what the form of a quern should be. Continuous and long-lived interactions between stone workers and this design of lava quern may have reinforced socio-cultural understandings relating to quern form, prolonging manufacture of this type. However, once the querns were relocated to a different environment, i.e., Britain, the option to reinterpret the design and meaning of a lava quern became possible. As vitally important domestic tools, pre-existing socio-economic and cultural practices existed in Britain regarding human-quern interaction, and these may have been very different to those of the people manufacturing the stones. Renegotiation of the form of the quernstones occurred to suit the user's preferred mode of operation and placement within the home. The quern was translated into a form that could embody the user's concept of what a quern should look like and how it should be interacted with.

Unfortunately, as these two modified querns are fragments, little can be confirmed with regards to how they were worn, as this can also be indicative of a fixed position quern stone. There are several examples of querns that have been found *in-situ* fixed into position this way (Cool, 2006, p. 74), such as a non-lava lower stone that was found set into the surface of a workbench at Vindolanda (Bidwell, 1983, fig. 29). It is highly probable that this practice was carried out elsewhere and perhaps adjustments needed to be made to a quern to allow for it to be appropriately fixed. However, a similarly chipped quern stone would need to be recovered from a use-context for this theory to be confirmed and more work needs to be done to examine whether other altered lower stone examples exist in other assemblages. The implications of having fixed position querns would indicate that grain processing was occurring at a scale where space could be permanently set aside for the activity to be undertaken. This might relate to centralised processing, commercial endeavours, or elite persons with large areas of space where domestic activities could take place.

Instances of reuse are difficult to identify in an assemblage that is so fragmented, but there are occasions where this is visible, both in terms of the effect on the object and/or the context of deposition. There were six objects that showed possible evidence for repurposing (table 7.4), all of which were recovered from within the town. Of these, four show signs that they were reused for other processing activities, such as for sharpening knives or blades. These were found to have surfaces that were not identifiable as being the result of original use as a quern, due to a change in texture, wear, or shape. For example, CAT 2566 was worn very thin and, if still part of a complete stone, would not have had the weight necessary for it to be used for milling. Both the surfaces were worn smooth and flat, and these could not be attributed to wear from rotary or oscillatory grinding. The fragment had also been reshaped (see figure 7.20 below).

The geology in and around Verulamium does not provide a ready supply of stone suitable for ground stone tools and stone brought into the town for other purposes is likely to have been reused regularly. Lava was clearly a popular and common choice for milling tools at Verulamium and its reuse is likely to relate to this. It is surprising that there are not more examples of reuse, as it would be expected that with the size of the assemblage this would similarly increase. However, the high level of fragmentation across

Figure 7.20. Examples of quern stones from Verulamium that show possible signs of reuse. a- complete lower stone that is shattered into fragments with swirled incisions on grinding surface CAT 2559 (© St Albans Museum), b- small thin fragment with two it has been reshaped for reuse CAT 2566 (photo by author © St Albans Museum Service), c- thin and curved fragment that has one very smooth surface CAT 2684 (© St Albans Museum).

the assemblage might relate to reuse and the deliberate breaking up of querns to accommodate this. Though there are fragments that could correspond to natural breakage from wear, several are more difficult to explain, such as body fragments that would not have been vulnerable to damage (figure 7.21). This type of fragmentation has often been correlated with ritual deposition, especially in the Middle Iron Age, and could explain the damage seen. However, many of these fragments lack the context of deposition needed to establish such interpretations.

One other of the reused lava quern stones has been summarised in the section on manufacture (CAT 2594), as it is unclear if the unusually shaped rynd chase is a production feature, a sign that the stone was mended for continued use or be related to a possible reuse of the stone for other functions. Unfortunately, it is not possible to correlate this possible adjustment to any specific type of reuse, though it appears that the shape assumed to be a rynd chase has been deliberately and carefully cut into the stone. The only other example of possible reuse may also be identified as a possible 'roughout' quern stone that was not completed and is between stages of production (CAT 2559). It is unique within the British lava quern dataset and shows some very interesting characteristics. The stone is clearly identifiable as a lower stone and has many of the standardised features associated with those manufactured at Mayen, such as the convex grinding surface and harped dressing on the grinding surface. However, the lower surface of the stone has not been fully hollowed out as would ordinarily have been done probably to make the stone less cumbersome. It appears that attempts had been made to do this, but the job had not been finished to the level that is normally observed on other querns. There are clear tooling marks on the lower surface to show where these attempts had been made (figure 7.22).

Tooling marks are usually very visible on the lower surfaces of lower quern stones, but the apparent incompleteness of this example is highly unusual. The only other evidence that could indicate that this was a quern roughout is the signs that show that the stone had been drilled to create the central perforation. Grooved marks in the centre of the stone show that a rotational motion had been used and this may also correspond to the swirled incisions found on the grinding surface of the stone. However, this evidence could also relate to the reshaping of the central hole so that the stone could be repurposed. Whatever the reason, the stone must have shattered because of the drilling, and the multiple fragments were deposited. There are clear signs

Figure 7.21. Body fragment of lava quern that may have been part of a quern deliberately broken as it does not correlate with normal breakage due to wear or other vulnerabilities. The left side if the image shows the grinding surface CAT 2571 (photo by author © St Albans Museum Service).

Figure 7.22. Underside of CAT 2559 showing the tooling marks where the underside had started to be hollowed out (photo by author © St Albans Museum Service).

of the original production and intention to use the stone as part of a quern, a purpose that from its thickness and lack of wear was apparently not undertaken. It may already have been rejected as a suitable stone, and the decision made to repurpose it. The 'overwriting' of the original dressing on the grinding surface to produce a swirled design may have been intentional, or it may have been a by-product of the process of repurposing.

7.7. Verulamium: Deposition

Of the earlier lava groups at Folly Lane, three (CAT 2623, 2624 and 2650) were recovered from shafts, ABZ, DKM and AAB. These shafts are within a larger group of similar features that have been interpreted as having a ritual function due to their frequency, placement and the finds recovered from them. Folly Lane has many of these 'ritual shafts' alongside ovens, which are also found at the temple sites in *insulae* VII and XV and are likely to have been used as part of ritual activities (Niblett, 1999, p. 100). Alongside the lava, the shafts also yielded large volumes of animal bone (including dog and horse), coins, glass, items of personal adornment, a 'deliberately' bent key, and a chisel. The pottery assemblages from these contexts are unusual, and it was found that fragments from shaft ABZ could be refitted with those from a 'paired' shaft nearby indicating that these were contemporary. They included fragments from complete vessels, all closed forms, such as jars, beakers, flagons, and face pots. Two fragments from separate Eifelkeramik vessels were found within shaft AAB, and it is interesting to note that these may have travelled to Verulamium alongside the imported lava querns (see Chapter 2). The contexts could be dated from the mid-second to early third century, when many of these ritual shafts were dug and then filled in (Niblett, 1999, p. 99). It is possible, therefore, that the lava recovered from the three shafts may also have been ritually deposited. However, the interpretation of these shafts and their contents as remains of ritualised activity is questionable. The 'ritual pits' at Silchester were subject to nuanced investigation with regards to the pit deposits, stratigraphy, and the finds recovered. It was found that there was no clear way to prove or disprove that there had been a ritual aspect to the placing of these deposits and finds in the pits involved (Eckardt, 2006). Fulford (2001) has also argued strongly for a more theoretically informed and evidence-based approach when identifying ritual deposits from rubbish disposal, and the 'ritual shafts' at Verulamium would benefit from more in-depth scrutiny of this kind with regards to the interpretations presented by Niblett (1999).

Unfortunately, the lava fragment groups recovered from the ritual pits are not described in the excavation report, but it would be interesting to see whether these comprised pieces of the same quern in the same way that fragmented pottery appears to have been split between two other shafts. The degree of fragmentation and the possibility for its destruction to have been a purposeful act would also be of interest. Deliberate fragmentation and deposition of querns in ritualised contexts is a phenomenon that is well-recognised in prehistoric archaeology with querns of indigenous stone types (e.g., Brück, 2006; Watts, 2014) but is less well-understood in terms of Roman era practices. The use of lava in such contexts implies that the material may have been less significant than the function of the object, and that it was selected for deposition for this reason. The frequent presence of ovens at ritual and religious sites in and around Verulamium indicates that food preparation and consumption may have been involved in some of the ritual activities carried out there. Querns and ceramic vessels may have been similarly employed and deposited afterwards (Niblett, 1999, p. 94).

A further parallel example to this depositional behaviour can be seen at the site of Mount Pleasant, which was a rescue excavation completed in the 1960s at an area 175m to the south of Folly Lane, where a continuation of the ceremonial landscape appears to have existed. Here, two deep shafts were uncovered, interpreted as possible wells (Hertfordshire HER number 14005 and 14006). These were of a similar chronology to those at Folly Lane, with the primary infill dated to the second century with a third century upper fill. The shafts contained coins, a fragment of lava, pottery and the disarticulated human remains from three individuals: a male, female, and an infant. The presence of lava in a third possible ritual shaft from a similar phase and location strengthens the possibility that lava querns had been ritually selected and deposited. However, details for this excavation were poorly recorded and remain unpublished making it impossible to establish where in the shaft the lava was recovered. Quantities of third-century domestic refuse has been found in sinkage hollows in the upper fills of shafts and pits in and around Folly Lane and without clear context and stratigraphic information, it may be that the lava constitutes intrusive rubbish. The weathered and undiagnostic nature of the lava found at the site indicates that this is a possibility.

Overall, the pits at the sites around Verulamium present an interesting possibility for lava to have been ritually deposited, possibly as part of practices that involved the preparation and consumption of food. However, there are many other potential explanations, and it is important to recognise the difficulty involved when interpreting such deposits. What classifies an object as part of a 'structured' deposit is a heavily debated topic (Hill, 1995; Fulford, 2001; Garrow, 2012), and interpretations often rely on more detailed stratigraphic information than that currently available for the specific examples mentioned here. Even then, connecting the meaning and intent behind the deposition of different objects within the same feature, possibly as part of different 'events', is a complex process that lacks any certainty regardless of how theoretically informed that process may be. This was found to be the case at Silchester during the detailed investigation of the late Roman pits completed by Eckardt (2006), as explored in the next section. Residual and intrusive material, alongside the recutting of features adds another layer of complexity. It is only possible to discuss the probability that lava was ritually deposited, which in the case of

Verulamium, appears highly likely due to the nature and context of the deposits involved.

7.8. Verulamium: Conclusions

Overall, the lava milling tools from Verulamium present some interesting results and can add to our interpretations of urban life. Verulamium was located within a region that both depended on imported milling tools and was able to access them easily via exchange routes on the eastern coast. Consequently, a significant amount of lava has been recovered from the Roman town, though the true volumes may not have been fully captured in this analysis due to the age and retention policies of some of the excavations. As a town that had good access to lava products, querns were deposited after their primary use was no longer possible, but when they were not heavily worn when compared to other querns from the full British dataset. Much of the dataset shows that the querns from Verulamium were of an average thickness prior to deposition, with some examples that were considerably thicker than the Romano-British average at the time of their disposal. However, many fragments were found to have been reused for other purposes, probably another outcome of the local geology that also was not able to provision whetstones, hones, or other ground stone tools.

Distributions of deposition have highlighted some possible activity areas related to quern use, such as the area around the *macellum*, but were generally associated with reuse and levelling contexts, with a large volume of probable residual material being used in this way. Ritual deposition of lava seems possible at Folly Lane and Mount Pleasant where late Roman pits appear to have been used for this purpose. There may have been a relationship between querns, pottery vessels, and the preparation and consumption of food as part of ritual activities, with these objects being deposited afterwards. More research and better stratigraphic information would be needed to explore this more fully.

In terms of the economy, Verulamium yielded a high number of millstones, indicating that centralised grain processing was occurring in or close to the town, probably after the period of the Antonine fire when lava querns cease to appear in secure primary contexts. The apparent transition from small granaries to a large tower granary also suggests that centralisation was occurring, and this probably involved the nearby villa estates that also provide evidence for large-scale storage of grain products. These changes could be related to changes in the status of the town and the development of more specialist economic and administrative activity. The centralisation of beer production provides one example of this, with the development of a brewery and malthouse appearing around the third century. This does not appear to have utilised lava querns in its operations, which may have been due to the impracticalities of using lava for this purpose. Overall, evidence appears to suggest that, over time, small-scale domestic grain processing using querns may have been rejected in favour of the consumption of ready-ground flour. The presence of a millstone in Antonine fire debris suggests that this transition had commenced by the mid-second century. However, it would be necessary to examine the non-lava quern data to establish this fully, as the current data may only relate to decreasing amounts of lava entering the province and reaching Verulamium.

7.9. Silchester: Introduction and Background

Calleva Atrebatum (Silchester) (Figure 7.23) is one of the most well-known and researched Iron Age and Roman towns in Britain (Fulford, 2021, p. 1–4). Extensive excavations that took place in the Victorian era revealed the full plan of the town grid and many of the querns from the dataset were recovered during these investigations. However, it is likely that these only represent a small number of those excavated (Shaffrey, 2003, p. 144–145). From the volume of Roman material that has since been recovered from the Victorian excavators' rubbish dumps in *insula* IX, it is clear that querns and bulk finds were selectively retained, with perhaps only the larger or most diagnostic examples kept. If any specific discard policy existed, this is not documented, and it is impossible to know what proportion of all recovered material was retained or why some objects were selected over others (Fulford, et al., 2002, p. 229). Furthermore, the early excavations were completed without an understanding for the importance of stratigraphic relationships and much of the data produced fails to address important research questions, such as those related to the Iron Age phases of the site.

Investigations have continued into the present day using modern techniques and approaches, such as environmental sampling and geophysics (Fulford, 2021, p. 8–19). These more recent excavations have predominantly focussed on *insula* IX (Fulford, et al., 2006; Fulford & Clarke, 2009; Fulford, et al., 2011; 2018a; 2020), with smaller-scale investigations conducted at the forum-basilica (Fulford & Timby, 2000), the temple in *insula* XXX (Fulford, et al., 2017), the amphitheatre (Fulford, 1989), and the bathhouse (Fulford, et al., 2018b; 2019), while an extensive geophysical survey completed by Creighton & Fry (2016) examined the town and the extramural landscape.

Silchester is in central southern England, in a region that is thought to have been within the tribal territory of the Atrebates. Initial settlement of the town has been dated to 20–10 BC from coin and ceramic evidence, though it is possible that other parts of *Calleva* might reveal earlier phases (Fulford, 2021, p. 32–34). It is likely that the inhabitants at Silchester had pre-existing contacts with the Roman world prior to the Claudian invasion, as evidenced by imported foodstuffs and commodities with connections to Italy, Gaul and Spain that have been recovered from pre-conquest contexts (Fulford, 2021, p. 36). This pre-exposure and connection to the Empire may have influenced the experience of the town's inhabitants during and immediately after the Roman conquest of AD 43 and is a pattern that has been observed at other southern

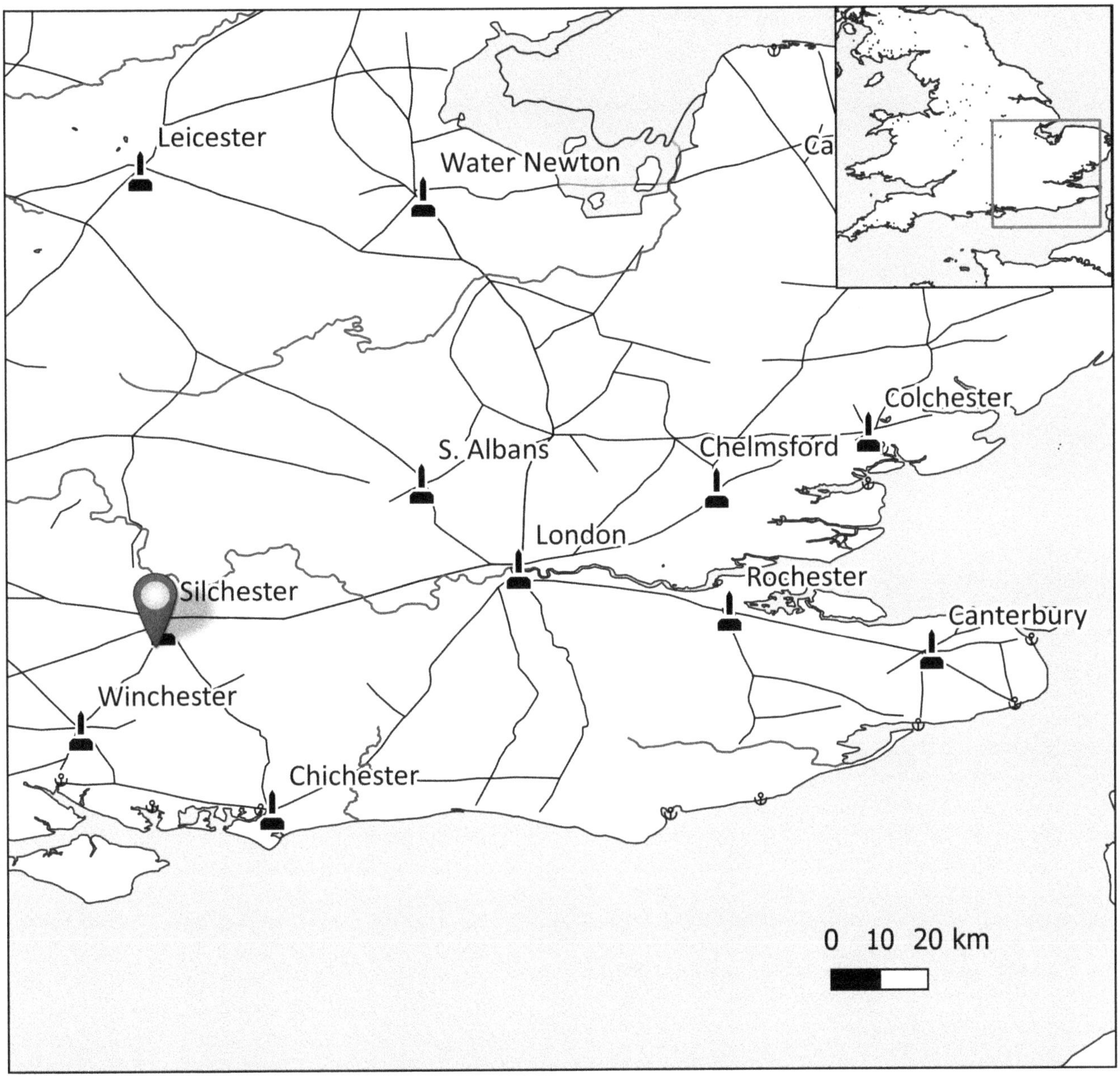

Figure 7.23. Location of Silchester Roman town (Background GIS data courtesy of the Ancient World Mapping Center).

urban/oppida sites (Creighton, 2006, p. 24). During this period, agricultural production involved the cultivation and processing of grain and there are also indications that animals grazed nearby (Fulford, 2021, p. 41). Most querns dated to this era were sourced from the Upper Greensand quarries at Lodsworth (Shaffrey, 2021, p. 7–8).

A possible military *principia* provides the only structural evidence for a potential military phase in the early conquest period at *Calleva* (Wacher, 1995, p. 272). However, this identification remains controversial for several reasons (Millett, 1990; Fulford, 1991; Creighton, 2006, pp. 72–75) and an assessment of contemporary finds evidence has not shown any obvious military or state associations (Pitts, 2014). However, as Fulford (*ibid*) and Boon (1969) argue, plentiful finds of early Roman military equipment and material culture associated with the military show that a presence is highly likely.

Around AD 85 the revised grid system for the town was set out, orienting the streets on a north to south axis from the earlier north-north-west to south-south-east alignment (Fulford, 2020a). This was a major undertaking and may have occurred as part of a single process of renovations, including the infilling of the inner enclosure ditch system and the establishment of the adjusted road network. The establishment of the forum basilica during this period was a significant development in that it reveals the formalised introduction of centralised self-governance to the town and its surrounds, while the possible *mansio* emphasises the importance of Silchester as a transport and communications hub (Fulford, 2021, p. 76–102).

The second century saw *Calleva* reach its pinnacle in terms of population, activity, and investment, (Fulford & Timby, 2000, p. 68). The economy of the town may have been largely supported by the traffic of people moving through on their way to other destinations, and that of the rural production of its hinterlands, including elite rural estates. Imported goods and foodstuffs were widespread during this period, indicating a culture of consumption for a population of relatively well-connected and wealthy inhabitants, while grain processing does not appear to have occurred on the small and localised scale seen in previous periods (Robinson, 2011, p. 283–287). Evidence suggests a process of economic change had occurred around this time and that it affected the range and scale of industry at *Calleva*, alongside the culture and practices associated with the preparation and consumption of food. This includes a marked decrease in the number of milling tools present, as the organisation and scale of grain processing appears to have experienced a similar change (Shaffrey, 2021, p. 26–27). Further building and refurbishment works continued at Silchester into the late third century.

Fourth-century occupation of the town saw some refurbishment and building was still taking place in *insulae* IX and III, where gravel-filled foundations show a town that was still changing and developing (Fulford, 2021, p. 158–163). Imported goods in the form of wine from Turkey, olive oil from Tunisia, and figs were still clearly available and consumed, while evidence for the consumption of more locally produced foods, such as spelt wheat, barley, beef, mutton, and hazelnuts occurs more frequently. Cultivation and animal penning appears to have occurred within the town itself and local provision may have generated much of the population's dietary needs, while grain would have been brought from the surrounding countryside. Grain processing evidence is, again, lacking, suggesting that this may have been completed using mills as a centralised activity, or simply carried out elsewhere (Ingrem, 2006, p. 188; Robinson, et al., 2006, pp. 216–218; Shaffrey 2021).

7.10. Silchester: The Dataset

The dataset used for this study was compiled by Ruth Shaffrey of Oxford Archaeology, who has generously shared her data to allow for an in-depth study of lava use at the site to be explored. It includes lava millstone or quern fragments from 62 contexts, with a further 548 fragments from milling tools of other stone types to provide a broader analysis for the use of lava at *Calleva* (figure 7.24). This data has been analysed previously by Shaffrey (2021) in an integrated approach to examine changes in grain processing at Silchester and its hinterland more broadly. The scope of this current work aims to specifically explore lava use by detailing the object biographies of lava querns and millstones at Silchester. The objects in the dataset were recovered from excavations dating from the Victorian era to the present day and finds records are, therefore, inconsistent in their detail depending on the era of excavation, especially with regards to context of recovery. However, all finds can be assumed to be associated with Late Iron Age or Roman activity as there is no significant medieval activity on the site. Some objects were recorded using published accounts, while others were recorded in-person by Ruth Shaffrey. The dataset presents the largest assemblages of querns and millstones from a Romano-British urban site.

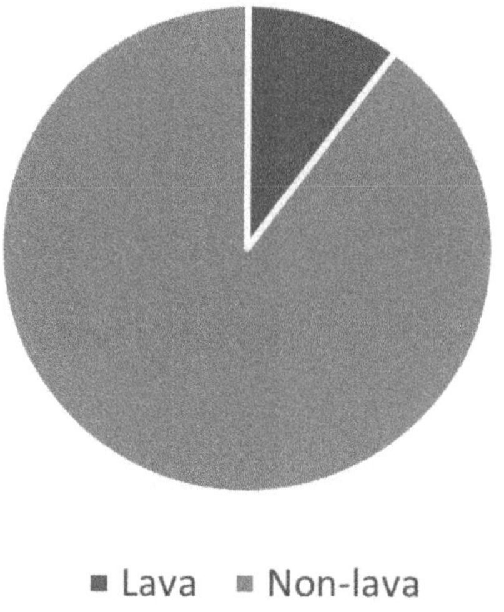

Figure 7.24. The number of querns or millstones of lava in the assemblage compared with those of other stone types (T= 610).

This provides a unique opportunity to fully explore the role of food processing, grain consumption, the stone trade, and changing uses of specific stone types for milling tools within a Roman period urban environment. Furthermore, it also allows for an investigation into the chronological changes in quern and millstone use during the Iron Age to Roman period transition.

Like other datasets in this study, the lava from Silchester was very fragmentary and it was not always possible to identify the features of a quern or millstone (figure 7.25). There were only five occasions when the fragments were at least a quarter of the full circumference of a quern or millstone, with most of the dataset comprising small or very small pieces. This has produced some interpretative issues, for example in identifying features from manufacture and use-wear. The chronology of the dataset is, however, excellent and provides a much-needed glimpse into how changes in lava use may have occurred at the site over time.

7.11. Silchester: Manufacture

Due to the fragmentary nature of the dataset, the manufacture features of these lava milling tools were less readily identifiable than they would have been for more complete examples. There were 12 upper, or probable upper and 12 lower, or probable lower stones within the dataset.

7.11.1. Millstones vs Querns

The diameters of the milling tools from Silchester are shown in figure 7.26, where the data from upper and lower stones has not been separated. The differentiation should not make a difference to the distribution of diameters due to the necessity of pairing stones of equivalent sizes. Millstones were identified from their size (500mm or larger), or because they possessed millstone features, such as holes for iron fittings for lifting (figure 7.27). Though the dataset is relatively small for Silchester, it is notable

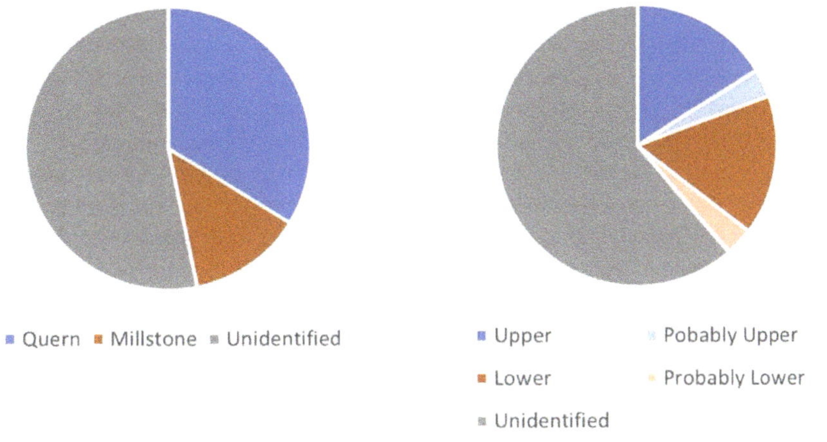

Figure 7.25. Number of lava fragments identified as derived from querns or millstones (left), and the number of fragments identified as upper or lower stones (right) (T=62).

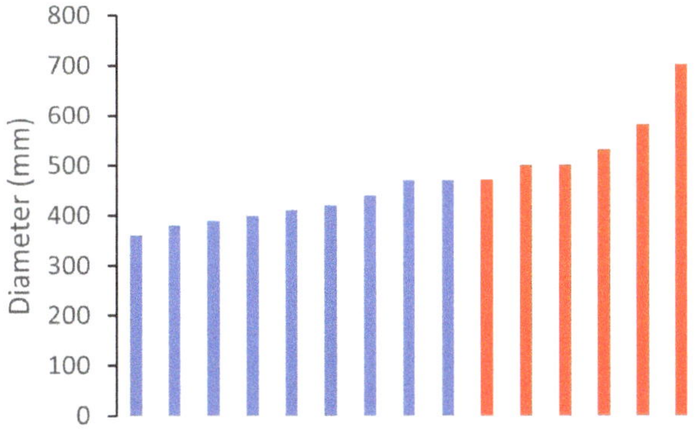

Figure 7.26. Diameters of both upper and lower quern stones and millstones from Silchester (T=24).

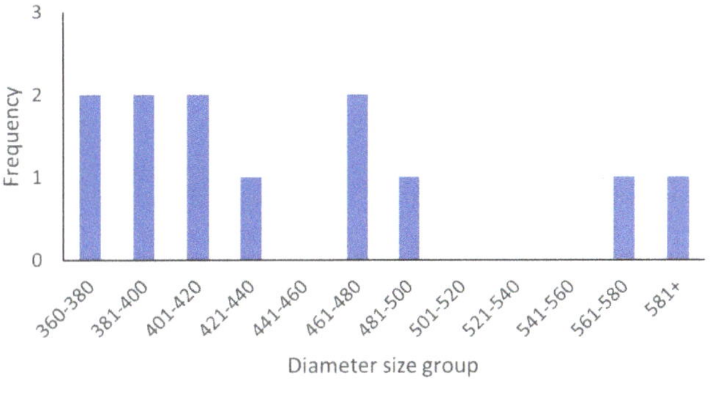

Figure 7.28. Frequency of mill and quern stones within different diameter ranges.

Figure 7.27. Lava millstone from Silchester showing holes that would have been used to accommodate iron fittings, possibly for lifting the stone as part of its function (CAT 0919) (photo by Shaffrey, 2021).

that there is little differentiation between the diameters of millstones and quern stones. At other Romano-British sites, a trend is seen whereby the millstones are significantly different in size to those of the quern stones. At Silchester there is not the normal 'leap' in diameters that would ordinarily occur, usually around the 500mm size.

Two examples of 'Pompeian' style milling stones, as identified by Allen (2012), have a relatively small diameter (estimated to be around 470mm and 500mm) when compared to the more typical disc type lava millstones and these have slightly skewed the results (CAT 0870, 0871). As this is a different style of millstone with a diameter that is variable depending on where the fragment is from on the stone, these diameters cannot be directly compared to those of disc type millstones (see Chapter 2). These objects have blurred the separation between querns and millstones and should be discounted from the general analysis of size. After their removal, it appears that there are three size categories present in the dataset: those clearly identifiable as querns in the smaller group, those clearly identifiable as millstones in the larger group, and a central group (figure 7.28). This varies from what is ordinarily observed in the size data, whereby querns and millstones are relatively easy to differentiate (see Chapter 5). In this case, the two objects in the 461–480mm range have been classified as querns, while a single object measuring 500mm in diameter has been classified as a millstone. Upon further investigation, it appears that the measurement for this single example was a minimum diameter and that the stone may have been significantly larger. This resolves the issue and shows that this method of differentiation still works well, so long as the data are carefully scrutinised.

It is also necessary to question the unusually high number of identified millstones in the assemblage, and whether this is representative of ancient practice at *Calleva*. A total of eight millstones were recorded out of 24 identifiable artefacts. However, it must be remembered that Silchester has been subject to decades of excavation, with many years undertaken using modern excavation techniques, where milling tools have been consistently recorded. A large area of the town has been uncovered when compared to other sites, and this has also improved the chances of recovering a higher number of millstones. However, this does not explain the percentage of millstones (25%) compared to quern stones (75%) present in the dataset from identifiable fragments, which is much higher than seen in the province-wide data (6% millstones and 94% quern stones). This may relate to the deliberate selection of more substantial lava quern examples for retention during the Victorian excavations. Only good-sized or distinctive fragments of any material appear to have been retained (Shaffrey, 2003, p. 145), which in relation to lava, are more likely to have belonged to the larger millstones. The fact that many of the fragments were analysed by a quern specialist may also help to explain this anomaly. It is probable that lava millstones are often misidentified in excavation reports and that there are far more in Roman Britain than have been accounted for. The Silchester data presents the possible scale of this issue and can be compared to Verulamium where in-person data collection has also revealed a relatively high number of lava millstones. This suggests that the export of millstones from Mayen to Britain may have been more significant than first anticipated, especially in relation to urban sites.

There is a greater disparity seen in the diameter ranges for millstones when compared to those for quern stones. If excluding the Pompeian style millstone fragments, the remaining four disc-type millstones showed a possible diameter range between 500–700mm. The range for the quern stones was much narrower, with diameters between 360–470mm and 50% of the querns falling into the range of 385–455mm. Although there needs to be some consideration taken for the estimation of diameters, this is a pattern that is seen elsewhere and is the result of a high level of standardisation for querns and a greater degree of flexibility in the sizing of millstones that were probably made to order (see Chapter 2). In the provincial-wide dataset, 50% of lava querns from all sites have diameters that lie within the range of 400–420mm, which is much narrower than that seen at Silchester. With such a small dataset for Silchester, it is difficult to determine whether this wider range should be interpreted as a meaningful disparity.

The question is: why would a reduced level of standardisation exist at Silchester, and how would this relate to the manufacture of lava milling tools at Mayen? The lack of any evidence for a watermill at Silchester (Shaffrey, 2021, p. 3) and the absence of a suitable body of water to power it may provide one possible explanation. This could have resulted in the need for larger milling tools that could still be operated by hand, or small-sized mills that could be powered by animals or humans. Possible millstone plinths have been recovered in the town in House 3 of *insula* XVIII, and it is likely that these are the remains of mills that relied on this method of rotation (Shaffrey, 2021, p. 21–22) (see figure 7.29 for location). The introduction of this technology would have generated a need for milling stones that were larger than the domestic quern, yet small enough to be powered for long periods of time, perhaps by donkeys or slaves. It, therefore, becomes a possibility that some of the larger 'querns' in the assemblage were operated mechanically, alongside the smaller millstones. A lack of diagnostic handle fittings means that it is impossible to identify the method of rotation for any of the fragments in the assemblage and querns have only been categorised if they were found to have diameters under 500mm. A lack of standardisation implies that Silchester may have been obtaining its relatively small millstones directly from the manufacturer, with stones made to order that may not have been significantly different in size from the more standardised lava querns.

7.11.2. The 'Pompeian' Style Millstones

As has been briefly mentioned in the previous section, fragments of a possible upper and lower stone of a Pompeian style millstone have been recovered at Silchester. These were identified and published by Allen (2012), though there remains some uncertainty with regards to this classification due to the extent of fragmentation (Shaffrey, pers. comm.). This type of millstone is uncommon in Roman Britain (Williams & Peacock, 2011), which may be due to identification issues but is more likely to indicate that this style was rarely used in the province (see Chapter 2). Animal and human-powered mills are likely to have been important tools for large-scale processing, especially in a town such as Silchester that had a relatively large population and no immediate means of harnessing waterpower to turn its millstones. However, the use of these millstones may not have necessarily been a functional response to the need to increase the scale of processing. It has been suggested by Williams and Peacock (2011) that these mills may have been used in 'patisserie' type bakeries used by specialist bakers to produce highly commoditised foods, the like of which would have been available in Rome and Pompeii. Evidence from the town suggest that the wealth and means to install such technology would have existed, and the range of imported foodstuffs evident in the diet of the townsfolk, even from the early Roman period (Fulford, 2021, p. 69–72), implies that they were a population open to exploring 'Roman' ways of eating and drinking, which strongly correlates with what is generally understood of Romano-British urban identity (Alcock, 2001, pp. 13–16; Cool, 2006, pp. 172–183). Furthermore, Pompeian mills are of a highly iconic style formed from a recognisably foreign stone and consideration should be given to its presence as an exotic symbol of Roman and Mediterranean culture in a town that was experiencing huge social and economic change. The millstones are likely to have been used around the mid-first century AD (Shaffrey,

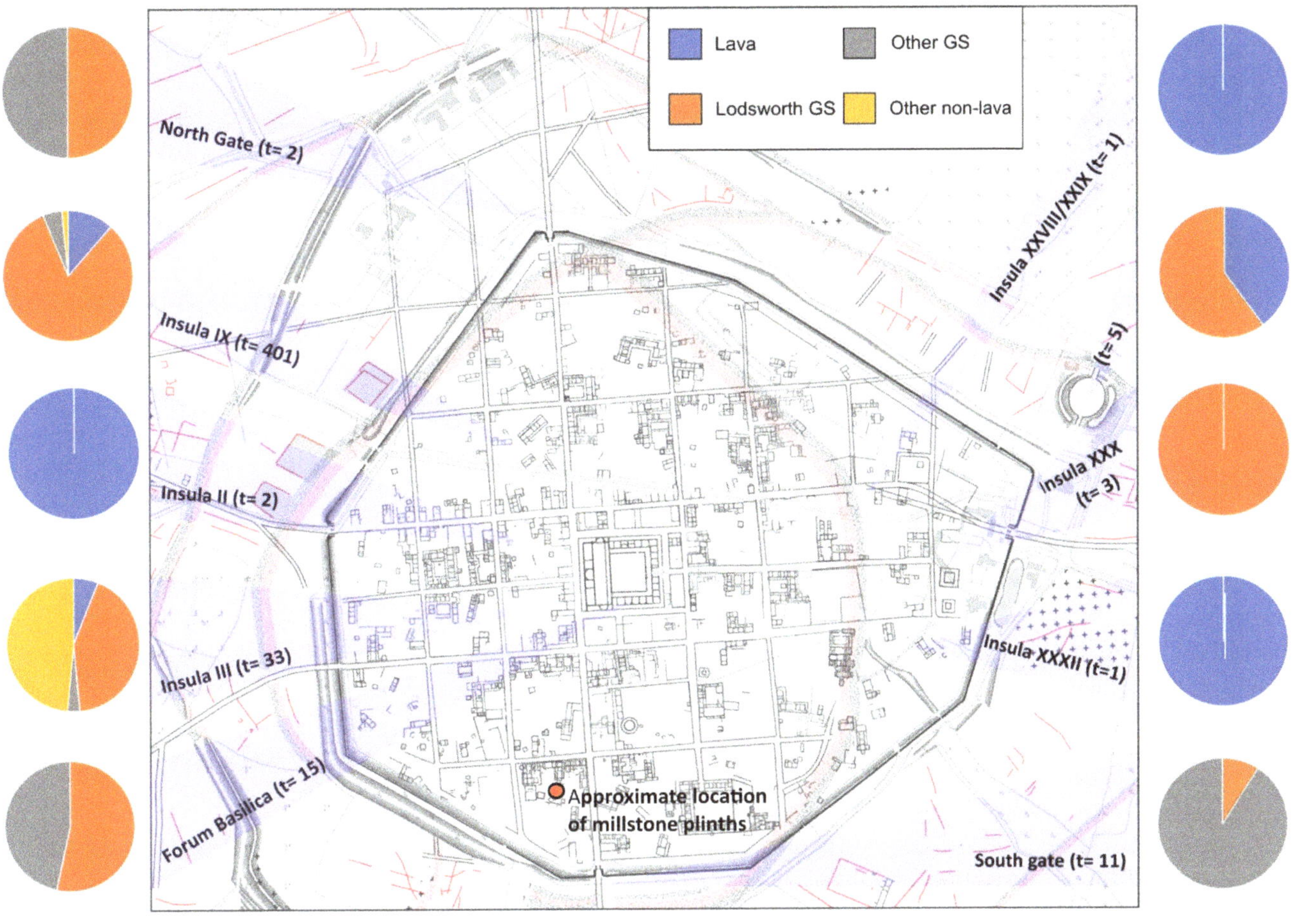

Figure 7.29. Distribution of milling tools of different stone types in Silchester (backdrop map Creighton & Fry, 2016).

2021, p. 21), at a time when long-distance connections to the wider empire and signs of Nero's imperial authority and influence in construction projects are visible at Silchester (Fulford, 2008a; Fulford, et al., 2017; Fulford, et al., 2019; Fulford & Machin, 2021). The possible use of a highly iconic and techonologically advanced 'Roman' mill may have been associated with this conspicuous messaging of advancement and imperial control.

In terms of provenance, geochemical analysis was undertaken on the lava of the possible Pompeian millstones from Silchester. They were found to probably have been sourced from Mayen (Allen, 2012, p. 266), though the analysis did not draw comparison with geological samples from other alternative lava regions, such as the Massif Central in France where Volvic lava was sourced. A broader approach is necessary with any geochemical analysis of this kind; sample data is now readily available for such comparisons to be made, and includes lava from the Volvic quarries, all known lava sources in the Eifel region (including some individual lava flows), and many of the Mediterranean lava quarries (Gluhak & Hofmeister, 2009; Gluhak & Hofmeister, 2011). A variety of lava sources are present in distributions for mainland north-western Europe, emphasising the need for wider analysis to be undertaken (Gluhak & Hofmeister, 2011; Reniere, 2018). This is not, however, to dismiss the possibility that Mayen was the source, as it is known that Pompeian style mills were produced from Mayen lava, but finished in Andernach (*contra* Williams-Thorpe, 1988b, p. 286; Mangartz, 2008, pp. 76–78). However, further work is necessary to eliminate other lava sources, such as the Volvic quarries of France, which is known to have been the source for a Pompeian lava mill recovered from Princess Street in London and, also, one from Corfe Mullen in Dorset (Williams & Peacock, 2011). The close relationship that Silchester appears to have shared with the south coast and northern Gaul at this time might make a French provenance more likely.

7.11.3. Identifiable Typological Features

Other than the range of diameters seen across the assemblage, there were few unusual features related to manufacture or typology present on the milling tools. The identifiable features of both querns and millstones generally conformed to the expected types, though the possibility for variation should not be excluded due to the heavy degree of fragmentation observed. One identifiable difference relates to the method by which the sides of the querns were formed. Vertical grooved tooling on the sides of both upper and lower quern stones is a highly identifiable feature that is strongly associated with Mayen lava products, and this was visible on nine quern stones from the assemblage, comprising both upper and lower stones (CAT 0868, 0883, 0900, 0910, 0915, 0917, 0918, 0921, 0922). Four other examples (CAT 0914, 0920, 0923 and 0924), varied from this usual style; the sides of three fragments were pecked, while CAT 0920 had visibly chiselled edges. Pecked or chiselled sides were only present on lower quern stones. As this is a small number of cases, it is difficult to draw conclusions as to what this variation might mean. Though it may demonstrate a difference in manufacture style, there is also the possibility that these lower stones were adjusted post-manufacture, removing the linear tooling associated with the production process. As has been seen at Verulamium, the sides of lower stones were sometimes trimmed and cut away, perhaps as a way of fitting the stone into a floor or other surface to permanently secure it in place when in use.

This may also have been a method used at Silchester based on a documented find from the Victorian excavations that is no longer within the museum collections. In the excavation report by Fox (1895), a pair of (non-lava) quern stones were recovered from *insula* X that appeared to be *in-situ*. The lower stone had been fixed onto a base of burnt brick approximately 2 inches (5cm) high (Shaffrey, 2003, p. 144). Such a feature would have enabled the stone to be fitted securely into the floor surface, whilst ensuring that the grinding surface of the quern was raised above the level of the floor. Other interpretations are possible, as the base may simply have been a means of adding more height to the stone. However, there is growing evidence that querns were used in this way, and that fixed position querns may have been common in a domestic setting.

7.12. Silchester: Distribution

Analysis of the distribution of lava and other milling tools at Silchester presents some problems due to the intensity of excavation at *insula* IX that cannot be paralleled with the smaller excavations conducted elsewhere at the site. It should be expected that the volume and types of finds would be different for each part of the town, but comparison is complicated by the difference in scale of investigation.

7.12.1. Distribution by Stone Type

Figure 7.29 shows the number and types of milling tool recovered from different areas of Silchester. These have been broadly categorised by stone type to take into consideration the main lithology groups present: Lava, Lodsworth Greensand, other types of Greensand, and the final group comprising all other non-lava types. This last group included a range of different sources including Old Red Sandstone, Hertfordshire, Worms Heath and French Puddingstone, and Millstone Grit (Shaffrey 2021 and unpublished data). Initial observations indicate that there was variety in the presence of different stone types depending on the area of the town. However, the extent to which this corresponds with the location of use for these different stone types is difficult to ascertain. Chronology must also be considered, as the town saw a long history of development, and access to different stone types would have varied over time. For example, lava was not readily available in the Iron Age. The dumping of material, including stone from deposited milling tools, may have occurred during specific phases of development or demolition. This would skew the volumes of particular types of stone that were more popular in specific phases within certain locations.

As a ratio of all querns and millstones recovered, lava was not as popular as is seen in more easterly settlements in Britain and is typically less present than examples of other stone types. Though there were some areas of the town that had only lava or high proportions present, these were generally small assemblages and do not match the trend seen elsewhere. This shows that although lava was used, it was clearly not as important as Lodsworth Greensand and other Greensand stone types that must have been utilised for most milling activities. It is possible that lava only saw a short period of use, as Lodsworth and other Greensand querns were already readily available at Silchester at the time of the Roman conquest of Britain. The abundance of Roman period Lodsworth querns seen in Sussex, the scale of their manufacture and the possible political and social associations between Chichester and Silchester may have made them much more accessible and desirable to the population of Silchester than imported types. As Silchester lacked a local supply of suitable stone for grinding and milling (Shaffrey, 2021, p. 7), the supply of querns (and probably other goods) from Lodsworth to Silchester would have been a well-established and vital exchange network by the Late Iron Age and it is probable that this continued to operate well into the Roman period (Shaffrey, 2021, p. 15). However, if lava was more popular in later periods, it may not be as archaeologically visible as the earlier stone types due to changing deposition practices or altered use of space. It has been noted that a decrease in milling tools occurs in the late first to early second century at *insula* IX, where the bulk of the Silchester assemblage was excavated (Fulford, et al., 2006, p. 254; Shaffrey, 2021, pp. 20–21). The chronology of lava use in the town provides some clarification, and this can be found in the following section.

Lava was completely absent at both gates (though assemblages there are very small) and the Forum Basilica, where only Lodsworth Greensand and other Greensand types were present. This is difficult to reconcile with what is known of the Forum Basilica, as excavations in this area have been relatively extensive and multiple phases of occupation have been investigated. As a location that may have seen a military phase (see 'background to the site' for discussion of this), the expectation would be for it to yield lava due to the close associations of this material with the military, especially in the early Roman period. Similarly, its use as an elite space should have improved the probability for imported milling tools to be present, in the same way that imported foods were found to be relatively abundant at this location (Fulford, 2021, p. 71–72). It should be noted that although the number of querns recovered from this area appears to be small, a larger volume of fragments had been recovered from the Forum Basilica than is illustrated here (see Fulford & Timby, 2000, pp. 386–387 for quantification by weight). However, due to complications relating to the quantification of fragmented milling tools and the fact that this data has been combined with that for the grinding and rubbing stones in the excavation report, it has not been possible to include all fragments in the analysis. In total, 36 fragments of quern were identified, but only 15 have been included here. None of these fragments included lava, which further confirms that this is unusual. However, environmental evidence at Forum Basilica indicates that no agricultural activity, including grain processing occurred there (Jones, 2000, p. 512), and the presence of querns, therefore, is probably more indicative of depositional practices (Shaffrey, 2021, p. 19). The context of the Forum Basilica and the gates are, perhaps, less likely locations for grain processing to occur as opposed to more domestic or agricultural settings within the town.

7.12.2. Chronological Distribution- Lava Volume and Comparison with Other Stone Types

Detailed examination of the changing use of querns and millstones has been undertaken by Shaffrey (2021) and this valuable research can be utilised and summarised here with a specific focus on the role of lava. As has been previously mentioned, changes in the volume and presence of milling tools at Silchester occurred over time. This is represented in figure 7.30, which demonstrates the extent to which a decline occurs from the late first and early second centuries in *insula* IX. A similar decrease is seen in other parts of the town, though the smaller datasets create some challenges in terms of making trends visible (figure 7.31). Reasons for this change could be related to a decrease in milling activity within the town in association with a decline in urban populations as the role of the town changed in later Roman Britain, changing deposition practices or more centralised methods of processing. It is likely that all three of these reasons impacted the chronological distributions seen to different degrees in Silchester. The presence of a relatively high number of millstones in the town suggests that large scale processing was occurring within specialist locations. This likely translates into there being fewer activity areas where grain processing occurred, but that this happened at much larger scales within those places. Some of this may have occurred within the town itself, though ready-ground flour may also have been brought into Silchester from the surrounding rural areas (Shaffrey, 2021, p. 26–27).

In terms of the changing numbers of lava milling tools over time, only data from *insula* IX could be used as there are too few from other parts of Silchester to realistically present any patterns or trends. Chronological information has been presented in figure 7.32 to display the minimum and maximum numbers of lava milling tools that could be assigned to any one period. The examples that could be clearly allocated to a specific date range are shown separately to those that had wider date ranges spanning two categories. As each milling tool is represented twice in the data, these values have been halved on the chart.

Although the volumes of dated lava milling tools is much smaller than the overall quern dataset, there is a clear trend that does not correlate with what is seen for other milling tools. The early post-conquest period and mid to late first century show the lowest number of lava milling tools at Silchester, with the peak occurring after AD 85. This continues at a relatively high level until AD 250, when a decrease begins to occur, coinciding with a period when the changing role of urban centres is most evident across

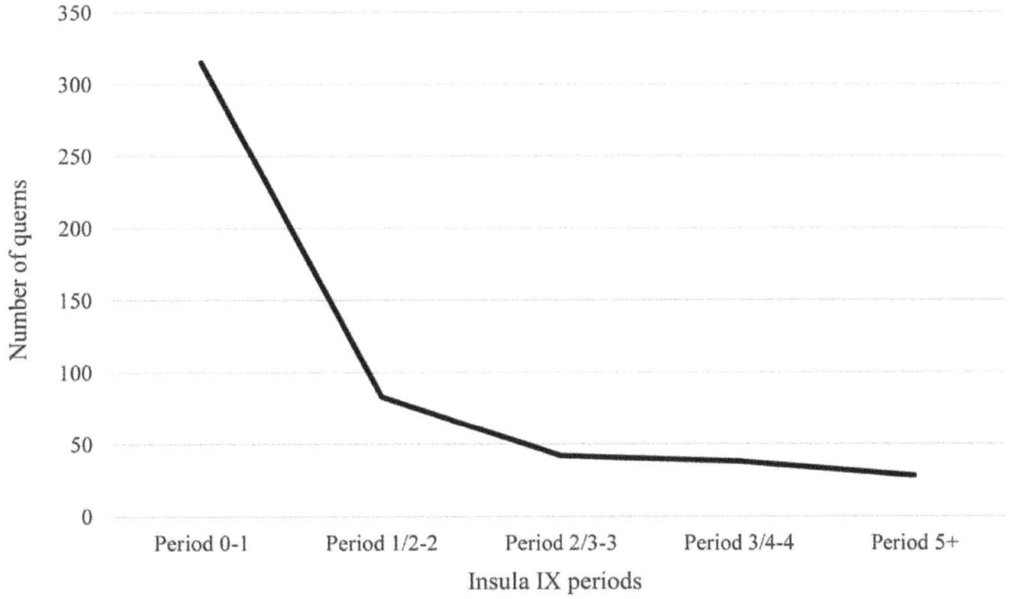

Figure 7.30. Change in number of querns recovered from insula IX by period (Shaffrey, 2021).

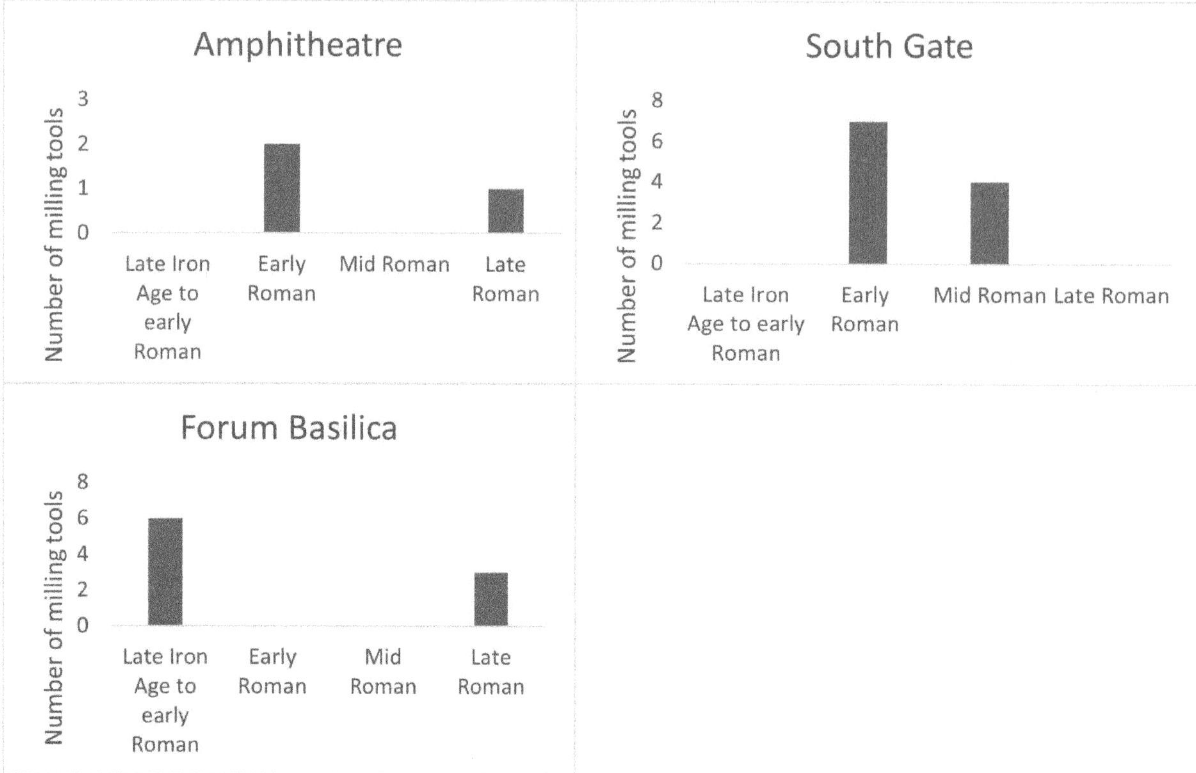

Figure 7.31. General trend showing decrease in milling tools numbers at other parts of Silchester with smaller datasets.

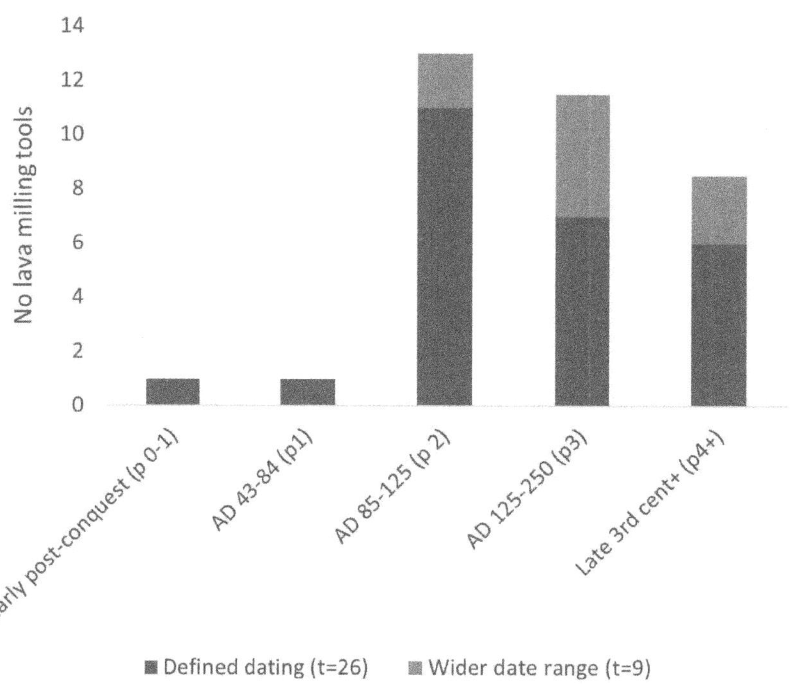

Figure 7.32. Change in the number of lava milling tools from insula IX over time.

the province (Woolf, 1999, p. 37–38). The changes seen do not directly correspond to the period of use for these objects as this only indicates when the lava was deposited. As objects with long use-lives and a high incidence of reuse, it is probable that there is a time lag before lava milling tools enter the archaeological record. As a type of material culture that is associated with the early Roman period, this is an interesting revelation and shows that *Calleva* was different to the eastern major urban centres, such as *Londinium* and *Camulodunum*, where lava is present from the early post-conquest phases (Buckley & Major, 1983, pp. 73–76; Dunwoodie, 2004, p. 51). It must be emphasised again the extent to which the exchange of Lodsworth querns appears to have impacted the adoption of lava milling tools, which occurs much faster in other areas of south-eastern Britain where suitable stone for milling tools was less readily available.

The probable military nature of the earliest settlement phases at London and Colchester must also have played a role in the early adoption and deposition of lava at these sites (Crummy, 1977; Dunwoodie, et al., 2015) and their surrounds, alongside the stronger connections they had to the North Sea trade system in their function as early Roman ports. The Boudiccan destruction that took place at Colchester, Verulamium and London must also be considered, as this would have been a cause for the discard/abandonment of unused or little used products long before they may have ordinarily been deposited. Furthermore, most of the period 2 lava at Silchester was incorporated into levelling deposits or dumped material, blurring the chronology of primary use. It is likely that some of the period 2 lava saw its final primary use in the period prior to final deposition and that mass dumping and levelling events have made this appear to have occurred more suddenly. Nonetheless, the data strongly suggest a later date for lava adoption at Silchester when compared to more easterly towns in the south of Britain.

The relative distance between Silchester and the east coast of Britain might be considered as a barrier to access to lava products, and it would be assumed that this would be a significant factor with regards to the volumes present at the town. However, it must be remembered that the inhabitants of Silchester were consuming a wide array of different imported goods, both prior to and after the conquest period, (Fulford, 2021, pp. 36–39; pp. 71–72). The absence of lava may suggest that in the early and pre-Roman phases of the town, these other imports were entering Silchester via ports on the central southern coast. These are not within the North Sea trade catchment area, and thus were not involved in the distribution of lava querns (see Chapter 2). This would make it more difficult for lava milling tools to 'piggyback' with other goods on their route to Silchester, and easier for Lodsworth stone querns, that were produced close to Chichester and Fishbourne, to be moved along these routes instead.

A transition occurs by the late first century, possibly fuelled by a change in the organisation and scale of processing activities. Shaffrey (2021) suggests that as Lodsworth stone was not often manufactured into millstones, it saw a natural decline in use at locations that had become more reliant on this new form of technology. This seems very likely considering the unusually high number of millstones present at Silchester. However, the increase of lava use in the town at a time where Lodsworth stone shows a rapid decrease in presence also indicates a change in the direction for the movement of goods, with increased focus on the east coast. This probably relates to the instigation of London as a major Roman port being slightly closer in distance to Silchester than Chichester, and possibly more accessible along the Thames valley than Chichester across the South Downs. Lodsworth stone still dominates the mid-to-late Roman assemblages in terms of proportions, though other exchange networks were probably creating a more competitive environment in terms of choice of milling tool. The fact that lava persists into the later periods may be a consequence of residuality but may also indicate that these wider exchange systems continued to operate throughout much of the life of Silchester as a Roman town, and that a connection to the redistribution centre of London had become an important social and economic association.

Changes in other industrial and economic activity within the town must be considered in correlation with changes in exchange networks. If querns were entering Silchester via a specific route, it should be assumed that goods were moving in the opposite direction also, perhaps as products of Silchester, the surrounding countryside, or other production/agricultural areas along the same route. The possibility that Silchester was moving goods to the southern coastal towns, or even exporting goods to the continent via the southern ports remains a means by which Lodsworth querns may have been transported, on the return journey, to Silchester. Examples may have included less archaeologically visible products, such as grain, hides and textiles, the former two of which were commented on by Strabo (Strabo, trans. 1923; Fulford, 2021, pp. 43–44). Textile production, and possibly leatherworking, appear to have been important activities in Silchester in the Late Iron Age and the early Roman period, evident in the number of tools recovered that can be associated with these crafts. Crummy (2020, pp. 279–280) suggests that this large-scale production may have been to supply military contracts or for consumers in growing civilian markets, such as London. These goods could also have been exported to the continent, or redistributed via other towns, such as the rapidly expanding Chichester, which is known to have held strong political and social ties with Silchester during this era. A swift decline in objects related to textile or leather working occurs after AD 85, around the time that Lodsworth querns begin to decline in numbers at Silchester (Shaffrey, 2021, p. 7–10). If fewer textiles were being produced, fewer journeys to the south coast would be required, resulting in fewer return journeys that would bring the Lodsworth querns to Silchester. Alternative stone types, including lava, would have become more important to fill the deficit as a result, and exposure to new types of milling tools may even have initiated a more rapid transition to the use of mill technology. Other crafts see a similar decline around the same time, and it may be that this avenue for personal economic advancement was halted due to state intervention, such as increased taxation or control of production. However, as is also discussed by Crummy (2020, p. 279), these changes in craft activity may simply have been the result of the changing status and economy of Silchester and the resulting shift to consumerism, as opposed to self-sufficiency. This would correlate with the transition to an increased use of mills and more prevalent centralised grain processing. The possibility that grain processing was occurring away from urban areas or in parts of the hinterland that are less likely to have been excavated is also a possibility. A combination of different factors is likely to have been responsible.

7.13. Sichester: Primary Use

Identifiable features of the lava fragments that can be used to associate them with their primary use as milling tools are difficult to establish due to the fragmentary nature of the assemblage. Signs of primary wear are especially difficult to identify due to the high incidence of reuse seen at Silchester, a subject dealt with in the following section. Traces of wear that are common for rotary querns are present on many of the fragments, including thinning of the area around the eye for upper stones and the creation of a 'lip' around the central hole for lower stones. There is nothing to imply that these were used in any way considered to be unusual. There are two consequences of primary use that can be explored using this assemblage; thickness of stone prior to deposition and dressing style on the grinding surfaces.

7.13.1. Stone Thickness

For Silchester, there were 6 upper and 9 lower stones that had measurements taken that were suitable for comparative analysis. As with Verulamium, this data has been plotted alongside that of the whole British lava quern dataset, excluding the querns from 1 Poultry. The upper stones were found to partially lie within the range of 'normal' thickness, between 48–72mm, corresponding with the thicknesses of 50% of all British upper lava quern stones (figure 7.33). Three stones, comprising half the dataset, fell below this range, although one of these (CAT 0883) was only 1mm short of this measurement. The other two (CAT 866 and 875) were not extraordinarily thin in comparison to this range, with the smallest measuring 35mm. A relatively standard level of thickness of upper stones was reached at Silchester prior to deposition. For lower stones, 50% of the British data (excluding 1 Poultry) lies between 32mm and 55mm, which also accounts for two-thirds of the Silchester assemblage (figure 7.34). Three examples fall below this range (CAT 0910, 0920 and 0924). Of these, CAT 0910 shows some signs of reuse, and so its smaller thickness may be a result of wear that occurred after its original function as a quern. CAT 0920 was an example showing chipped edges that may correspond to it being reshaped to set it into the ground in a fixed position or general reuse. As a dataset, the thicknesses are on the smaller end of the scale overall and are comparable with those seen from Verulamium. All were deposited once they had reached a thickness between 20–40mm, like that seen at Verulamium. However, the thicker examples of both upper and lower stones that were seen at Verulamium are completely absent at Silchester. This implies that lava was more heavily used in a consistent way at Silchester prior to deposition, which may relate to the availability of lava milling tools in a town more distant to the centralised distribution centres of London and Colchester. The lack of suitable local supplies of stone for grinding, milling, and processing activities would also have been a significant factor, as querns and millstones would have been more difficult to replace and were probably more likely to have been reused for alternative purposes once the primary function ceased to be possible. This is a pattern that is also observable on non-lava quern and millstone fragments from Silchester (Shaffrey, pers. comm.)

7.13.2. Dressing Style

The style of dressing on the grinding surface of a quern or millstone might be considered to fall under the category of

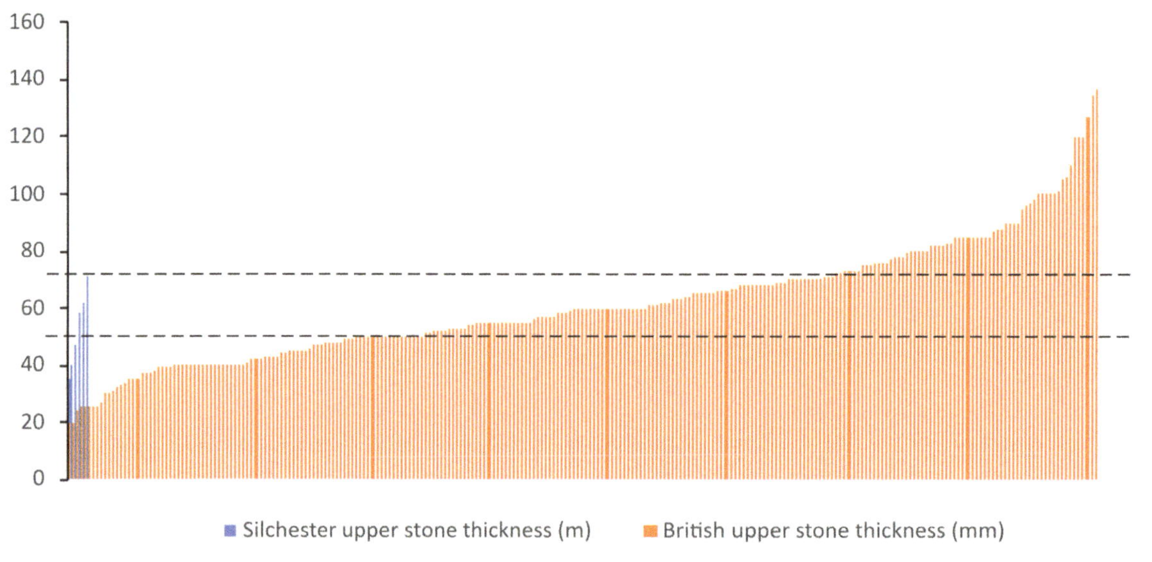

Figure 7.33. Graph showing the edge thickness of upper stones from Calleva compared with those from all Roman Britain. The 25th and 75th percentiles are indicated using dotted lines.

Figure 7.34. Graph showing the edge thickness of lower stones from Calleva compared with those from all Roman Britain. The 75th and 25th percentiles are indicated with dotted lines.

typology, in that dressing would be added to a milling tool in the manufacturing stages of its production. However, milling tools would have been redressed at multiple stages of their use lives due to wear and be more representative of society, culture and the availability of stone working skill. There were very few dressed examples within the Silchester assemblage, and those that had identifiably dressed grinding surfaces were often 'pecked', as opposed to the typical radial or harped style of dressing that would be expected (table 7.5). Around half the assemblage was pecked, which presents a significantly higher proportion than is seen across the dataset from Britain. It must be noted that pecking is difficult to identify on versicular lava and is much less likely to be commented on than radial or harped dressing. This has skewed the results substantially. However, it is worthwhile considering why the more commonly used method of harping was not more consistently applied to lava at Silchester.

This pattern may relate to stone working skills, as pecking would be easier to apply to a stone and could be accomplished easily by any person with a chisel. Alternatively, pecking may have been used to replicate the familiar style of dressing associated with Lodsworth Greensand querns, becoming a cultural or social choice. The extent to which this might be true is debateable, as the grinding surface of a quern would not be immediately visible and may not have been significant. Nonetheless, querns of indigenous stone types have replicated harped

Table 7.5. Comparison for volumes of pecked and harp-dressed querns in the Silchester and British datasets

Dressing style	Silchester Quern	British Quern
Pecked	5	30
Harped	6	179

styles of dressing, sometimes with the inclusion of other features reminiscent of lava. The reverse of this process appears to have occurred at Silchester, with the adjustment of new forms of material culture to resemble more traditional ones. This may have made lava querns more identifiable and relatable to a culture of people who were more resistant to change.

7.14. Silchester: Reuse/Modification

Drawing out the specific aspects of secondary use is a complicated process, but it appears that Silchester experienced a high level of stone reuse throughout the Roman era. This is apparent in the whetstone assemblage, which largely comprises old building material and quern stones from a variety of different stone sources (Allen, 2014). Silchester is located within a geological region with little in the way of immediately available stone suitable for ground stone tools, and these needed to be brought from further afield (Shaffrey, 2021, p. 7). Once at Silchester, it is logical to assume that as much use would be made of such stone as possible prior to deposition, regardless of intended primary function. This may explain why the lava assemblage from Silchester is so significantly fragmented, as lava milling stones may have been broken up to serve other functions after primary use was no longer possible. Although it is highly likely that lava was reused, very few incidents can be identified from the fragments available for study, and even these identifications are tentative. Unlike other stone types, lava does not always preserve well under certain soil and exposure conditions, creating sometimes irregularly shaped fragments on which wear traces can get lost.

In total, four lava fragments could be associated with incidents of reuse: CAT 0867, 0894, 0895 and 0910. Of these, CAT 0867 was identified as having been used as a whetstone and was recovered from an unstratified context at the amphitheatre (Allen, 2014, p. 86). Unfortunately, no details have been provided in relation to this example or how we can tell that it was used specifically as a whetstone. However, this means of reuse seems like a realistic scenario and it would not be surprising if other lava fragments were reused in such a way. The other three fragments can only be associated with reuse due to the presence of unusual regions of wear that cannot be attributed to the normal movements used during milling. For example, CAT 0894 was worn on both the upper and lower surfaces. The grinding surface was used as a rotary quern, while the upper surface was used to sharpen blades within the same stage or at a later stage of the quern's lifecycle.

By far the most common and visible way that lava was reused at Silchester was within levelling deposits, flooring and to create hard surfaces. Lava was not the only quern material that was used in this way, and this may be another reason for the high degree of fragmentation present. The reuse of stone, including that of querns, for this purpose was a common practice in Roman Britain (see Chapter 5). This has been discussed in the next section in relation to deposition.

7.15. Silchester: Deposition

Deposition of lava at Silchester has only been explored for *insula* IX as this was where the bulk of the lava querns were recovered and because excellent context information exists. Location information relating to the feature of recovery alongside categorisation of the feature type was available, and this allowed for a more detailed investigation into deposition than has been possible for other sites in the dataset. It should be noted here that findspots have been used to explore deposition as opposed to distribution. With quern stones, location on an intra-site scale is more likely to correspond to where the object was deposited/disposed of than to where it may have been used. Alternatively, general distributions, such as looking at location data for the whole town (see figure 7.29 above), can suggest vicinities of more intense use. This works on the general assumption that querns are unlikely to have travelled far prior to deposition, though they may have been reused, redeposited, and moved about within the localised environment.

The deposition of lava milling tools in *insula* IX shows some interesting patterns (figure 7.35). In the early Roman period, lava was predominantly recovered from occupation and dump deposits, and in postholes. In the mid-Roman period, levelling deposits are the most common form of deposition, with pit deposition occurring more often in the late Roman period. Although it is tempting to see this as a change in deposition practice, it is likely that much of what is seen here relates to the types of development that were occurring at *insula* IX at each chronological phase. The customary use of timber buildings in the early Roman period can be equated with a higher frequency of postholes and a higher likelihood of lava being deposited into these. Most of the early period lava querns were dated to the period after AD 85, when large parts of the town were renovated, and timber buildings replaced with new timber structures (Fulford, 2021, p. 76–102). Similarly, there were very few pits at Silchester between AD 85–250 (Fulford, pers. comm.), meaning that rubbish disposal may have occurred in controlled locations or according to different traditions than that of later phases of occupation.

Querns were sometimes reused for post-packing during the Roman era (see Chapter 5). It is possible that the incorporation of querns into such deposits relates to some form of social or cultural practice, as has been identified for other periods and locations (Brück, 2006; Watts, 2014). However, this is difficult to identify or define as broken quern stones may simply have provided a useful source of suitable stone and deliberately deposited to fulfil a functional requirement that was related to their material properties instead of any specific meaning. They may also have been deposited into middens to become

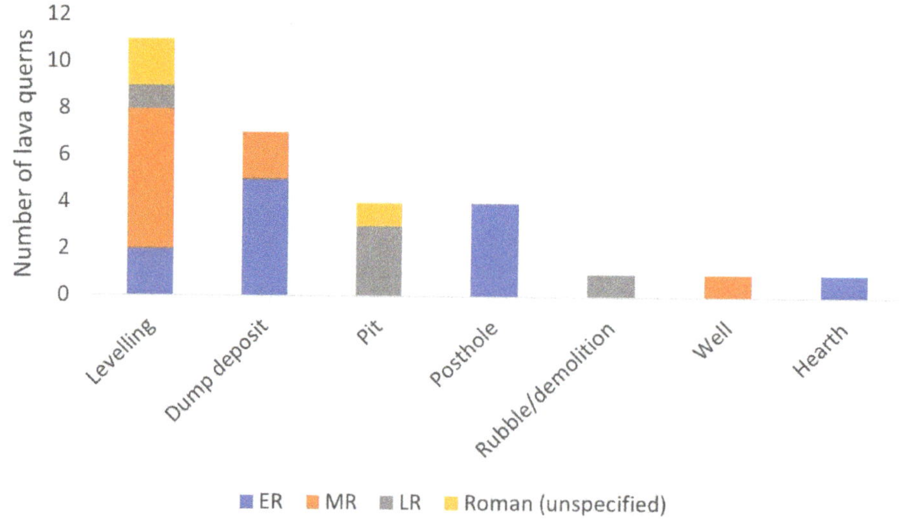

Figure 7.35. The types of feature where lava milling tools were deposited classified by chronological period.

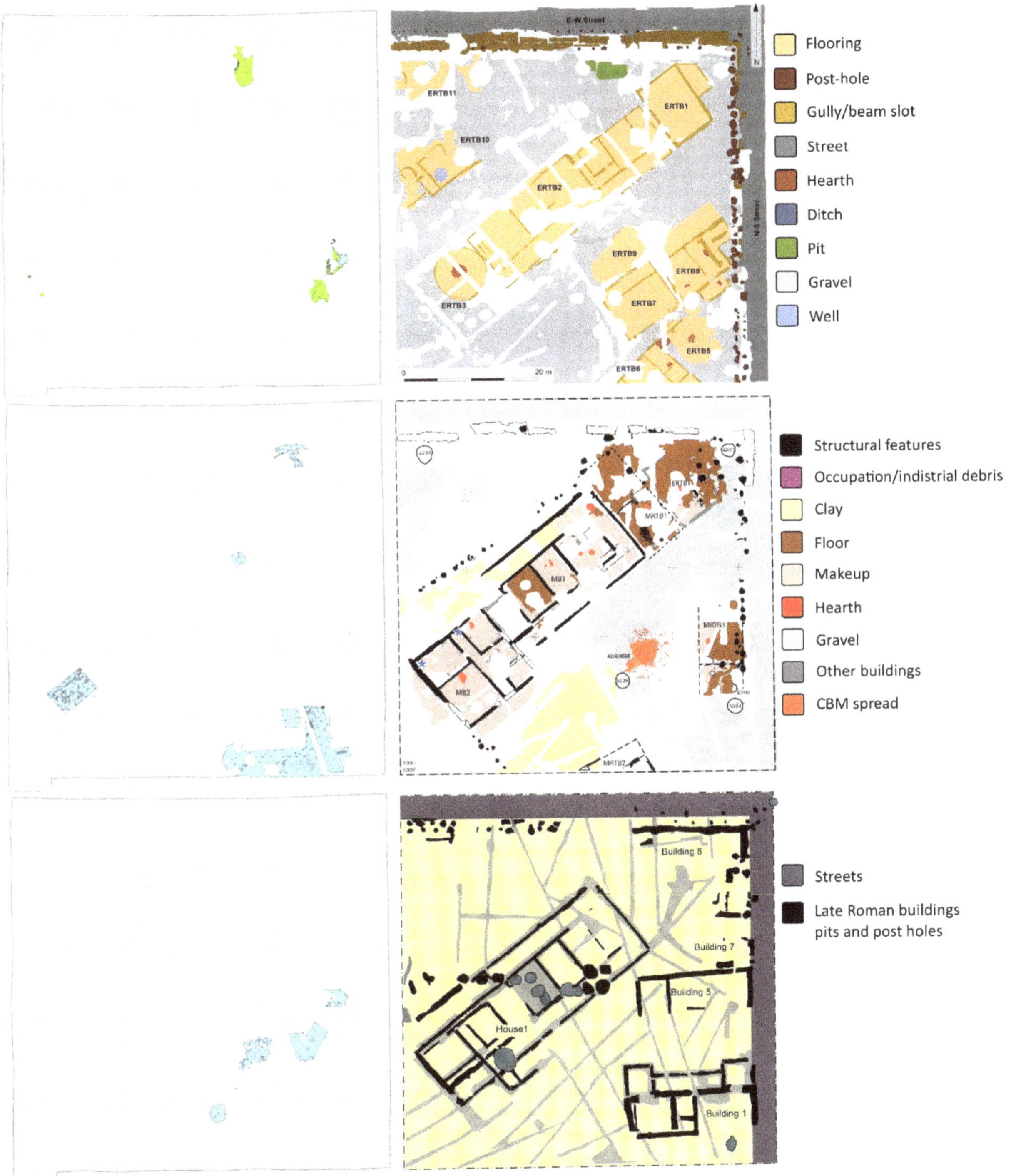

Figure 7.36. Plans of excavated area in insula IX showing the features where lava was recovered on the left, and corresponding excavated structures and features on the right. The top two plans relate to the early Roman period 2 (AD 85–125/50) (Fulford, 2021, p. 94), the mid-Roman period 4 (AD 200–250/300) (Fulford & Clarke, 2011) and the bottom to the late Roman period (AD 250+) (Fulford, 2021, p. 144). A key is provided for each of the site plans, as these do not use the same system for feature identification (access to IADB kindly provided by the Silchester Project).

incorporated with household waste that was incidentally used as backfill. At Silchester, two of the four lava milling tools were retrieved from a series of postholes associated with a rectilinear structure or fence line that replaced an earlier roundhouse (Fulford, et al., 2011, p. 3), which may have had ritual significance (Fulford & Clarke, 2009, pp. 9–10; Fulford, 2021, pp. 97–98) (figure 7.36). This group of postholes contained a range of finds that would be classified as household rubbish such as pottery and animal bone, alongside possible demolition debris including iron nails, CBM and stone tesserae (information taken from Silchester IADB). Though ritual deposition cannot be ruled out, there is nothing to signify that this had taken place or that the lava querns had played a role, and the replacement building does not appear to have held the same significance as the one that had preceded it.

Nonetheless, the distribution of deposited lava in the early Roman period can be predominantly identified as either incorporated into postholes and thus into the structure of buildings (CAT 0870, 0871, 0901 and 0907), or deposited within buildings themselves (CAT 0869, 0891, 0905, 0906, 0890 and 0892), except for one fragment that was recovered from an external gravel spread (CAT 0908). One fragment was recovered from a charcoal-rich hearth deposit and though this may not mean it was directly related to the use of the hearth; the deposition of milling tools in hearths and hearth structures has been discussed in more detail in Chapter 5. This provides an interesting pattern of deposition and suggests that quern stones had strong domestic associations, being used, and deposited within areas of habitation where other domestic refuse was allowed to accumulate. Other finds recovered alongside these internal deposits include material that would be defined as domestic waste, such as animal bone, pottery, oyster shell and charcoal. That quern fragments were included within such deposits indicates that they were disposed of in similar ways and were probably used close to the home.

The high number of querns associated with levelling deposits in the mid-Roman period probably relate to construction and renovation activities that took place in the early mid-second century when many timber buildings were reconstructed with masonry foundations. The formation of the large town house in *insula* IX was a major building project (Fulford & Clarke, 2011; Fulford, 2021, pp. 112–113) and relevelling of the ground would have been an important step in the reuse of any plot of land that had been previously utilised for construction purposes. Deposition occurred both in and around the structures, wherever a hard standing floor area would be required. As a result, distribution is more widespread and randomised. Finds recovered from the levelling deposits were also much more wide ranging: iron and copper alloy fragments, lead sheet, pottery, animal bone, oyster shell, iron tools and fittings, glass beads and vessels, iron working waste, daub and CBM, among a variety of other small finds were recovered alongside lava and other quern fragments. The material is not only domestic in nature, with evidence of industrial and craftworking waste being present, alongside demolition debris. It is probable that this represents material that had been collected from further afield and dumped to provide a level ground surface. It, therefore, becomes more difficult to associate the querns specifically with domestic refuse and use, and the degree to which residuality exists in the dataset is likely to be significant due to the high probability that large-scale redeposition had occurred.

The late Roman period shows a higher propensity for deposition of lava into pits, which does not appear to have occurred in previous periods. The distribution, including that of the demolition rubble and levelling layer where other lava quern fragments were recovered, is relatively localised, being isolated to a group of features outside of buildings 5 and 1. These pits may simply have been dug as a way of disposing of household and industrial waste generated within *insula* IX, an activity that is much less common in the earlier periods of occupation at the site. Alternatively, they may have been utilised for ritual deposition, with the lava quern fragments forming part of those ritual deposits. One of the pits where lava was deposited, [3235], also contained human remains from a minimum of two individuals and a wide range of find types including a brooch, coin, and finger ring. Though these could be considered as unusual within this context, it is very difficult to ascribe these as specifically associated with ritual activity. All three of the pits were found to contain a variety of finds, with some objects that would be considered as personal items and, possibly, of social or cultural significance. However, in a town that had seen several centuries of occupation, perhaps it would not be too unusual for there to be an accumulation of such objects that were no longer useful or wanted (Eckardt, 2006; Fulford, et al., 2006).

7.16. Silchester: Conclusions

The excellent contextual information from the modern excavations at Silchester have provided a unique opportunity to examine quern and millstone use in the town in a significant amount of detail. The already established popularity of the more locally available Lodsworth querns is likely to have had an impact on the initial adoption of lava milling tools at the town, and this seems to have occurred more slowly than in the more easterly towns. However, several factors may have been responsible, including the possibility that the southern ports and towns were more involved in the economy of Silchester in the early Roman period, probably a residual effect of the social and economic ties that existed in the Late Iron Age. An increase in lava use is seen after AD 85, which despite the chronological delay in deposition due to the long use-lives of querns, coincides with several other major developments in the town. This appears to have been a period of social and economic change at Silchester. Changes in exchange routes and greater involvement with the eastern towns and ports is likely to have been one factor.

As a location that is geologically poor in suitable stone for milling tools, it is not surprising that lava querns were heavily worn prior to deposition. However, it also shows that access to lava may have been more constrained due to the distance of Silchester from east coast distribution centres and full use needed to be made prior to disposal. Reuse of lava is also visible, both on fragments from wear patterns and in the contexts of deposition. Extensive reuse of lava milling tools occurred in levelling and flooring deposits, and they were also reused as posthole packing. The deposition methods make it impossible to determine whether lava presence can be related to activity areas where grain processing may have been carried out.

The number of millstones recovered from Silchester suggest that large-scale food processing was occurring, either in the town or in the close vicinity. This may have been conducted predominantly using animal or slave mills, which were probably smaller than their water powered counterparts. Lava millstones appear to have been manufactured to specially ordered sizes, as the range of diameters for the querns and the differentiation between the millstones and quern stones are not as standardised as would be expected for Mayen products. The decrease in overall quern numbers at Silchester suggests that this method of milling may have been introduced relatively early and initiated the decline in Lodsworth products.

7.17. Urban Sites: Conclusions

As has been the case throughout this research, interpretations relating to the use of lava milling tools is largely dependent upon the level of detail that can be discerned from the archaeological reports generated at the time of excavation. Roman urban sites that are situated on undeveloped land, such as Silchester and Verulamium, tend to have been subject to early excavation and exploration, resulting in poor recording and missing records. More recent investigations at both towns have, however, provided much needed clarity and context to some of the earlier excavated material.

Overall, there are some interesting comparisons that can be drawn between the two urban case studies. With regards to lava volumes, direct comparison is complicated by numerous factors, such as differing scales of retention. Nonetheless, it is worthwhile to note that Verulamium yielded the greatest number of lava fragments, followed by Silchester. It is highly likely that Verulamium did make more use of lava for milling tools and that the volumes are indicative of greater lava use when compared to Silchester. The proximity of Verulamium to the east coast ports, especially London and Colchester, coupled with the deficit of suitable local stone sources for milling, made Verulamium a significant consumer of lava products. Similarly, the theory that Verulamium had an early military phase (Pitts, 2014) is also supported by the lava consumption observed at the town, which is comparable with other urban/military sites in the south. In comparison, Silchester was probably making better use of the southern ports for its imports and was utilising the well-established Lodsworth quarries for its milling needs, negating the requirement for lava imports. By the time that lava started to become more accessible and popular at Silchester, sometime around or before AD 85, the local economy had shifted in favour of more centralised processing and lava querns failed to infiltrate consumer choice in the same way that it had at Verulamium.

Another way of exploring this comparison is to look at the relative volumes of lava at the different urban sites in comparison with querns of other stone types. Although only lava milling tools were recorded during data collection for Verulamium, the online collections might be used as a rough guide for what is present within the full archive, as this has catalogued 119 querns, most of which have been identified by lithology. Data collected from Chichester can also be included, and this can be collated with the extensive work already completed by Shaffrey (2021) on quern supply to Roman towns, adding a further two urban sites to the wider dataset, which had already included Silchester and Dorchester-on-Thames (figure 7.37).

As has been discussed by Shaffrey (2021), the Thames appears to be the defining border between the towns that that had good access to lava, and those that did not. Lava querns and millstones were clearly transported to locations west of London via the River Thames. Proximity to this major transport route must have heavily influenced the degree to which urban populations had access to these products. The addition of Verulamium and Chichester as towns north and south of the Thames has added some further detail to this interpretation. Verulamium, though not located on the Thames, was well connected to London by road and it may also have been accessible by water via the River Ver. The presence of Old Red Sandstone at Verulamium indicates that the east-west route that brought lava to the main urban centres along this main Thames corridor, also operated in the opposite direction. Interestingly, Chichester has a similar profile to that seen at Silchester, though without the presence of Old Red Sandstone. In comparison with Neatham and Winchester, more lava found its way to Silchester and Chichester, indicating that the status of the town may have had a role to play in its access to imported lava products. It appears that multiple factors influenced the types of milling tools that were used at each town, and it would be very useful to be able to identify chronological changes that may have occurred in these proportions as a further way of exploring the Roman stone trade and wider economy.

In considering the role of urban centres on the centralisation of grain processing, Silchester, Chichester and Verulamium provide some interesting comparative data (table 7.6). At Chichester, four millstones were recovered, with twelve at Silchester. For Verulamium, only the lava millstones were recorded and the online collections provide no further examples of millstones of other stone types. For these three towns, the number of lava millstones is extraordinarily high considering that there are only 76 recorded in the whole of Roman Britain. These appear to have been at least as popular as indigenous millstone types and show that urban populations may have had better access and the means to import specialist milling equipment, while consumer and cultural choice may have driven demand. The need to feed growing urban populations must have been a significant driving factor for the transition to mill technology and the fact that so many millstones were recovered from each of the urban sites should not be surprising. However, possessing the stone working skills necessary to create a working custom-design millstone and the suitable stone to form it must have provided some challenges to a province that had, hitherto, only been used to hand operated rotary

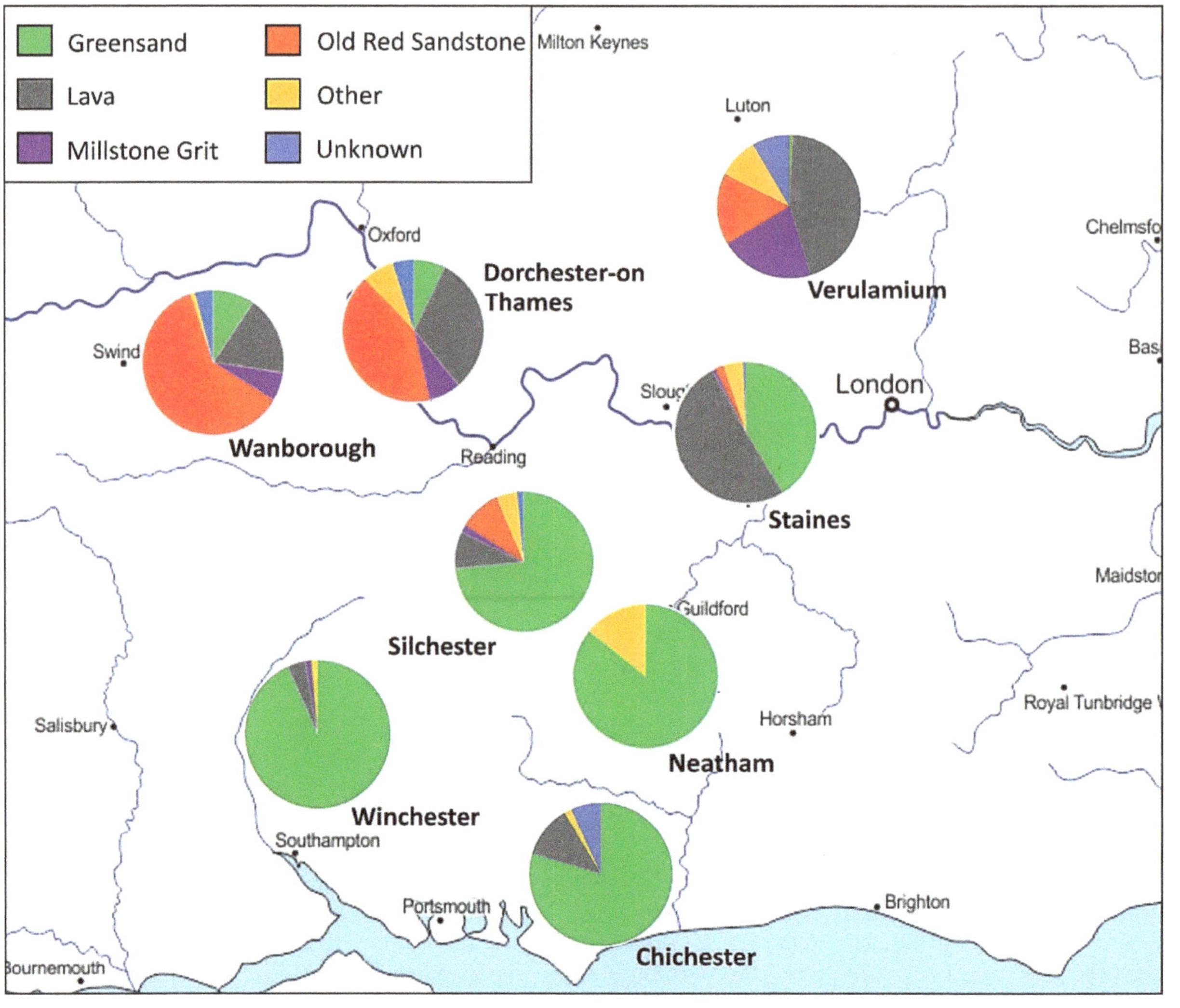

Figure 7.37. Map of central southern Britain showing the breakdown of stone types used for querns in the Roman period at various urban sites (after Shaffrey 2021, with inclusion of new data from Chichester and Verulamium).

Table 7.6. Summary of the number of millstones recovered from urban case study sites

	N° lava millstones	N° non-lava millstones	Total
Chichester	2	2	4
Silchester	8	4	12
Verulamium	6	?	At least 6

querns. Lodsworth stone and conglomerate are both very difficult to work and the fact that these do not often appear in millstone form indicates that they may not have been suitable for this purpose. Lava may have provided one of the best early choices for the introduction of mechanically driven mills as the skills and materials for production were already available at Mayen.

The high overall number of millstones at Silchester also raises some questions despite the difficulties involved in drawing direct comparison across the three sites. This is because of the probability that watermills existed at Chichester and Verulamium, while Silchester is more likely to have made use of animal and/or slave mills. Watermills are required to be situated near to a suitable head of water, placing them outside of the towns, while animal or slave mills could be located anywhere. The possible millstone plinths in *insula* XVIII at Silchester suggest that these operated in the town itself (Shaffrey, 2021, p. 53), which could explain why more millstones were recovered nearby. Centralisation of grain processing may, therefore, have looked very different across these separate urban landscapes and have been organised in diverse ways. This has implications for understanding how grain and flour entered towns for civilian consumption, the organisation of labour and the distinct ways that urban societies adapted to sustain the population of the town. However, it must be remembered that such organisation need not have been static, and that change would have occurred over time. It is not known whether the evidence seen here relates to civic and state-led provision, or whether the introduction of mill technology was a commercial enterprise. It is difficult to reconcile the presence of millstones with a specific source of capital. However, it can be agreed that a great deal of wealth and probable high social status would have been required to install mills at these sites.

The difference seen between Silchester and Verulamium for the extent of use of lava querns prior to deposition provides another noteworthy comparison. Greater extent of wear was visible on the Silchester querns, while those from Verulamium were more likely to be discarded when they were slightly thicker. This is likely to relate to the extent to which lava milling tools were replaceable at each of the sites, and Silchester being further away from the main distribution centres probably forced greater longevity of use. Both sites showed relatively high levels of reuse of lava which, again, is probably the consequence of there being a poor local supply of suitable stone for ground stone tools.

A final observation is the similarity seen in the deposition of lava in late Roman pits at both Silchester and Verulamium. Both examples might have ritual connotations, though the complexities associated with such interpretations creates a barrier to any positive identification of ritualised behaviours involving lava. This is a pattern of deposition that has not been observed at other site types, though the prevalence of lava at urban sites increases the probability that lava would be accidently introduced into 'ritual' contexts. Nonetheless, the possibility for specifically urban cult activity involving the preparation and consumption of food using milling tools should be considered. To fully explore the similarities between the two towns, a more detailed examination of individual ritual pits should be undertaken for Verulamium, in a similar manner to that carried out by Eckardt (2006). This might provide greater opportunity to identify patterns in urban deposition that might relate to ritual actions.

8

Milling at the Edge of an Empire: Lava Quern and Millstone Use at Northern Military Sites

Lava querns have long been associated with the Roman military, mainly due to the large volume of lava milling tools that were uncovered in some of the earlier excavations of military sites on Hadrian's Wall, such as at Newstead (Curle, 1911). Although the data produced for this research emphasises that lava use was not restricted to military communities alone, concentrations of lava milling tools at military sites show these objects must have played a significant role in everyday military life. There is also strong evidence in continental Europe that Mayen lava querns were initially part of Roman army equipment (Gluhak & Hofmeister, 2011, p. 1616), and this has led to the hypothesis that they may thus have been introduced to Britain at the time of the Claudian invasion (Watts, 2002, p. 33; Peacock, 2013, p. 153). Distribution analysis in Britain has shown that the chronological change in the spatial distribution of lava follows what we understand historically of military campaigns and movements (see Chapter 5).

However, this is not where investigations into the military use of lava milling tools should end. More recent examinations of material culture from military sites have shown the potential that ancient objects have for revealing the lived experiences of those in Roman military society. The most notable of these was completed by Allison (2006) who explored the concept of gendered spaces by comparing known gendered objects and activities at Pompeii with finds evidence from Vetera I, Ellingen and Oberstimm forts on the Roman frontier in Germany. However, the application of spatial distribution analysis in this study has since been criticised as it does not consider the extent to which redeposition may have affected interpretations, while the degree to which specific objects or object types can be classified as 'male' or 'female' has also been subject to debate (Campbell, 2010; Hodgson, 2014). Aside from spatial distribution (for example, Van Driel-Murray (1995; 2001), and Alberti (2018)) and the highly significant Vindolanda tablets (Bowman & Thomas, 1983; Evers, 2011; Greene, 2013), investigations that have involved any detailed examination of material culture from military sites have been relatively limited.

Currently, there are a wide range of different approaches that are yet to be applied to military site assemblages, though these could potentially inform our understandings of life on the frontier and in associated settlements. For example, lifecycle and object biography approaches can evaluate the human-object relationship at various stages of an object's existence (Gosden & Marshall, 1999); use-wear analysis can identify how tools can become extensions of the body and how embodied practice might vary according to changing traditions and identities (Harman, 2002; Olsen, 2010, pp. 68–75); typological analysis can help explore how innovation associated with object design can change according to the social, political and economic conditions (Bijker, 1997). The aim of this chapter is to incorporate some of these different approaches and to use these to answer questions relating to the lived experiences of people associated with the military, and to address these within the broader perspectives of frontier life, Roman Britain, and the wider empire.

The case studies examined in this chapter are Corbridge, Chesters, Housesteads, and Vindolanda. These can all be classed as northern military sites, but with very different characteristics and functions: Chesters and Housesteads were fort sites that formed part of Hadrian's Wall; Corbridge may have been part of the original 'Stanegate' system, probably becoming a military storage and supply depot after the establishment of Hadrian's Wall that provided protection at a major river crossing point for Dere Street, and later becoming a town; Vindolanda was also part of the earlier 'Stanegate system' of defences that saw continued use at other periods of Roman military occupation in the north. Despite the variety in the site characteristics, they would all have belonged to the same military network of supply, consumption, and society. How this varies or correlates across the sites may give some indication of how uniform or individual everyday practice involving food processing activities may have been. As a wide range of diverse cultures are represented within the Roman military community, it is interesting to investigate how food consumption practices and treatment of lava milling tools may have been dictated by the uniting factor of being a soldier or differed to reflect individual/group cultural practices.

This chapter reflects on lava quern and millstone use at each of the case study sites in turn, examining the artefactual and site evidence for manufacture, distribution, primary use, reuse or modification, and deposition. The analysis commences with Corbridge, followed by Chesters, Housesteads and Vindolanda. The largest dataset has been produced from Vindolanda, which has been subject to modern standards of excavation and recording.

8.1. Corbridge: Introduction and Background

The site of Corbridge to the south of Hadrian's Wall, also referred to as *Corstopitum* or *Coria* as its possible Roman name, is in modern day Northumberland (figure 8.1) and may have belonged to the earlier fortifications of the 'Stanegate system' (Breeze & Dobson, 1976, p. 20–26).

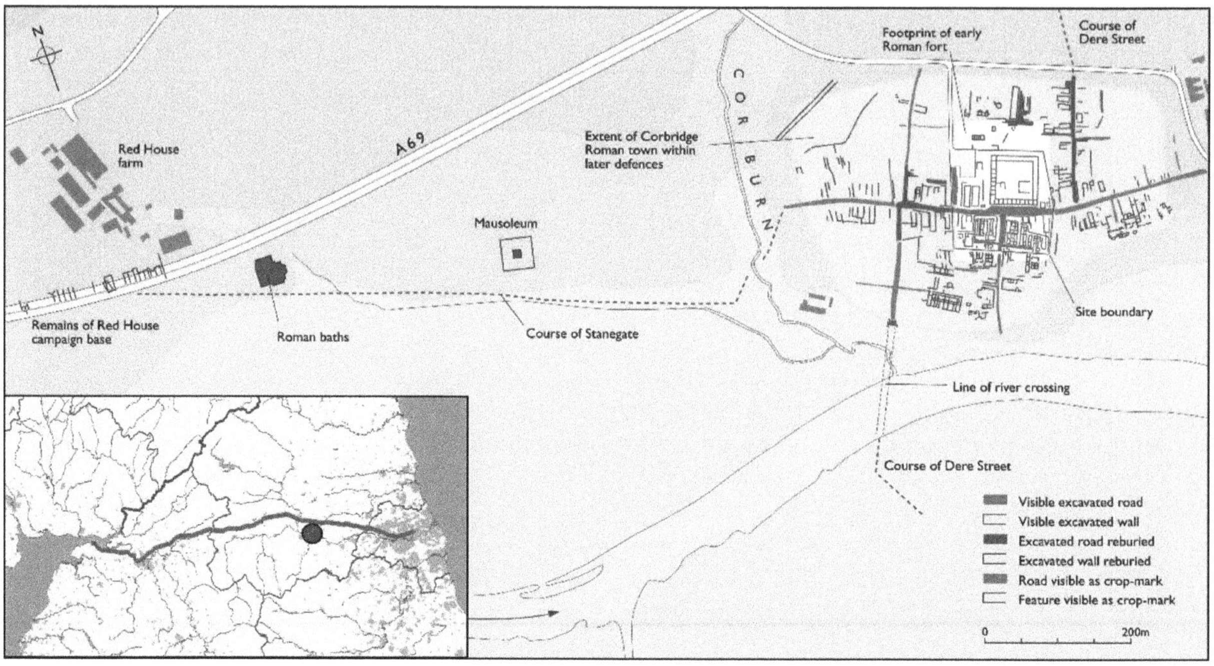

Figure 8.1. Location of Corbridge on Hadrian's Wall and with respect to the site at Red House and the River Tyne and bridge crossing (plan of Corbridge ©English Heritage Trust; map of Hadrian's Wall, Ordnance Survey OpenData).

Initial military occupation in the area appears to have been established at the site of Red House, which was a military base approximately half a mile to the west of *Corstopitum* (Bishop & Dore, 1988, p. 126). This earlier fort, probably dating to the Flavian period, has been interpreted as a possible supply base with open-ended buildings used for storage (Hanson, et al., 1979). There has been more recent discussion as to whether these interpretations should be revisited, as site plans suggest that this area was within the extra-mural settlement and may not constitute a separate site (McIntosh, pers. comm.). From epigraphic evidence, it is possible that troops stationed at this early fort, and maybe the succeeding relocated fort, were the elite cavalry regiment *ala Petriana* (Collingwood & Wright, 1965, RIB 1172). Military diplomas indicate that they were previously posted in Mogontiacum (Mainz), Germania prior to redeployment to Britain (Spaul, 1994, pp. 180–182; Clauss, et al., 2022).

Occupation at the site known as Corbridge was established around AD 86 in a timber and daub fort (Phase 1) after the dismantling and abandonment of the buildings at Red House (Hanson, et al., 1979, p. 1). Estimates of its extent indicate that it was larger than a standard cohort fortification, but smaller than a legionary fortress (Bishop & Dore, 1988, p. 129), often referred to as a vexillation fortress (Frere & St Joseph, 1974, pp. 6–7; Goldsworthy, 1998, p. 126). A second phase dating to around AD 103 (Phase 1b) where some of these buildings were modified or repaired has also been identified. Structures associated with this phase were destroyed by fire (Bishop & Dore, 1988, p. 126–129).

The second main phase, occurring around AD 105–122 (Phase 2) saw the construction of the secondary fort, also a timber and daub construction. This was more like the traditional style of cohort forts in its layout and form. The format and structures were not fully removed at the end of this period, though modifications that occurred to the site in the next phase of occupation dating to around AD 122–139 may have coincided with a change in stationed troops (Phase 3). Following this around AD 139–158, the first buildings in stone were constructed within the central range of the fort, although a complete conversion to stone was not carried out and some remained in timber form (Phase 4a). Twin stone granaries in their earliest configuration are likely to belong to this phase. Historical records suggest that this work may have been partly carried out by *legio II Augusta*. The changes continued into the next phase (Phase 4b), with the reconstruction of the barracks with stone footings around AD 158–163 (Bishop & Dore, 1988, p. 135–137). This phase was the last one associated with a fort, which was demolished after AD 163, and the function of the site appears to have changed by the middle of the second century.

The post-fort archaeology has been difficult to reconcile due to the disjointed nature of the investigations that have captured these later phases and the ambiguous nature of the excavation recording. More recent work has revealed that Corbridge became a military base and supply station after the fort had ceased to operate between AD 160–185 and was subject to large-scale reconstruction in stone. New buildings included the magnificent twin granaries, a large warehouse structure and *macellum* (Hodgson, 2008). The previously timber bridge was also reconstructed in stone at this time, emphasising the continued importance of the crossing site. The function of Corbridge would have been to store and supply goods for the forts on Hadrian's Wall and along Dere Street, and to manage administration tasks associated with this. It housed detachments from *legio*

XX Valeria Victrix, whose main garrison was at Chester, and *legio VI Victrix* from York (English Heritage, 2016). However, it appears that many buildings were incomplete at the time of a large fire, which was possibly the result of attack. Rebuilding appears not have taken place immediately afterwards (English Heritage, 2016).

After the restoration of stability at the northern frontier in the early third century, Corbridge remained an important military base and saw the construction of two stone walled compounds to house detachments. It is probable that the granaries were rebuilt in this phase, as the continued role of Corbridge as a supply base would have necessitated the storage of grain on a large scale. Around the same time, a civilian settlement grew up around the military activity areas, which gained recognised status of *civitas* by the third century. Continued military presence seems likely but the character of the later site is considered as urban in its nature (English Heritage, 2016).

8.2. Corbridge: The Dataset

Data for Corbridge was recorded in person, which allowed for measurements and records to be taken that are consistent with the aims of this research. The dataset comprises 36 lava milling tools, with no millstones identified in the assemblage. From this 20 were recorded in person, 4 were taken from the excavation report for the site at Red House, and 12 were compiled using information stored on the archive database for objects that appear to have been discarded at the time of the excavation. Most of the assemblage comprises undiagnostic fragments or objects that were only briefly described in publication or database information. There were 16 quern stones with diagnostic features, and it was possible to identify 6 upper and 7 lower stones from these (figure 8.2 and table 8.1). This seems slightly unusual when compared to the data from other sites, where upper stones tend to dominate lava quern assemblages. However, the preservation of lava at Corbridge was much better than has been observed for other sites, and this is likely to have contributed to a more equal representation of quern parts. However, querns in the assemblage are unlikely to be the full number excavated from the site, and we have little record of what the retention policy may have been over the course of the many decades of excavation work. A great many more quern stones would have been at the site, and those recorded here probably represent a small sample of what was present. Other retained querns are kept at the English Heritage stores at Helmsley, but these were not accessible at the time of data collection. Unfortunately, as these have not been identified by specific stone type in the archive database, it has not been possible to include them in the dataset at this time. Interpretations must, therefore, be mindful of this.

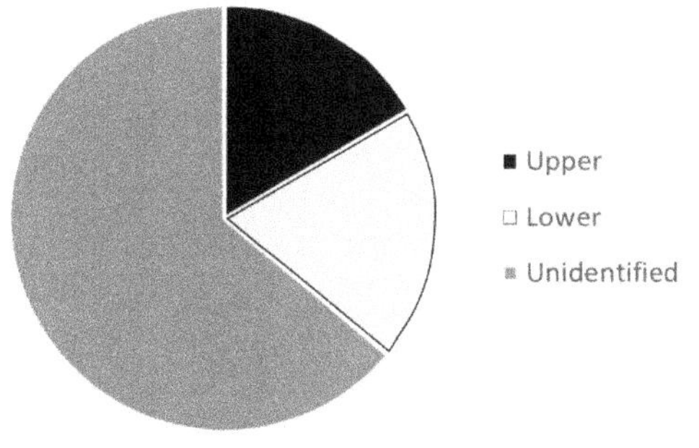

Figure 8.2. Chart showing the number of upper and lower stones in the dataset, alongside those unidentified (T= 36).

Table 8.1. Number of upper and lower stones in the dataset

Stone part	Number of objects
Upper	6
Lower	7
Unidentified	23

Unfortunately, chronological and context information is patchy, which might be expected considering the age of some of the excavations and the lag between the time of excavation and the production of the excavation report. Further to this, only the querns from the site at Red House have been published, as none were listed in the main Corbridge excavation monograph. However, all lava querns in the assemblage can be confidently assigned a Roman date due to the nature of the site, and due to typological characteristics of Roman period lava querns being present on most of the stones themselves. Nonetheless, data analysis focusses on the data that is available, namely detailed examination of the objects themselves and distribution analysis.

8.3. Corbridge: Manufacture

The lava querns from the archived collection at Corbridge were well preserved. This might be indicative of selective retention, as most examples retained diagnostic features that are typologically distinctive. Unfortunately, no descriptions or details were provided for the querns that must have been discarded. The presence of these was noted in the excavation documents and utilised in the dataset (16 objects), so for this data there is a record for context of recovery and a general idea of volumes, but nothing pertaining to physical characteristics. Therefore, this section focuses on those lava quern stones that were examined in-person (20 objects), with consideration for the fact that this denotes a sample of the full assemblage. Evidence for manufacture considers: the diameter of the stone, presence and sizes of kerbs, handle fittings, rynd fittings, and dressing patterns.

8.3.1. Stone Diameter

There were eight stones with measurable diameters, and these can be seen in figure 8.3, which also presents the diameters of querns from the whole dataset for comparison. The results show that all the stones with measurable diameters can be classified as querns. There were no identifiable millstones in the collection, and this was also the case in terms of diagnostic features associated with millstones. The Corbridge querns all appear to be of

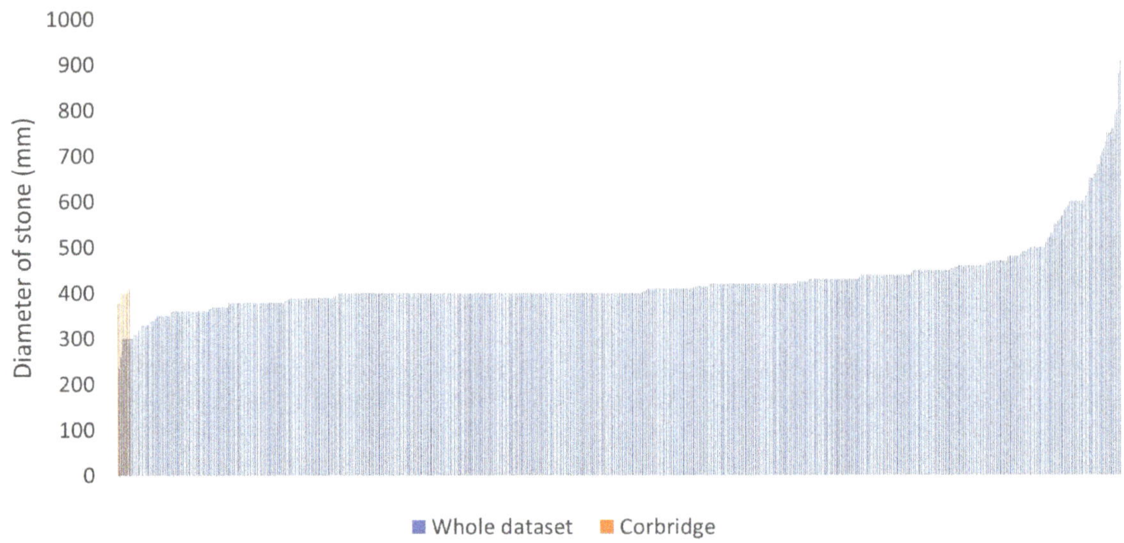

Figure 8.3. Corbridge quern diameters compared with the diameters of milling tools for the whole of Roman Britain (T=8).

the standardised size and fit very well within common dimensions seen across the rest of Roman Britain. The smallest example was 380mm, and the largest was 410mm, with no unusual sizes within the assemblage. This shows that these querns are likely to be Mayen and from the same location as the bulk of the Romano-British lava querns.

8.3.2. Kerbs

Of the seven upper stones six had kerbs present. Two of these kerbs were very faint and not clearly visible, while the remaining four were more prominent. One of the faint kerbs was from a stone that was heavily weathered, and deterioration of the lava affected visibility of the kerb. However, the second example was a complete upper stone that was well preserved, and the faint kerb appears to have been an original feature of the stone (figure 8.4) (CAT 2179). As discussed in Chapter 2, the kerb may have played a role in reinforcing the edge of the stone for the elbow handle fitting. This example with the faint kerb does not have any evidence for a handle fitting, despite obvious signs of wear from use. It may be worth considering whether an iron band may have fixed the handle to the stone in this instance, evidence for which would be absent if the metal had been removed from the stone prior to deposition. Such a handle fitting would have made the functional purpose of a kerb redundant, but it may have remained in later examples as a visual reference to previous styles. Unfortunately, as the chronological information for the lava querns from Corbridge is poor, it is impossible to determine whether this change occurred over time or was simply a contemporary variation of the standardised lava quern in use across the province.

8.3.3. Handle Fittings

Further to what has been observed with regards to CAT 2179, two stones did have elbow handle sockets present

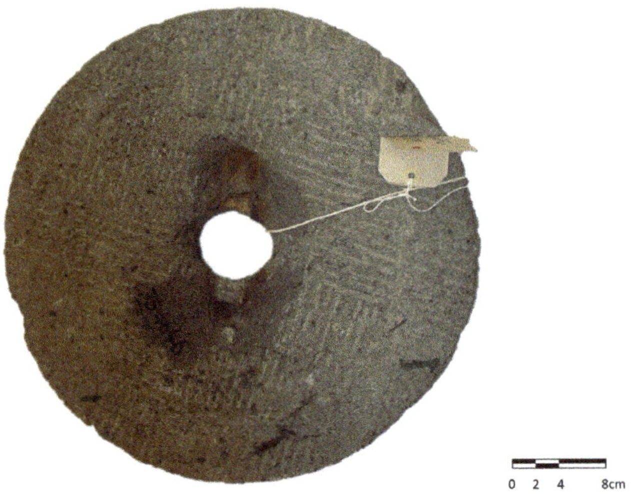

Figure 8.4. Complete upper quern stone showing a faint kerb and no sign of the standard elbow handle fitting (photo by the author, ©English Heritage) (CAT 2192).

(CAT 2257, 2188). These were visible at broken edges for both, which is a normal occurrence at a place where the stone is most vulnerable. Both these objects had kerbs, and these were prominent; though other features that would be expected on a standardised upper stone, such as the upper surface dressing and rynd fittings, are not visible due to heavy weathering of the surfaces and breakage. The elbow handle fittings are like those observed at the quern finishing sites at Mayen, which are included in the quern typology produced by Röder (1955). It is thought that these were added to the object at Mayen, and that this explains the prominent level of standardisation seen across lava quern assemblages. The method of adding an iron band does not, however, appear in this typology. One explanation is that quern roughouts recovered from Mayen without elbow handle fittings have been assumed unfinished, as opposed to representing a different type. This method may have produced stones that had a longer use life to those with the traditional elbow socket, as it removed the vulnerability that such sockets created. The iron band would also generate external pressure to the stone to reinforce stability for a brittle stone type that appears to have been sensitive to breakage as the stone thinned. In a comparable way to the French Burr stones, it may also have served as a method for consolidating several pieces of lava to form a complete stone (Tucker, 1982) or provided a means for fixing a fissured stone. The iron band may, therefore, have been a technological innovation and show development of the original type, or have provided a means for extending the use-life of the stone.

Another possibility is that handles were added nearer the site of consumption. The addition of elbow handle sockets to mended querns at Vindolanda shows that the ability and tools to do this existed in military societies, though there is an argument that these were not particularly successful attempts (see section 8.2). Adding an iron band may have been a less intrusive and potentially less damaging way of creating a handle fitting for a community that was not habitually used to working with lava stone. Further to this, as previously discussed, there may have been some advantages to this alteration. However, a greater level of innovation and variation would be expected of quern handle fittings from non-military contexts if this were the case, and elbow sockets are dominant across the province (see Chapter 5). This is not to say that iron band handle types do not occur more frequently than has been recorded, as these would not be identifiable on fragmented upper stones. However, evidence appears to suggest that the iron band handle fitting is a typologically different quern to those with the elbow fittings.

8.3.4. Rynd Fittings

Two upper querns from the assemblage had remains of a rynd chase (CAT 2192, 2179) (figure 8.5), one which is the complete upper stone previously discussed. Both are of the standardised rynd chase fitting types, being recessed into the top surface of the stone across the central hole in the form of a rectangular iron bar fixed with lead. Remnants of iron and lead remain in the chase for CAT 2192 showing that the rynd was fitted in this way, while CAT 2179 retains only the rectangular shaped slot that would have accommodated the metal fixture. There is nothing on any of the other upper stone fragments to indicate that other forms of rynd fitting were in use at Corbridge, and evidence suggests that the standard types were utilised.

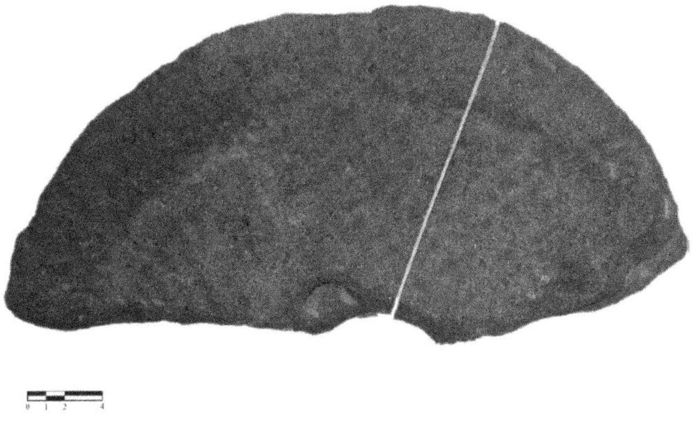

Figure 8.5. Upper quern stone showing recessed rynd chase in the upper surface. Scale is in cm (photo by the author, ©English Heritage) (CAT 2179).

8.3.5. Surface Dressing

Dressing for both upper and lower stones typically occurs on the grinding surface and the sides, while for upper stones, the upper surface is also dressed (see Chapter 2). Although much of the Corbridge lava quern assemblage had been subject to weathering, evidence for these types of dressing were apparent on 13 stones:

- Well-executed and clear 'harped' dressing on one upper and one lower stone (CAT 2179, 2255),
- Poorly executed harps on one lower stone (CAT 2254),
- Unidentified grooved patterns on eight upper lower and unidentified stones (CAT 2180–2, 2186, 2188, 2190–1, 2256),
- Clear radial dressing on one upper stone (CAT 2192),
- Pecking on one lower stone (CAT 2193).

As with other sites, there does not appear to be any correlation between whether the stone is an upper or lower and the dressing type on the grinding surface; a variety of different dressing types are present. As wear would require redressing of grinding surfaces, these patterns are not associated with the manufacture process unless the stones are little used or unworn. Traditionally, the lava quern is associated with harped style dressing, but they were not always redressed with this design. The variation suggests that people with differing skills in stone working were responsible for redressing the quern stones, and that it may not always have been preferable or possible to replicate the more complex harped design after wear had erased it.

The vertical tooling was present on six of the twenty querns recorded in person and is the most common characteristic of a standardised quern from the archive (CAT 2179, 2181, 2188, 2192, 2255–6). However, edges were missing from many of the stones in the assemblage, and so evidence for this feature is absent due to this. Possible reasons for the broken edges has been discussed in the section on reuse/modification. The upper surface dressing was visible on three of the six upper stones (CAT 2179, 2180, 2192), while weathering appears to have removed the evidence for this on the other three.

8.4. Corbridge: Distribution

Location data was available for 20 querns. A further 4 querns are known to have been retrieved from the site at Red House, which are not included in the distribution analysis but are worth considering in association with the other site data. Most findspots could be reconciled to specific contexts and their locations. Those from the 1980 excavation season must have been recovered from the location of the current visitor centre, which is where digging took place that year, while those from the 1970 season, though recovered from the spoil heap, must have been from the area north of the officers' house. Unfortunately, 12 querns could not be allocated a site location, and these have been omitted from the analysis. It must be emphasised that the distribution does not reflect a full dataset and that it is highly probable that evidence for other milling tools was recovered and not recorded. Locations have been plotted onto a site plan that also includes the known phasing for the excavated areas (figure 8.6). However, it should not be assumed that the objects necessarily follow these dates as these relate to the structural evidence.

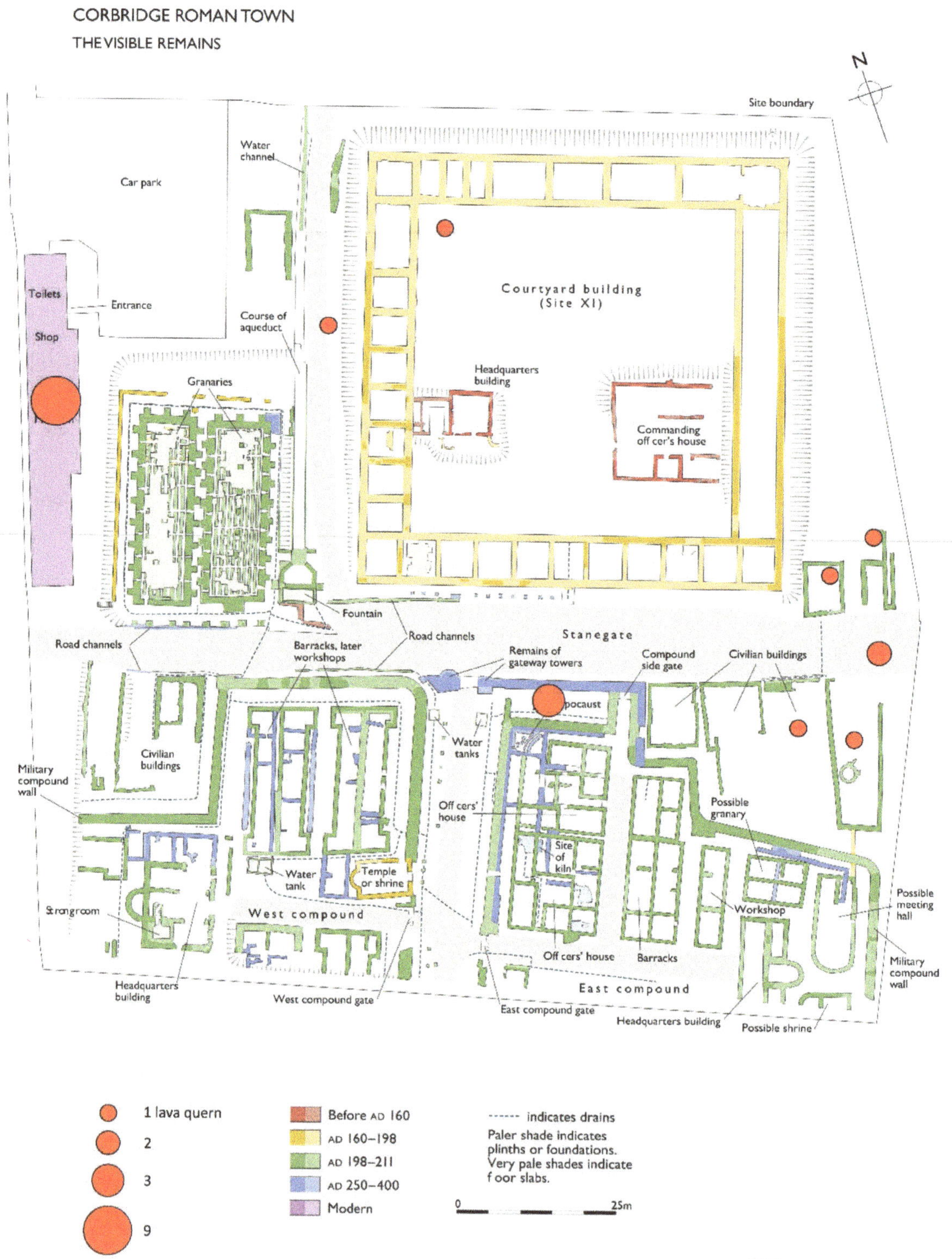

Figure 8.6. Plan of the site at Corbridge showing the distribution and volumes of lava querns by findspot (amended site plan ©English Heritage Trust).

The results show a cluster of lava at the area of the east gate and on either side of the road. These are external to the compound walls that are dated to the late second/early first century and close to an area of civilian buildings belonging to the same phase. A small group of three querns were found within the compound wall and near to the officers' house. Two other lava querns were retrieved from the north-western area of the courtyard building, one external and one internal to the structure. The final cluster occurs in the area where the modern-day museum and visitor centre sits and comprises the largest group of the material recovered from the site.

A total of nine querns were found during excavations of the visitor centre area, referred to in the excavation monograph as site 9, and this was the excavated area that showed the greatest amount of post-fort activity. It seems highly probable that the lava was recovered from archaeology of this phase and therefore constitute redeposited material. Evidence at site 9 indicates occupation in timber buildings and the presence of two hearths and an oven. Only one quern had information relating to the archaeological feature it was recovered from; this had been reused within a cobbled surface and cannot be considered as being from a primary use context. Similarly, the three querns from the 1970s excavation in the area near the officers' house were without context and any association with location of use is impossible to identify. However, this group is close to the main cluster of querns near and at the east gate area. This region is referred to in the excavation monograph as the 'inter-compound' area (Bishop & Dore, 1988, p. 110–125). This is the area where the earlier phases of the fort were accessible, while the east gate excavations were similarly dated to the earlier military phases (Bishop & Dore, 1988, p. 88–93). It is not possible to determine whether the lava querns were uncovered from these phases as the third-century civilian area overlies it, but it does produce some questions with regards to the concentrated distribution in these areas and the association of lava with early military activity. However, as the region is outside of the third-century compound wall, it may simply have been a suitable refuse disposal location external to the main occupation area. Similarly, large areas of road surface were uncovered in this area and the lava may have been reused within this context, creating a greater concentration of deposition.

8.5. Corbridge: Primary Use

As has been shown from the analysis of the whole dataset, comparison of the thickness of upper and lower stones presents the best opportunity for exploring primary use of the querns in the assemblage, which mainly comprises fragments that cannot be examined for method of use or rotation. Stone thickness can be easily compared with the stone thickness of querns from the whole dataset, allowing for the identification of especially thick or thin stones that may represent more, or less extensive wear and possible meanings associated with this. Comparison has been completed for the six upper and seven lower stones separately.

8.5.1. Upper Quern Stone Thickness

The upper lava quernstones from Corbridge are thicker than those recovered from elsewhere in the province (figure 8.7). Of the six upper stones recovered, four were in the upper quartile for thickness when compared with the full provincial dataset. Though this may seem significantly high, it must be remembered that only larger fragments

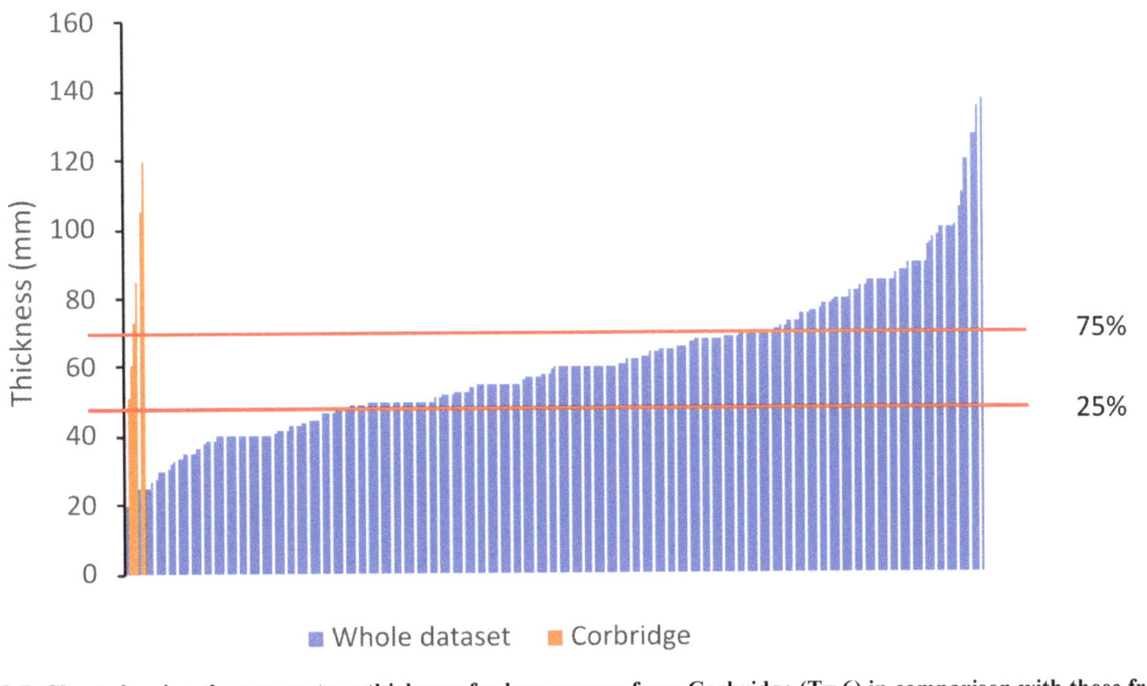

Figure 8.7. Chart showing the upper stone thickness for lava querns from Corbridge (T= 6) in comparison with those from the rest of Roman Britain (T= 248). The querns from 1 Poultry have been excluded from the analysis.

Table 8.2. All sites with upper stones thicker than 100mm

Site Location	Site Type	Number of querns
1 Poultry	Military/Urban	89
Carlisle	Military/Urban	4
Colchester	Military/Urban	3
Corbridge	Military/Urban	2
Segedunum	Military	2
Castleford, North Yorks	Military	2
Chesters	Military	1
Holditch, Staffs	Rural	1
Strageath, Perthshire	Military	1

may have been retained on excavation, and that thinner and smaller fragments that may have existed in the site have not been included in the analysis. However, three of the thickest fragments are significantly thick, with two possibly being unworn, or very little used prior to deposition (CAT 2180, 2192). This is also the case for the lower stones (next section).

To emphasise how unusual these thicker querns are, table 8.2 names the nine sites where upper stones were found to be at least 100mm thick. The majority were from the site at 1 Poultry, London, where their presence has been taken as evidence for London being a centralised distribution hub for imported lava milling tools. The few other examples occur more frequently on military/urban sites, and occasionally on military sites. This may be an indication that lava was redistributed via military/urban sites, as querns that were broken in transit are most likely to have been deposited on arrival. As Corbridge appears to have had a significant role in storing goods and supplying the northern frontier, the analysis may suggest that querns were part of this supply chain and were entering northern military circulation at this location, amongst others. This is a theory that is worth investigating further. This could be done by taking the lava quern data from more recent excavations at northern military sites that were also involved in supplying forts along Hadrian's Wall, namely Wallsend, Carlisle, and South Shields. These could be compared with the quern thickness data for Corbridge to see if any similarities in quern stone thickness can be identified and if these differ from lava quern 'consumer' sites on the Wall.

8.5.2. Lower Quern Stone Thickness

As with the upper stones, the lower stones from Corbridge show a similar trend in being generally thicker at the point of deposition than querns from the wider province (figure 8.8). Of the seven lower stones from Corbridge with known thickness measurements, four fall within the upper quartile for thickness when compared with the whole British dataset. The reasons for this are likely to be similar to that seen for the upper stones, and the choices relating to discard or retention at the time of the excavations provides one possible explanation for this. As with the upper stones, one lower stone also falls into the category of being a possible unused or little used stone (CAT 2190). The remaining three lower stones in the Corbridge dataset fit comfortably into the range that half of the British lava querns do for thickness, showing ordinary extent of wear prior to deposition.

8.6. Corbridge: Reuse/Modification

The extent of weathering on the surfaces of many of the quern fragments makes it difficult to determine whether any of the surfaces had been reused for purposes other than for grinding and milling, and in many cases, wear associated with primary use could not be established either. Although perfect preservation cannot be expected, the surface weathering observed could be indicative of lava that has been exposed to frost, suggesting reuse or deposition in an

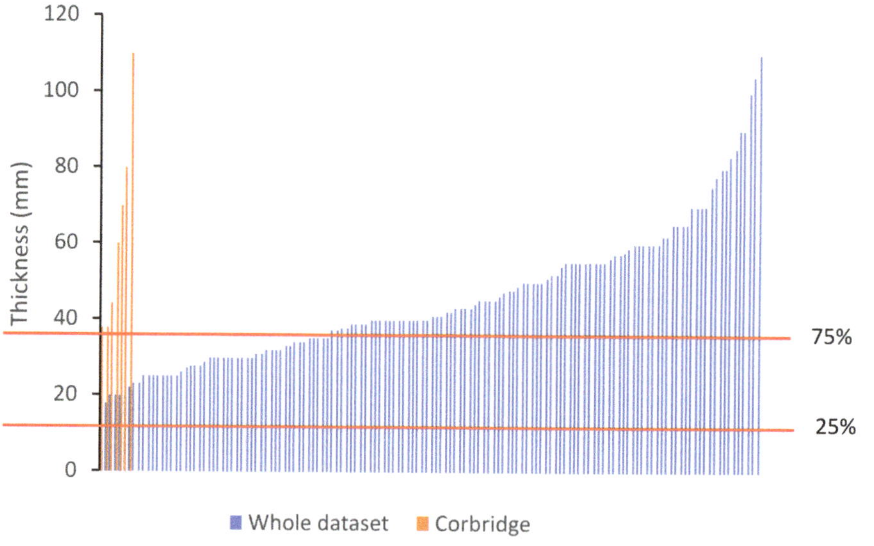

Figure 8.8. Chart showing the thickness of lower stones (T= 7) compared with the lower stone thickness of data from the whole of Britain (T= 138) excluding the querns from 1 Poultry.

exposed environment, such as for a road surface. Similarly, the extent and way that the lava was broken may give some indication that reuse occurred, or that deposition of lava took place after it had been deliberately broken. Several of the fragments were broken in a way that is unlikely to have occurred accidentally, with thick pieces broken in places that are not usually vulnerable to damage. One example includes CAT 2180, which shows thick areas at the side of the quern stone broken or chipped off (figure 8.9). Unfortunately, it is not possible to differentiate between reuse or deposition in terms of the treatment of the lava. As lava is often reused in floors, surfaces and roads, the breaking of querns and frost damage could relate to this. Reuse as building material is also possible.

The trend seen for modified lower stones by reducing the sides towards the bottom of the stone, such as seen at Verulamium (Chapter 7, CAT 2605 and 2592), is also apparent in the Corbridge assemblage with CAT 2190 (figure 8.10). The extent to which this had been carried out has changed the profile of the stone, angling the sides inwards where it has been reworked. The sides have been roughly removed, cut irregularly, and show tooling marks associated with this process. The expected standardised vertical tooling is also absent. Meanwhile, the full shape and size of the grinding surface has been retained meaning that there has been no loss to the area of the grinding surface. This shows that the modification was not intended to affect the use of the stone as a milling tool. Thus far, there has been no evidence for similar modifications associated with upper stones, showing that the alterations probably relate to the functioning of the lower stones only. It can be compared to an indigenous lower quern stone from the Clayton collection at Chesters, which has been fashioned in a similar way (figure 8.11). Placement into a floor surface is one interpretation for this form of reworking, and it may have been a customary practice that is only identifiable on more complete and less worn stones. What is interesting about this stone is that this example is especially thick at 110mm, and may constitute a new unused quern, or one that has seen little use. Such a situation might be interpreted as a quern roughout, but this is unlikely as the grinding surface has been dressed, a process that would have taken place at the later stages of quern manufacture (see Chapter 2). Why this stone was not used or was little used cannot be explained, but it was broken in half prior to deposition. It is possible that the intended modification failed, that accidental breakage occurred, or the quern was rejected for another reason.

One other lower stone in the assemblage shows similar features and reworking, though wear has made this more difficult to identify clearly (CAT 2254) (figure 8.12). This

Figure 8.9. Upper quern stone with broken thick edges that may have been fragmented deliberately for the purposes of reuse (CAT 2180) (photo by author, ©English Heritage).

Figure 8.11. Lower quern stone of indigenous stone that has been similarly worked to round the lower surface, possibly to fix it into a floor (Acc No CH266) (photo by author, ©English Heritage).

Figure 8.10. Modified lower stone (above and right) that has had its sides removed near the lower edge (CAT 2190) that is similar to reworked lower stones from Verulamium and Silchester (photos by author, ©English Heritage).

Figure 8.12. Possible modified lower stone that has been more heavily worn and has a decreased diameter (CAT 2254) (photo by author, ©English Heritage).

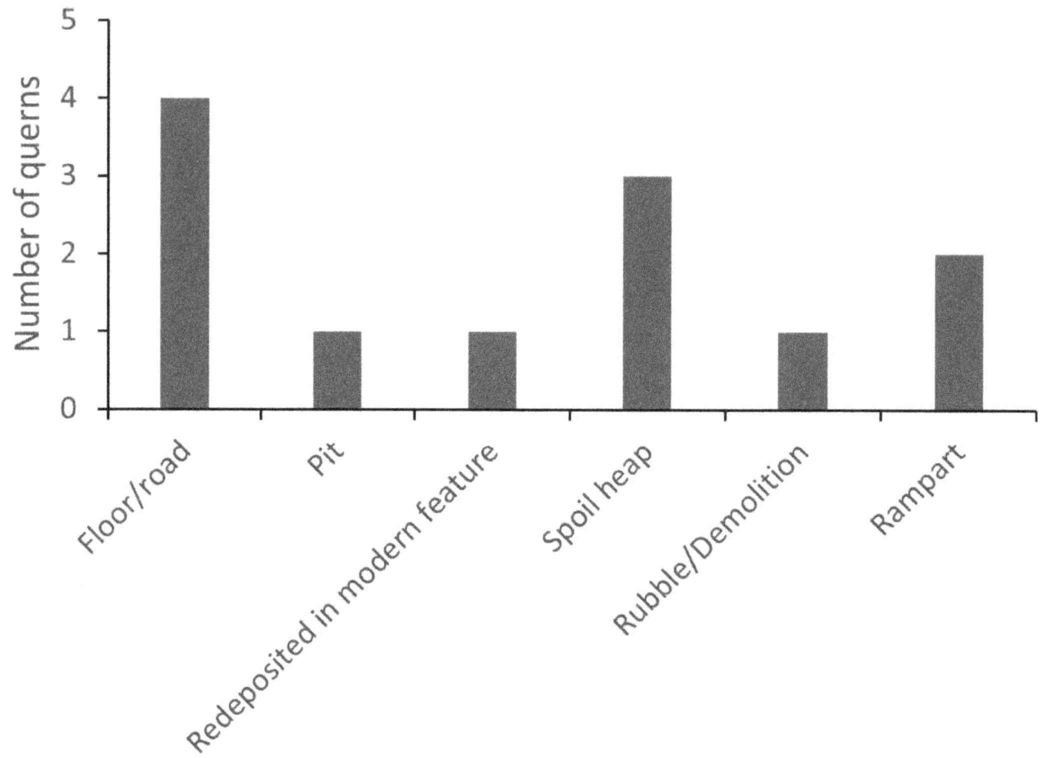

Figure 8.13. Chart showing the feature type where lava querns were deposited at Corbridge (T= 12).

stone is 60mm thick, and so is more worn than the previous example. The sides have been clearly reworked, though these have been smoothed and tooling is not visible on the sides. What is also notable is that the diameter is much smaller than the standardised lava quern at 350mm, and it must have been reduced during the modification process. One reason could relate to wear, as removal of the sides has caused a gradual decrease in diameter at the lower levels of the quern. This may have impacted the efficiency of the quern, and deposition may have occurred as result. The fact that the lower stone would no longer have fit and paired with the upper stone may also have posed some problems.

8.7. Corbridge: Deposition

Unfortunately, specific archaeological features were not identified in relation to the deposition of lava querns in the Corbridge archives, and there is little that can be discerned from the available information. What is known has been presented in figure 8.13. This includes the four groups of lava fragments from Red House that had context information included in the relevant publication. Of these, three were incorporated into a cobbled floor surface and so the Red House querns have influenced this category significantly. This is a type of deposition that is known for lava at many other sites in Roman Britain and appears to have occurred most often at military sites. However, it is not possible to comment on the prevalence of this deposition type as there is not enough data to make comparison possible.

The treatment of the lava querns prior to deposition has already been commented in the previous sections, but it should also be noted here that the deliberate breaking of querns may have been part of a practice of ritual deposition. In order to establish if this were the case, it would be necessary to complete a detailed examination of context type and associated finds recovered alongside the lava, which is not possible for Corbridge.

8.8. Chesters: Introduction and Background

The Roman fort at Chesters is in modern day Northumberland on the western bank of the River North Tyne, which is bridged by Hadrian's Wall at this point (figure 8.14). The fort was one of the earlier constructions established as part of Hadrian's Wall. This required adaptation of the original design of the wall defences, resulting in alterations to the structure itself and associated ditch system (Breeze & Dobson, 1976, p. 45). The fort was garrisoned by a cavalry regiment, evidenced by the barrack stables and inscriptions that identify the troops stationed there (Hodgson & Bidwell, 2004, p. 139; Tomlin, et al., 2009, pp. 467–480). The first of these regiments was the *ala Augusta ob virtutem appellata* – 'the cavalry regiment styled Augusta for its valour,' though the date of this initial occupation is not known and can only be estimated to have occurred sometime after the construction of the wall, which commenced in AD 122. The regiment is thought to have been previously stationed at Calunium (Lancaster), and possibly in Germania prior to that (Spaul, 1994, p. 55–57). Little is known of the use of the fort during the mid-second century when troops were moved to the Antonine Wall, but activity appears to have continued to some extent. Building work and inscriptions have been recovered from

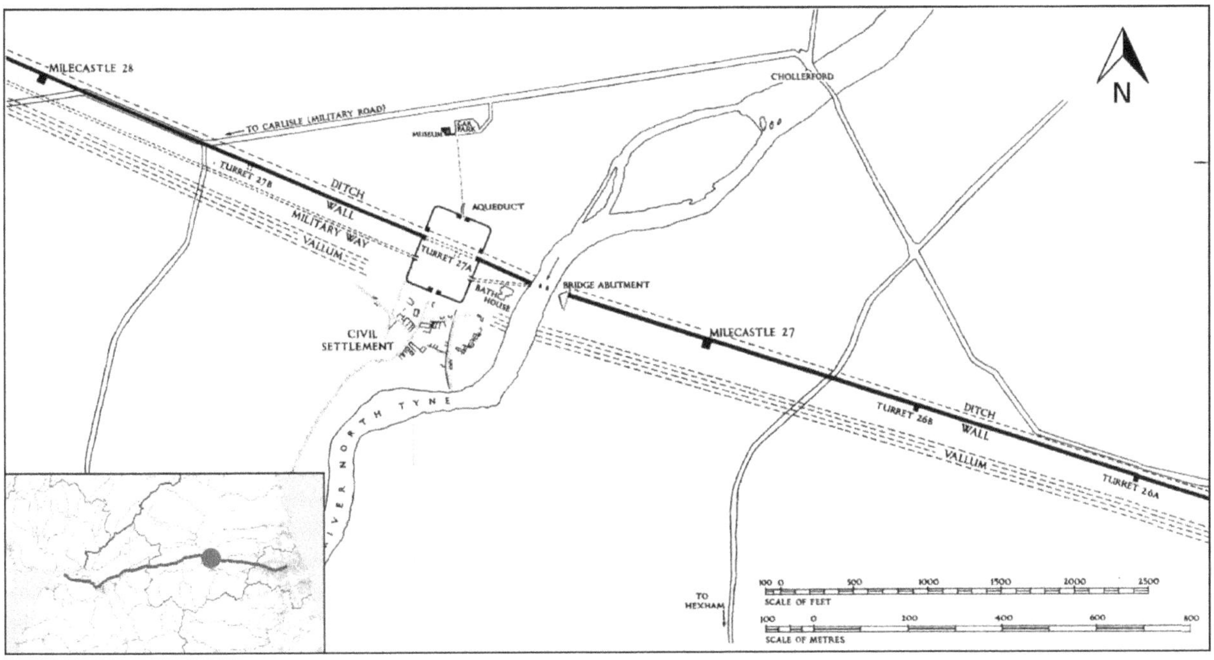

Figure 8.14. Plan of the location of Chesters Roman fort (background map, HMSO, 1960; map of Hadrian's Wall, Ordnance Survey OpenData).

the site that date to this period, which suggests that the fort was being maintained and used, even if this was not necessarily on a permanent basis. Permanent occupation by the military continued by AD 178–84 with the arrival of the *ala II Asturum*, 'the Second Asturians', another cavalry unit who remained at the fort until the end of the Roman period. It is thought that they had previously been stationed at Ribchester in Lancashire from a dedicatory inscription (Vanderbilt, 2020). Their arrival initiated large-scale reconstruction and refurbishment works to accommodate the specific needs of the garrison (Hodgson, 1999).

Excavations of the fort were completed by the Clayton family during the nineteenth century and the early nature of these investigations means that the chronological development of the fort is not well recorded due to a lack of stratigraphic data. Further to this, only specific finds were saved and very few of these retain context information (McIntosh, 2019). More recent geophysical survey of areas of the site that remain unexplored have been subject to geophysical survey, which has revealed previously unknown structures, such as a large granary and more barracks (Hodgson, 2009, p. 109). The vicus settlement has not been subject to excavation, meaning that all retained finds were recovered from the main fort area. However, aerial photography and geophysical survey has been used to identify its extent and layout (Bidwell, 1999, p. 116; Burnham, et al., 2004, p. 273).

Separate to the main fort area and highly relevant to this study are the remains of a possible mill structure at the eastern abutment of the Roman bridge over the River North Tyne. Excavations of the abutment took place at the same time as those of the fort buildings and are described by Clayton and Bruce (see summary by Simpson, 1976, p. 44). These descriptions in association with the remaining evidence and comparison with the more substantial remains of a Roman mill at Haltwhistle Burn have been used to identify the site as a potential watermill (Simpson, 1976, p. 44). The proximity of the structure to the fort and the sizable population of the garrison makes any possible mill a probable military structure. Further to this, the technological skills, resources, and labour to establish such an innovation would be more easily met by the military community. The mill at Haltwhistle Burn was a third-century construction and it might be assumed that any other mills on the Wall were similarly dated. Millstones were recovered in association with the Haltwhistle Burn mill, which still had part of the wooden structure preserved intact. None of these millstones were of lava, though a range of stone sizes were present; probably to cope with the seasonal changes in waterflow and, thus, power to turn them (Simpson, 1976, p. 36–42). The absence of lava is interesting and might relate to the change in lava consumption seen more generally in the province during the third century (see Chapter 5). The millstones were made from a locally quarried sandstone (Simpson, 1976, p. 36), allowing for the specific dimensions of the mill mechanisms to be factored in at the time of stone extraction and fashioning.

8.9. Chesters: The Dataset

The dataset for Chesters is the smallest of all the military case studies, comprising only seven stones, four being lower, two upper and one unidentified. As explained in the site introduction, none of these objects has associated context information and the findspots have not been recorded. However, it is certain that they were excavated at Chesters and are of the Roman period, meaning that they can confidently be included in this research. Throughout the analysis, the unknown finds discard policy must be considered alongside the high probability that the stones retained were the best examples recovered from the whole site. It is notable that the lava in the archive at Chesters is well preserved and consists of very large fragments or complete stones. Their condition implies that the retention of querns during the excavation may have been more selective than for other sites and that these objects do not constitute a random sample of what would have been present overall. It is also worth considering the dominance of lower stones in the dataset, as this is highly unusual in the province where upper stones tend to dominate.

Figure 8.15. CAT 2172, lower quern stone showing all characteristics of a Mayen lava product (photo by author, ©English Heritage).

This does not reflect an ordinary assemblage, and this must be acknowledged in any interpretations relating to it. The gaps in the excavation data limit the types of investigations that can be carried out on these milling tools in terms of object biography, and data analysis predominantly focuses on what can be learnt from the objects themselves, while making clear any bias that might be inherent in the data. In the absence of context data, distribution and deposition cannot be explored, while only a limited view of primary use can be investigated. This section, therefore, looks at the typological characteristics of the stones relating to manufacture, signs of primary use, and any reuse or modification visible on the objects.

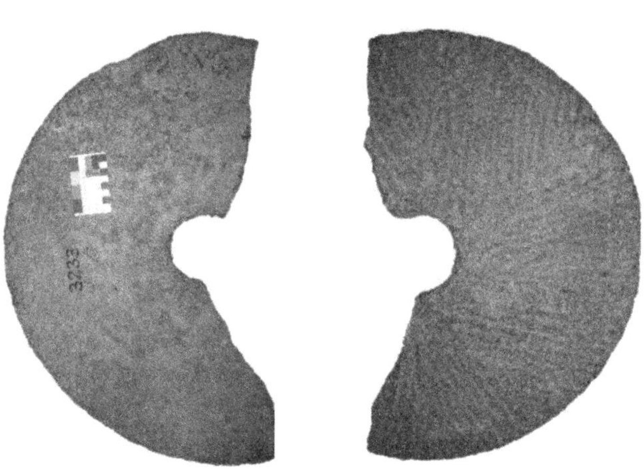

Figure 8.16. Part of the complete lower millstone showing top and bottom view (CAT 2176) (photo by author, ©English Heritage).

8.10. Chesters: Manufacture

8.10.1. Lower Stones

The features of the lower stones from the assemblage are largely compatible with the standardised lava querns that are known to have been produced in the Mayen region. Weathering or wear has removed signs of dressing for CAT 2173, but this stone retains the convex profile, straight vertical sides and hollowed lower surface, all typical characteristics for Mayen lava lower stones. CAT 2172 presents these features but is a better-preserved example that also has the harped dressing on the grinding surface, which aligns with the vertical striae dressing on the sides (figure 8.15). CAT 2174 and 2176 are complete lower stones, the latter being that of a millstone with a diameter of 600mm. The grinding surface for the complete lower quern stone (CAT 2174) does not have any signs of dressing present and has been worn and weathered in a comparable way to CAT 2173. There is a possibility that the grinding surface was pecked, but this cannot be identified with any certainty.

The complete lower millstone (CAT 2176) has broken into two pieces and is of the same style as that of the lower quern stones, only larger (figure 8.16). As there is no evidence present on lower stones that would be indicative of methods of rotation as they are not active in the millstone mechanism, the stone has been identified as a millstone based on its size alone. The profile is of a convex grinding surface with a hollowed out lower surface. The sides are vertical and straight, and it has a circular central hole that fully perforates the stone. The grinding surface is dressed in harps that are clearly visible, but there are no vertical striae present on the sides. It has been surmised by Moritz (1958) that bigger querns/mills of this type were used by larger units of men, such as 'centuria' or 'turma' to process greater volumes of grain to accommodate their needs. The suggestion is that these needed to be operated by groups of soldiers or turned using animal power, though operation using a turning crank has also been proposed by Jacobi (1914).

However, the association between specific unit sizes is based on the presence of a limited number of inscriptions found on millstones in Germany and this theory has not been tested using a larger dataset with consideration of chronological change. Nonetheless, the millstone does demonstrate that larger scale food processing was occurring at Chesters and that lava was utilised in this activity. Whether this was in association with the potential water mill at Chesters is also a matter for consideration, though the absence of lava in the millstone assemblage at Haltwhistle Burn dating to the third

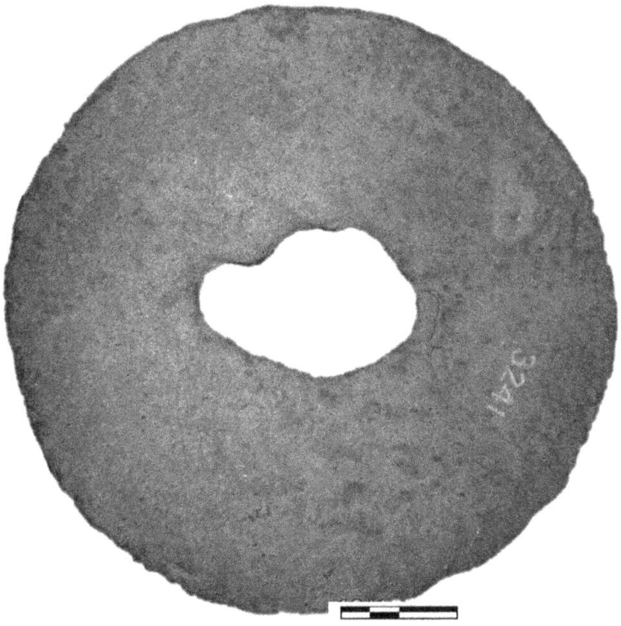

Figure 8.17. Complete upper stone that has been weathered and heavily worn around the central hole. A small part of the rynd chase remains on the right side of the central hole. No handle fitting exists on the stone (CAT 2177) (photo by author, ©English Heritage).

century indicates that watermill technology at Hadrian's Wall may have been introduced at a time when lava imports decreased in Roman Britain.

8.10.2. Upper Stones

The two upper stones were similarly less fragmented, comprising one half stone and one complete example, both of which were weathered. The complete stone (CAT 2175) is heavily abraded from weathering, and it is not possible to identify any surface features, such as dressing or a kerb, except for a small trace of vertical tooling on the sides. The central hole is very heavily worn, showing the probable reason for deposition, but a small amount of a rectangular rynd chase remains (figure 8.17). Although the standardised features are minimal, those present suggest that this was manufactured to the same specifications as other Mayen quern stones. There is, however, no handle slot and can be compared with CAT 2179 from Corbridge where it is likely that an iron band handle fitting was employed. As another example of the use of such a fitting, evidence indicates that this type may have been more prevalent than first thought, at least in the context of military sites.

The second example (CAT 2177) has retained more of the characteristic features of a standardised upper stone, with a well-pronounced kerb, part of the rynd chase and dressing on the upper and grinding surfaces. The dressing on the grinding surface appears to be harped and has worn almost through to one of the holes for the mechanism fitting. This stone appears to have been modified, distinguishing it from the other upper stone and this has been examined in more detail in the section on reuse/modification.

Figure 8.18. Small fragment of unidentified lava stone that has been pecked on the grinding surface (CAT 2178) (photo by author, (photo by author, ©English Heritage).

Table 8.3. Left, Diameters of the lava milling tools in the assemblage with their associate catalogue numbers. Right, values of the upper and lower quartiles for lava quern diameter at Chesters and for the complete British dataset

Cat No	Diameter (mm)
2172	360
2173	390
2174	400
2175	400
2176	400
2177	600

	Lower quartile (25%) Upper quartile (75%)	
Chesters	375	400
British data	400	425

8.10.3. Unidentified Stone

The unidentified stone is the smallest lava fragment in the Chesters assemblage and does not retain any features that allow for it to be determined as an upper or lower stone (CAT 2178). Some faint dressing marks on the non-grinding surface could suggest that it is an upper, but these are not clear enough to confirm that this is the case. What is interesting about this stone is the dressing on the grinding surface itself, which has been clearly pecked (figure 8.18). The wear of the stone suggests that this is not associated with the manufacture process, and it may have been redressed in this style. Its presence in the assemblage suggests that this may have been considered as an unusual example at the time of excavation or may have been the best sample of this type of dressing style.

8.10.4. Stone Diameters

Diameters were measurable for six out of the seven stones in the assemblage, and the five querns show a smaller range to the standardised range of diameters when compared with the whole British dataset (table 8.3). However, the dataset is small and more susceptible to distortions as the result of unusually large or small stones. The one example measuring 360mm is likely to have been the cause of this and as a single exception of an unusually small stone, cannot be considered as representative of the dataset. Overall, the diameters agree with the standardised sizes for lava querns. Similarly, the lava millstone is of a diameter that occurs regularly elsewhere and is not unusually large or small.

8.11. Chesters: Primary Use

As has been noted in the discussion on the dataset, the lava milling tools from Chesters are likely to have been retained at the time of excavation because they were well-preserved and more complete. A comparison of the thickness of the lava querns with the rest of the British dataset is, therefore, not be representative in terms of showing the range of thicknesses present and the degree of wear that was normal at the site prior to deposition (table 8.4). It does, however, show the probable upper range of thicknesses that are likely to have been present. The results show that there is one significantly thick upper stone, with the second being within the normal range for thickness as that of 50% of the British lava data. Similarly, the lower stones were all within this normal range. The outcome is unexpected but could signify that at Chesters the querns experienced a greater degree of wear prior to deposition when compared to other military sites in Britain. The reasons for this are not clear, but it is possible that Chesters was continually occupied, as shown by dated inscriptions during the period when troops were stationed at the Antonine Wall. Continuous occupation would enable prolonged use of milling tools and more extensive wear prior to deposition. Another possibility is a lack of or changing supply of lava to the site, possibly relating to the function or chronology of the fort.

8.12. Chesters: Reuse/Modification

Reuse was not visible on any of the stones in the assemblage, due to the extent of surface weathering on most of the milling tools, but modification was present on one. This is the previously mentioned upper stone, CAT 2177. This has

Table 8.4. Upper and lower stone thickness for the querns from Chesters (left), and the upper and lower quartiles for the lava querns from the whole British dataset (right)

Cat No Upper stone thickness (mm)	
2175	100
2177	50
Lower stone thickness (mm)	
2172	50
2173	45
2174	40

	Lower quartile (25%)	Upper Quartile (75%)
British dataset upper	48	73
British dataset upper	31	55

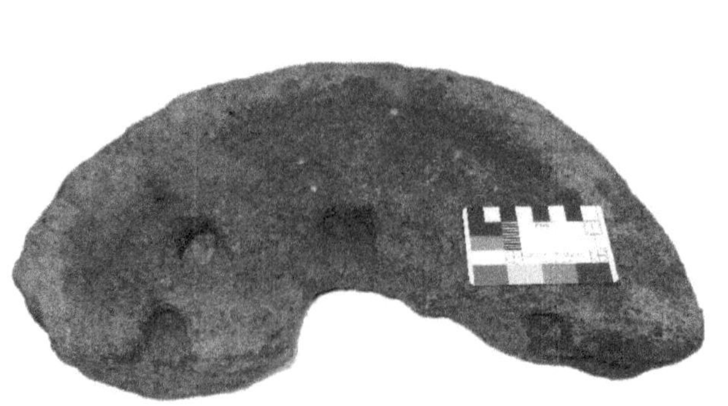

Figure 8.19. Upper quern stone from Chesters (CAT 2177) that shows evidence for modification in the form of holes that would have accommodated iron fittings not associated with manufacture (photo by author, ©English Heritage).

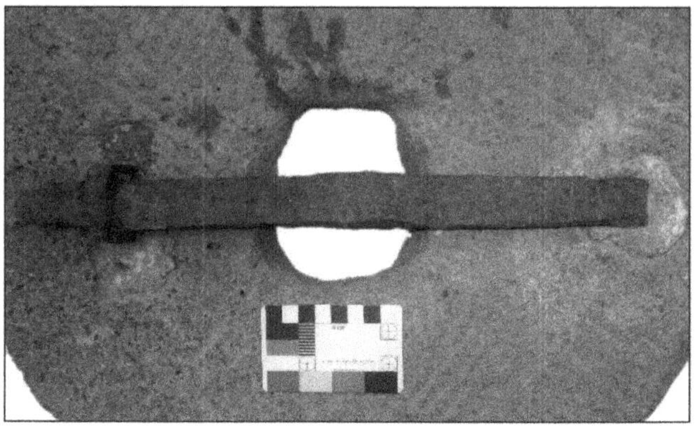

Figure 8.20. Upper quern stone from Housesteads (CAT 2284) that retains the iron fittings and provides a parallel for the incomplete stone from Chesters above (photo by author, ©English Heritage).

many of the standardised features of a Mayen lava upper stone, with the addition of two parallel holes adjacent to the central hole (figure 8.19). These are located at a right angle to the remains of the rynd chase a few centimetres away from the central hole. A more complete parallel to this quern exists at Housesteads that retains the iron fittings that match the holes in the Chesters upper stone. In this example, an iron bar braces the central hole and is fixed into one side of the upper surface of the quern via an iron loop fitted into two holes using lead. On the other side, the iron bar is fed into the upper surface and fixed with lead (figure 8.20). Part of the kerb has been removed, indicating that this addition was an adjustment to the original design and not related to the manufacture process. The function for this bar is not clear, but interpretations could relate to its use as a handle to lift the stone, or as a tentering mechanism to adjust the gap between the upper and lower stones. This has been discussed in more detail for the quern from Housesteads.

8.13. Housesteads: Introduction and Background

Housesteads Roman fort, also referred to as *Vercovicium*, is in modern day Northumberland and located in the central region of Hadrian's Wall on the Whin Sill escarpment

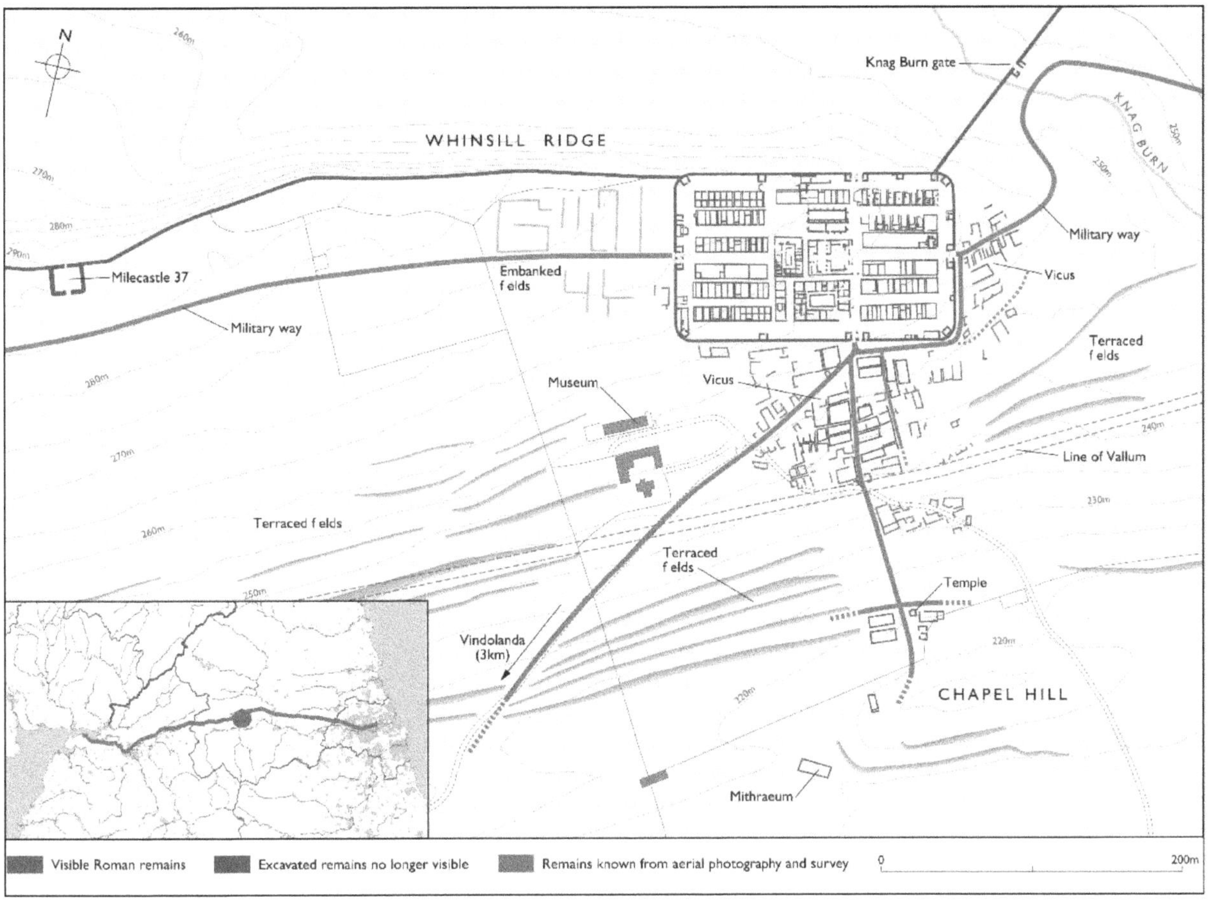

Figure 8.21. Location map of Housesteads Roman fort (plan of Housesteads fort © English Heritage, location map Ordnance Survey OpenData).

(figure 8.21). The site has produced some of the best-preserved Roman military archaeology along Hadrian's Wall, much of which has been excavated using modern excavation methodologies to yield detailed insight into the lives of those who were stationed at the fort and who lived in the associated *vicus* settlement (Birley, 1961, pp. 178–184; Rushworth, 2009, p. 1). Preliminary investigations focused on the curtain wall and Hadrian's Wall, with some work also done to uncover the granary structures. At the turn of the century, excavations were carried out on the *principia*, while a fuller plan of the structures on the site was also 'traced', with little consideration of the chronology of the buildings. Further explorations followed with several small-scale targeted excavation to examine specific areas of the fort interior. As excavation methodology improved during the mid-to-late twentieth century, greater focus was given to identifying chronological development of structural evidence, and to investigate the *vicus* settlement (Rushworth, 2009, p. 1–8).

The fort was built within a decade of the initial construction stages of the wall, after AD 122 (Crow, 1995, p. 17–21). Unlike Chesters, the fort lies parallel to the wall and does not project beyond it into the north. It was home to a large regiment of infantry soldiers, comprising auxiliary troops and numbering approximately 800–1000 men (Crow, 1995, p. 27). The first garrison may have consisted of the first cohort of Tungrians, as recorded by inscriptions dating to around AD 200, but this is not certain (Crow, 1995, p. 57–59). They were a cohort initially from Belgic Gaul (Birley, et al., 2013, p. 291–293). On the movement of troops to the Antonine Wall, continued occupation at Housesteads is evident, though probably by a caretaker garrison. The full complement of the first cohort of Tungrians returned to Housesteads by the early third century, presenting an unusual continuity of the cohort's occupation at the fort (Crow, 1995, p. 59; Rushworth, 2009, p. xiii).

Most buildings of the fort were initially constructed in stone within the stone curtain wall defences. The layout of the central range included the *praetorium*, *principia*, and the granary, alongside a storage building of unknown function. Further to this was a hospital and possible baths for the commandant, while to the east and west of the central range lay the barracks (Crow, 1995, p. 44; Rushworth, 2009, p. 271). A programme of rebuilding and refurbishment occurred on the return of the full garrison after the return to Hadrian's Wall around AD 162, with more extensive changes occurring around AD 300, when the barrack blocks were converted into individual 'chalets', reflecting a change in the way that soldiers were accommodated within the fort complex (Rushworth, 2009, p. 119–127; 134). Though suggestions have been made that this was to allow soldiers to live with spouses and children, finds evidence does not prove the presence of women or children in these areas of the fort and difficulties remain in interpreting the evidence this way (Hodgson, 2014). The final identified phase involved modifications to the 'chalets,' ramparts, and other fort structures, alongside road resurfacing that occurred around the fourth to early fifth centuries (Rushworth, 2009, p. 136–177).

In terms of structural evidence relating to the preparation and consumption of grain, the granaries and bakehouses have provided interesting details. The west gate, known as the *porta decumana* was the main vehicular access to the large granaries for the fort (Crow, 1995, p. 34–35). These were within the central range of the fort. The earliest granary built during the Hadrianic period consisted of a single structure with two aisles, a raised floor, and stone piers that supported a double span broad roof with eaves, reinforced with buttresses. The entire structure was designed to keep the grain within cool, dry and vermin free (Crow, 1995, p. 51), and would have been one of the more important buildings within the complex due to grain being a staple part of the military diet and representing the bulk of the rationed allowance of foodstuffs provided by the state (Groenman-van Waateringe, 1997). By the Severan period, adjustments had been made to the granary with the addition of two internal walls to replace the pillars, effectively splitting the building in two. The changes reduced the storage space significantly and could signify a contemporaneous change in the garrison size. It has been suggested by Crow (1995, p. 52) that the structural changes may have occurred in two phases, with the addition of the walls as separate occurrences that reflect a decrease and subsequent increase in garrison size. The first wall being designed to decrease the size of the granary initially, and a second added to enable use of the other half of the building later with the return of the fort to full force (Rushworth, 2009, p. 271).

Bakehouses have been uncovered in the east (figure 8.22) and west ramparts and in the angle towers of the fort, which were in use in the early occupation phases of the fort (Rushworth, 2009, p. 38–39; 42). The way that these have been spread out and the frequency at which they occur seem to suggest that use of ovens was divided between the military population and individual ovens may have been allotted for the use of set groups of soldiers. This correlates with the evidence that we have for quern use, in that inscribed querns from military sites indicate that food preparation was undertaken as a communal activity within units of soldiers (see section on Vindolanda, this chapter). The oven in the commandant's kitchen is of a similar style to that of the commercial ovens in bakeries at Pompeii (Crow, 1995, p. 36–37). The provision of specialist equipment of this type possibly indicates that a professional baker was employed to prepare delicacies for the military elite and their guests when stationed at the fort.

8.14. Housesteads: The Dataset

The dataset from Housesteads is complicated by the long chronology of excavations that have taken place at the site. It has resulted in inconsistent data standards due to varying

Figure 8.22. Oven at Housesteads in the bakehouse in the eastern rampart area of the primary fort. The layout and structure changes considerably over subsequent phases (Rushworth, 2009, p. 39).

levels of detail in the recording process, and different finds retention and reporting policies across the multiple projects undertaken over the years. It would appear the earlier excavations only retained the larger lava quern fragments, or those that had peculiar or distinctive diagnostic features. Disappointingly, these probably represent only a small number of those uncovered from the site, as comments made by Bosanquet (1904, p. 285) refer to a large number of querns excavated during the 1898 excavation season. It is probable that some of these have been retained and archived at the stores in Helmsley, which was not accessible at the time of data collection due to Covid restrictions. As these are not identified by specific stone types in the archive database, they could not be added to the dataset at this time. The lava milling tools recorded from these early excavations lack any context information or findspot data and may only be examined as objects known to have been recovered from the fort site, and highly likely to have been retrieved from Roman contexts. These comprise 8 recorded lava milling tools.

The bulk of the dataset are of lava objects recovered during the more recent archaeological investigations at Housesteads, most being found in the north-east area of the site where detailed recording was undertaken and comprise 17 objects. These have location data relating to the findspot of each quern, though some of these are more general areas of the excavation zone and do not relate to specific contexts or archaeological features. There were 10 querns that had associated context data, 6 with general location information, and one unprovenanced object. Many of the querns from this data group were heavily fragmented and non-diagnostic showing that most material must have been retained, while also providing a picture of the volumes that may have been discarded and not recorded in earlier excavations. They also allow for a general view of lava quern distribution of a sample of the site area to be examined. Of these, 12 were recorded in person at the archive, while a further 5 were taken from the published excavation monograph for the site compiled by Welsby (2009), which provides detailed descriptions for the lava querns recorded in the catalogue. These were cross-referenced with the lava recorded in person to ensure that none were duplicated.

Overall, the data includes 6 lower stones, 7 upper stones and 12 unidentifiable groups of fragments, totalling 25 (table 8.5). Of these, 3 might be identified as millstones/large querns, 11 as querns, with a further 11 that could not be identified as either due to an inability to accurately measure or estimate the diameter of the stone.

8.15. Housesteads: Manufacture

The manufacture stage for querns and millstones from Housesteads was assessed through analysis of their functional features and dimensions. This included measuring stone diameter, observing kerbs and handle fittings, and surface dressing. The results of the analysis are presented below.

8.15.1. Diameter of Stone

Stone diameter could be measured or estimated for 13 stones in the dataset, 3 of which could be identified as either small millstones or large sized querns due to their diameters, with 10 clear examples of querns. The possible millstones did not retain any features that might be related to the method of their use. However, as the three milling tools all have a diameter of 500mm or more, they have been treated separately and assessed alongside the millstone data for Britain. The Housesteads quern diameters can be seen in figure 8.23, which also shows the diameters for all

Table 8.5. Breakdown of the data in terms of stone part and milling tool type

	Upper stone	Lower stone	Unidentified
Quern	5	5	1
Millstone	2	1	0
Unidentified	0	0	11

Figure 8.23. Chart showing the diameters for lava querns from Housesteads (T=10) compared with those from the whole of Roman Britain (T=553). The upper line indicates the upper quartile, the lower indicates the lower quartile for diameters in the British dataset.

Table 8.6. Upper and lower quartiles for quern diameter at Housesteads compared with the full British dataset

Quern Diameter (mm)
400
400
430
430
450
450
460
460
460
470

	British dataset	Housesteads
Lower quartile	400	423
Upper quartile	420	460

British lava querns with the upper and lower quartiles for this dataset.

As has been discussed in Chapter 5, there is a high level of standardisation in the size of lava querns imported into Roman Britain and this is visible in the proximity of the lines that show the upper and lower quartiles for the British dataset. The Housesteads data presents a clear deviation from this, being larger than the normal range. The extent of this deviation is shown in table 8.6 which examines the differences in the upper and lower quartiles for the two datasets. It shows that for the British dataset, 50% of querns are sized within the range of 400–420mm, while at Housesteads the figure is for 423–460mm. Some consideration needs to be given to the size of the dataset from Housesteads, as occasional large values are more likely to skew the results upwards. However, there are no unusually large diameters represented. Only two of the querns fit into the normal range for British quern diameters, with the majority larger than 420mm.

Though the sample size is small, the results need to be reconciled with the pattern of diameter size seen at other northern military sites. There are practical reasons as to why querns may have been larger at military sites in general, and at Housesteads specifically. As communal objects that are likely to have been shared within a unit of soldiers, they were required to process large volumes of grain. This differs from the assumed traditional role of the household quern, used for everyday domestic food preparation. A larger surface area would increase the rate of flour production, and the large garrison size at Housesteads in the earliest phase of occupation may have secured the largest querns for its troops. Whether the larger diameters observed here would have made a significant difference when compared with those 30–70mm smaller is a question worth addressing using experimental archaeology, yet the clear variability across site types suggests that there was a reason this occurred.

The two smaller millstones in the assemblage are only 30mm larger in diameter than the largest quern in the assemblage. This means that the clear differentiation between millstones and quern stones that exists within the wider British dataset is more difficult to identify at Housesteads. This is like the pattern seen at Silchester (Chapter 7), where it is possible that larger lava querns, or smaller animal powered mills were in use to enable large scale processing without the use of watermills. This research has defined lava millstones as predominantly being 500mm in diameter or larger (see Chapter 5). Therefore, the 'millstones' at Housesteads are some of the smallest in the whole British dataset, being 500mm, 500mm and 510 mm. Considering the relatively restricted space inside the fort at Housesteads due to topography, and the difficulties associated with water sources at the fort, small human or animal powered millstones may have provided a solution to the problem generated by a large population and limited centralised processing capabilities. One would imagine that the military community had privileged access to forced labour or beasts of burden if these were necessary to keep the garrison functioning smoothly. However, as there is no current evidence to suggest that these were operated as millstones, either on the stones themselves or in terms of structures associated with mills, an interpretation that these were larger sized querns should be considered most likely. In depth analysis of lava milling tools at other northern military sites needs to be carried out to see if similarly larger sized stones present handle fittings, as these would prove that these were hand operated. This would suggest a different standard of quern existed at northern military sites when compared to southern (non-military) sites.

8.15.2. Kerbs and Handle Fittings

Kerbs and handle fittings only occurred once each across the assemblage and as these features occur together on one example, it is best to discuss them together. Both a kerb and handle fitting are present on CAT 2284, which is the best-preserved upper stone recovered from Housesteads. The kerb is wide at 70mm and is also very shallow standing only approximately 1mm proud of the upper surface. Although the stone has been weathered, the upper surface dressing is still clearly visible meaning that the shallow kerb is unlikely to be the result of post depositional damage and was part of the original design of the stone. There is no elbow handle socket, and so the combination of a shallow and wide kerb with the absence of the elbow handle appears to coincide with what has been observed previously, specifically that the kerb may not have been required on this style of quern due to the absence of the elbow handle fitting that required extra reinforcement of the edges (see Corbridge, this chapter). The kerb lacks the properties needed to make it a functional characteristic, and its continued addition may have been a reference to previous styles, or other quern types that were in contemporary use.

Figure 8.24. CAT 2284 showing the faint shallow and wide kerb on the upper surface with upper surface dressing (left), and iron corrosion product where the iron band would have been fitted around the circumference of the stone. This can be seen as a wide band above the scale, culminating in a cluster of corrosion product where the handle itself is likely to have been located (right) (photo by author, ©English Heritage).

The handle fitting is instead formed from an iron band wrapped around the circumference of the stone near to the upper edge (figure 8.24), which is still clearly visible as corrosion product. The handle itself is not distinct, but a concentrated area of corrosion product could indicate where this may have been and analysis using x-ray would have the potential to further identify the feature. This example of an upper quern stone with an iron band handle fitting and no evidence for an elbow handle socket provides further evidence for a previously unidentified lava quern type that is archaeologically less visible than the types previously recognised. There can be no question that when CAT 2284 was finished, that the intention was to fit it with an iron band handle. The continued addition of a non-functional kerb and the upper surface dressing on CAT 2284 suggest that these querns were still produced at a centralised location, probably at Mayen. Therefore, questions remain as to whether this type of quern can be identified at the quern finishing sites that have been examined in Germany and whether a chronological development can be seen to occur. This could have implications for how we date lava querns from Romano-British contexts and how we understand the lava quern manufacturing process; as a continuous centralised practice, or one that became devolved over time, allowing for alternate types to be produced.

8.15.3. Dressing

Surface dressing on the grinding surface was present on 8 stones: 4 upper, 3 lower and 1 unidentified. Of these, all were of the harped style, except for one upper millstone, which was pecked. There is, therefore, some relative consistency in the dressing style for the grinding surfaces of the lava milling tools from Housesteads, perhaps showing that a prominent level of stone working skill was present at the fort and that there was a uniform understanding of how querns should be dressed. The pecked millstone is an exception, which could relate to its method of operation as both other identified millstones had no visible dressing present despite being well-preserved. However, pecking is much harder to identify on lava due to its vesicular nature, especially on an assemblage that has experienced any surface weathering, as is apparent at Housesteads.

Vertical striae on the quern sides were present on 7 stones, including one upper stone that has been identified as a small millstone. These tooling marks are typical of lava querns manufactured in the Mayen region and could suggest that this example was also produced there, as opposed to the separate millstone finishing area at Andernach (see Chapter 2). The other upper millstone lacks these striae and has well-shaped but rough sides, markedly different to the other example, but has upper surface dressing; another feature of a milling tool finished at Mayen. The reason different combinations of features are visible on each stone might relate to post-depositional effects of weathering or soil conditions, or the way that they were operated or modified during their use lives. The upper surface dressing is visible on 4 of the upper stones in the assemblage in total.

8.16. Housesteads: Distribution

The distribution of lava milling tools at Housesteads could not be established for any objects that had been excavated prior to the 1974 excavations due to an absence of location data. For the 1974–81 excavations in the north-east area of the fort, site zones had been allocated for 12 lava milling tools, while a further 3 had general zones that still allowed for some form of mapping to be carried out. Those without specific zones have been marked differently on the plan. Therefore, 15 lava milling tools have been mapped, which constitutes all objects in the dataset that were recorded for Housesteads and from the 1974–81 excavation years (Crow, 1995). This presents a good sample of data for investigation within a focal area of the site. It should be noted that each object mapped presents the central point within the excavation zone from which the milling tools were recovered and does not relate directly to findspot data. The results allow for a general view of distribution to be visualised (figure 8.25). Locations of the bakehouses, ovens and granaries have also been added to this plan, as these are structures and archaeological features that relate directly to the preparation and storage of grain and may help with interpretations relating to the distribution of milling tools.

As has been previously discussed, the bakehouses and ovens for the fort were in the area of the ramparts and

Figure 8.25. Phased plan of the north-east area of Housesteads Roman fort showing the placement of bakehouses, ovens, and granaries. This shows the general locations where lava milling tools and non-lava milling tools were recovered (plan after ©English Heritage).

close to the curtain wall. It is assumed that the strategy was to avoid the possibility of setting fire to buildings within the fort by minimising burning in built up areas and is a common arrangement seen at other forts across the empire (Hajnalová & Rajtár, 2009, p. 204; Hatherley, 2020, pp. 29–30). From evidence elsewhere at Housesteads, the pattern seen in the north-east corner of the fort with bakehouses and ovens on either side of the angle tower is repeated at the other corners of the curtain wall, showing that the layout was designed and probably related to policy. Though the precise arrangement and number of ovens and bakehouses did not remain static throughout the period of the fort's occupation, each renewed layout was relatively similar to the one that preceded it. The area of the ramparts must have been an area where food preparation occurred daily, and the role of this area continued throughout the life of the fort.

Using this information, the distribution of lava milling tools at the site suggest a correlation with these food preparation areas. Care must be taken when interpreting findspot information, as milling tools may simply have been deposited in this area as it provided a convenient location at the outskirts of the occupation area. Secondary deposition is also highly probable. However, the distribution analysis suggests a strong association between quern deposition and the locations where grain products would have been cooked. Of course, querns need not have been used so close to ovens as flour can easily be moved. Nonetheless, the possibility that food preparation, including grain processing, took place in the area of the ramparts should be considered. This possibility has been noted at other fort sites in Britain, for example at Doune, Scotland, where high numbers of amphorae and mortaria were found in the rampart areas compared with other parts of the site, which also indicate that this was an area for food processing (Hatherley, 2020, p. 19–20). The fort at Housesteads had restricted occupation space due to the topography of the site, despite having a large population of troops when fully garrisoned. It is likely that grain was being ground almost continuously to accommodate the needs of the troops that probably worked in shifts and relied on grain as a large part of their diet. Everyday domestic activities, such as grain processing, may have required designated areas to maintain discipline and manage the limited space available. Being away from the main routes through the fort and the central range, the ramparts may have been the best place for such activities and conveniently close to the ovens and domestic areas to similarly reduce traffic and movement across the fort.

As a further way to explore the deposition observed for lava, the 19 published non-lava milling tools were also plotted on the same plan to see if a similar pattern existed. Interestingly, the non-lava show a very different distribution, being mainly clustered in and around barrack 13 where very few lava querns were recovered. Several of these were found consolidated into floor surfaces, one was reused in a wall, and one in a threshold. Very few milling tools of any type were recovered from barrack 14, and this absence provides a second feature of the data that requires some thought. Initially the matter of chronological change in quern types should be considered. If lava is assumed to be the earlier choice of milling tool, the non-lava querns are indicative of querns that were in sole use during the later periods of occupation, but may also have been used simulataneously alongside the lava in earlier periods. Table 8.7 shows the chronological distribution of the lava and non-lava querns. To visualise the data better, the chronology of deposition for different stone types is presented in figure 8.26. This suggests an earlier start date for the depoistion of lava, but a peak in all querns being deposited in phases III and IV, these being the chalet phases of the fort. The evidence suggests that this may have been the result of redeposition, causing the mass movement of broken querns to suitable reuse or disposal locations at a time when major refurbishment works were being undertaken. This helps to explain the reuse of non-lava querns within structures and floors in barrack 13, but does not help interpretations relating to the different way that lava and non-lava milling tools were treated in terms of the location of their deposition.

Table 8.7. Chronological distribution of milling tools of lava and non-lava types

Site phase	No. lava milling tools	No. non-lava milling tools
I	0	0
II	1	0
III	3	6
IV	4	6
Post-Roman	1	5
Undated	1	5

Possible explanations for the difference in the distribution between the lava and non-lava milling tools vary. For example, at the time when the barracks were remodelled into the chalet format all rubbish from the inside of the buildings, including broken and discarded lava querns, could have been removed and dumped onto the nearby rampart area. If lava was not considered as a suitable building material due to its physical properties, it may not have been reintroduced into the areas of habitation as recycled stone. This may also reflect different or changing deposition practices and meaning attached to different stone types, or to milling tools in general if indigenous stone types overtook the use of lava in later periods. It is also possible that the chalet phase of the barracks is indicative of the introduction of greater privacy and separated living conditions for soldiers to that provided by the preceding barracks. Previously communal activities, such as grain processing and food preparation, may also have become more of an indiviudal affair, occurring in the privacy of individual chalets instead of the rampart areas.

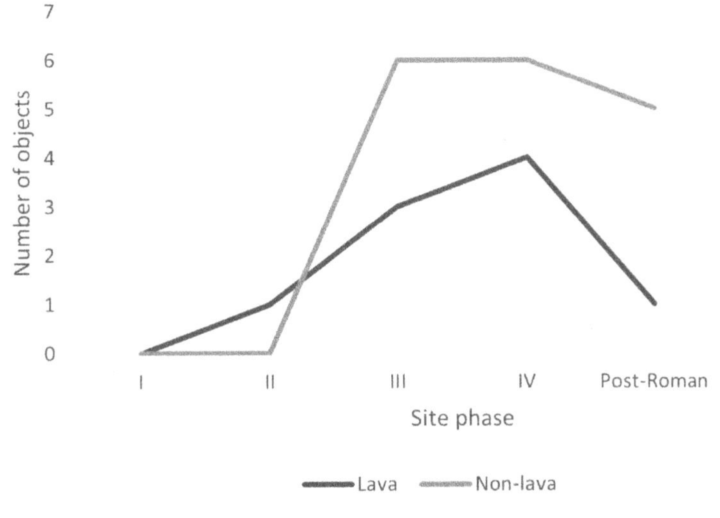

Figure 8.26. Chart showing the chronological change in lava and non-lava milling tool deposition in the north-east area of Housesteads fort (T=26).

This would result in different use locations for the lava and non-lava milling tools and possibly result in different deposition sites as a result. However, none of these possible interpretations explains the prevalence of milling tools deposited in and around barrack 13 and their absence at barrack 14.

8.17. Housesteads: Primary Use

As with the other sites in this study, primary use has been explored by looking at upper and lower stone thickness to measure extent of wear. The results are detailed below.

8.17.1. Upper Stone Thickness

The thickness of only 4 upper stones could be measured, providing a small sample for analysis (figure 8.27). Two of these were from the earlier excavations and were probably selected for retention due to their larger size. The other two were from the 1974–81 excavations, and although this might make them more representative of quern wear at the fort, they were the only upper stones that had survived as more than tiny fragments. It is likely that the most heavily worn quern stones deposited at the site have not survived, leaving only the larger and thicker examples that may not be characteristic of typical wear. Overall, the analysis does not produce any conclusive results because of the small sample size, but it is interesting to note that two of the querns are thicker than that of 50% of the upper lava querns from the whole British dataset. As the other two examples fall in with the normal range for quern thickness, it is not possible to state that the assemblage is particularly unusual. However, there has been a continuing pattern of thicker lava querns recovered from military sites, and this is also visible in the data for the upper stones from Housesteads.

8.17.2. Lower Stone Thickness

There were 4 lower stones from the Housesteads assemblage that had measurable thicknesses, one of which was from the 1974–81 excavations and 3 from the earlier projects. A similar issue to that with the upper stones is, therefore, also present in this dataset. The results of the analysis can be viewed in figure 8.28. The outcome of the analysis presents similar issues and results to that seen with the upper stones, with two of the stones being thicker than 50% of the lower lava stones in the British dataset, and two that are within the 'normal' range for thickness. Similar conclusions can, therefore, be drawn and there is some indication that Housesteads follows the same pattern for stone thickness as seen at other military sites.

8.18. Housesteads: Reuse/Modification

Reuse of lava querns, other than that associated with deposition, can only be speculated on for one example, CAT 2269, and this interpretation has been drawn from its context. This was a complete upper stone that was found next to a tipped-over stone mortar and may have been used as a lid (Welsby, 2009, p. 356; p. 355, fig 12.9D) (figure 8.29). This stone was not available for examination at the time of data collection and was included in the dataset using information taken from the site monograph. Unfortunately, no description of the object has been published as it was not seen by the author, Welsby (2009), at the time that the quern catalogue was complied. This means that if any modifications had been made to the stone to facilitate its reuse, they have not been identified. Such observations might support the evidence for reuse; large mortars often had a wooden lid with a hole, which would secure the position of the large wooden pestle. Reuse of a quern in this format would work well as a replacement lid and serve a similar purpose (Shaffrey, pers. comm.).

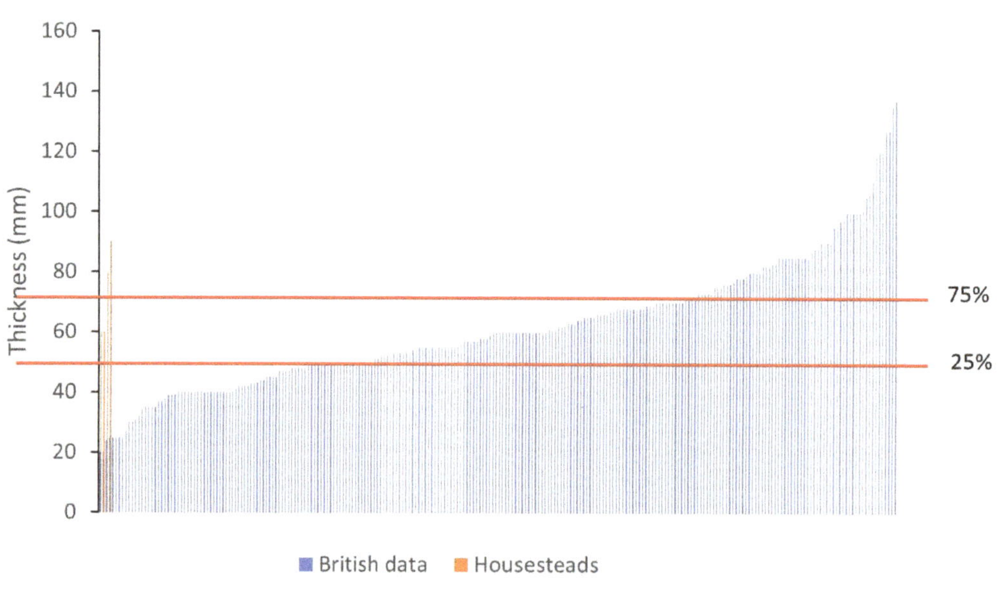

Figure 8.27. Chart showing the thickness for upper lava querns from Housesteads compared with those from the whole of the British dataset excluding those from the site at 1 Poultry. The 25th and 75th percentiles for the British data are marked with red lines.

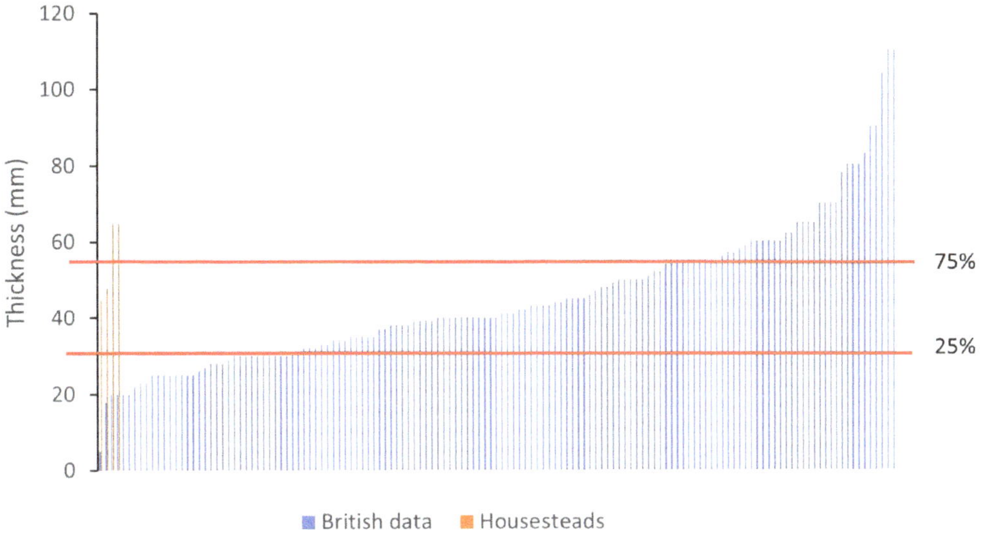

Figure 8.28. Chart showing the thickness for lower lava quern stones from Housesteads compared with those from the whole British dataset excluding those recovered from the site at 1 Poultry. The 25th and 75th percentiles are shown with the red lines.

Figure 8.29. Site photo of mortar, with a glimpse of the upper lava quern CAT 2269 that has been interpreted as reused as a possible lid (Rushworth, 2009, p. 355).

In terms of querns that had been modified, one lower stone (CAT 2288) may have been altered at the sides and lower surface in a comparable way to that of lower stones seen at Corbridge (this chapter) and Verulamium (Chapter 7). The edges on this stone were rough and irregular with no sign of vertical striae, showing that they may have been reworked. However, this is less convincing than some of the other examples identified. There were two upper stones that may have been modified from the original manufactured specifications, CAT 2284 and 2267. The first of these is the best-preserved upper stone in the assemblage and has been discussed in the previous section on manufacture. It retains an iron bar that has been fitted into the upper surface bracing the central hole; one side of the bar has been bedded into the stone, while the other side is fixed into position using an iron loop that is fitted into the surface with lead (figure 8.30). Addition of the bar has interfered with the kerb, indicating that this is a later addition to the

Figure 8.30. Upper stone CAT 2284 that has an iron bar fitted into the upper surface as a possible modification of the original manufacture design. The upper image shows the widened central hole that may have removed the original rynd chase. The end of the iron bar that is fully bedded into the stone has interfered with the kerb. The opposite end is held in place with an iron loop that allows for the bar to be adjusted and may have altered the angle of the upper grinding surface. The iron loop is fixed into the stone via two holes in the surface which have then been filled with lead (photo by author, ©English Heritage).

stone and occurred post-manufacture. The central hole itself has been widened and this has not occurred through wear suggesting that it was also adjusted deliberately. This process has removed the rynd chase, suggesting that the iron bar was added to replace the rynd component of the mechanism. The bar must, therefore, have functioned as the rynd and provided the pivot for the spindle. As the bar is moveable on one side within the iron loop, this may have been designed to alter the angle of the grinding surface as a method of tentering the stone. Although this provides one possible explanation, experimental work and further examination of the upper stone may produce alternate interpretations. After seeing a complete upper stone with the iron bar fitting *in-situ*, it makes it easier to identify other stones that may have been similarly modified. One example includes CAT 2267, a small fragment of upper stone that has a 'shallow oval depression' in the upper surface that may have accommodated one side of an iron loop like that seen on CAT 2284.

8.19. Housesteads: Deposition

Context of deposition was recorded for 7 of the lava querns recorded from the 1974–81 excavations in the north-east area of the fort. Of these 3 were recovered from rampart deposits and had been incorporated into the feature. A further 2 were in deposits of rubble, while the last 2 were incorporated into road surfaces near the eastern rampart. The querns included in rampart deposits may have been redeposited and discarded at the outskirts of the occupation area or relate to activity areas where food processing took place. The number of lava querns recovered from the area of the ramparts suggests that they had been deposited in this part of the site deliberately, but any meaning is difficult to extrapolate from such evidence. The querns were not subject to any specific treatment that would suggest ritual practice. Lava querns have been frequently associated with deposition in road surfaces at military sites and the two occasions of this at Housesteads correlates with the wider dataset of Roman Britain.

8.20. Vindolanda: Introduction and Background

The fort of Vindolanda is one of the most famous Roman forts in the country, with some of the best preservation conditions known from within the empire (Bidwell, 1983; Birley, 2009). The fort is, in fact, probably nine different forts, each constructed in the same location with slightly different layouts and to suit slightly different purposes. Associated with this is an extramural settlement that is thought to have had official *vicus* status from the third century AD (Birley, 2009, p. 153; Birley, 2016, pp. 225–226). Initial construction pre-dates Hadrian's Wall and is estimated to have occurred *c.*AD 85. The site is slightly south of the later Hadrianic frontier, though it probably formed part of the first defended frontier in the region as delineated by the Stanegate (Breeze & Dobson, 1987; Edwards & Breeze, 2000; Historic England, 2019).

The location of Vindolanda highlights why the site had continued significance throughout the Roman period, even after the movement of the frontier to Hadrian's Wall; situated almost halfway between Newcastle and the North Sea to the east, and Carlisle and the Irish Sea to the West (figure 8.31) (Birley, 2009, p. 13), the fort was probably a strategic midpoint for the centralised distribution of people and goods. Food and supplies may have been transported more easily by sea, but these would then have to make

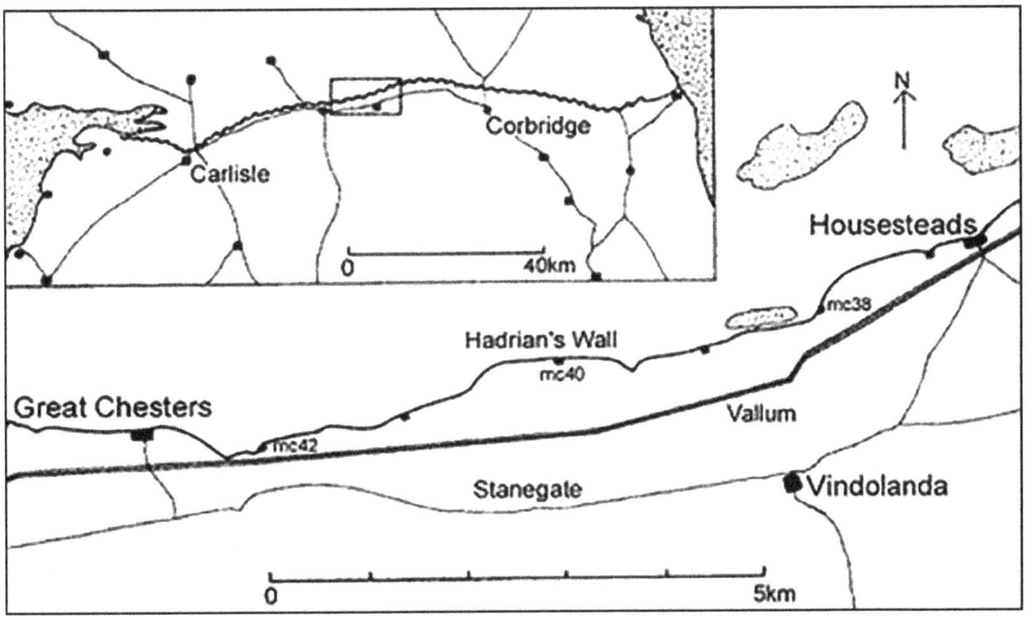

Figure 8.31. Map showing geographical location of Vindolanda and its relative position to other Roman period frontier fortifications (map ©Vindolanda Trust).

their way inland by river and road to ensure continuous supply to those stationed away from the main navigable waterways. Vindolanda was well-connected by water and road, and was centrally located to serve eastern, western, and northern supply routes. That the site was located away from the frontier in later periods would also have provided greater security; there would be less chance of supplies or back-up troops being compromised by incursions of 'barbarians.'

Although current interpretations identify the earliest phase of occupation at Vindolanda to date from *c.*AD 85, the possibility of an earlier Flavian fort has been speculated on from aerial photographs that show a possible fort structure to the northwest of the later forts (Birley, 2009, pp. 42–45). The first confirmed fort, dating from *c.*AD 85–92 was constructed in timber and was primarily occupied by troops from the *Cohors I Tungrorum* during this period (Birley, et al., 1993, p. 4–9). Demolition of the early fort and reconstruction work on a second timber fort occurred *c.*AD 92–100/103, which coincided with the backfilling of pre-existing ditches to allow for expansion of the fort's site (figure 8.32). This period probably saw continued occupation by the first cohort of Tungrians, who were followed by the ninth cohort of Batavians (Birley, 2009, p. 45–62), who were originally from regions in and around Belgic Gaul and Germania.

The third timber fort dates to *c.*AD 97–105 and was probably a reconstruction of the previous fort through refurbishment and modification. The end of the period saw the reassignment of the entire cohort to Dacia with the subsequent demolition and destruction of the site and its contents. Large volumes of equipment, administrative and personal writing tablets, tools, and other items were dumped or burnt when the occupants of the fort departed. There is, therefore, a wealth of finds relating to this period, with written evidence to further enhance interpretations of these finds (Birley, 2009, p. 63–90). The hurried departure of the cohort has preserved many finds *in-situ*, and it has, therefore, been possible to determine the probable functions of buildings and rooms using the objects recovered from them. For example, small room VIII was likely a kitchen; it had two small bread ovens and tablets discovered in a rubbish pit contained recipes and crockery inventories (Birley, 2009, p. 77). This provides a unique opportunity to potentially explore the socio-economic roles of objects within their context of use. However, not all finds were deposited in use locations, as many were intentionally gathered and relocated to be dumped or burnt (Birley, 2009, p. 72–73).

The *praetorium*, and possibly the entire site was deserted for a few months after the departure of the Batavians until the first cohort of Tungrians returned to Vindolanda later in AD 105. A new timber fort was constructed to a different plan to the last layout and may have been considerably larger. The fifth timber fort was likely to have still housed the first cohort of Tungrians with construction commencing after *c.*AD 120, though it is very probable that there were also legionaries and other wall builders stationed there. After AD 128, the Tungrian cohort were likely moved to the newly built Housesteads fort, and Vindolanda was empty for a few years (Birley, 2009, p. 64–91). A primary stone fort, stone fort 1, that saw the return of troops to Vindolanda can be dated to *c.*AD 130–213. A military bathhouse and the western temples have also been assigned to this period, along with an elaborately designed and ornate *principia* with similarly elaborate extramural stone buildings. Most of these major changes are thought to have occurred around AD 184. Nearer the end of the period, stone-built circular huts were constructed. These were clustered in groups of 10, and though their precise function remains unknown, they may have had domestic functions from the hearths discovered within them (Birley, 2009, p. 135–140). These have been interpreted by Bidwell (1985, pp. 28–31) as being self-built structures by forced labour, a suggestion that has been supported in a discussion of the evidence by Webster (2005, pp. 171–172).

Stone Fort 2, which forms the most visible remains at Vindolanda, must have taken several years to complete and was in use from *c.*AD 213–300. The change from a southern-facing headquarters, to one that faced northwards also indicates an alteration in the role of Vindolanda as a frontier outpost, instead of overseeing the Tyne Valley below. The end of this period may have seen a few years of disuse for the fort. Reconstruction works occurred around AD 300, when the size of the garrison may have declined, while the extramural settlement seems to have been abandoned by the fourth century. Reconstruction occurred again in the last quarter of the fourth century, with the granaries converted to living quarters after AD 400 (Birley, 2009, p. 141–168).

8.21. Vindolanda: The Dataset

There are 174 querns from Vindolanda, of which 59 are of lava. Of these, 48 were examined and recorded in-person, while the remaining 11 were added using publications; two from Collingwood & Wright's (1992) section on worked stone in the 'Roman Inscriptions of Britain', with the remainder extracted from the Bidwell (1983) excavation report. Of the 115 non-lava querns, 80 were recorded in person by Ruth Shaffrey, one was taken from Collingwood & Wright and 34 from Bidwell. It should, therefore, be noted that there is some disparity in the detail of recording, as the published accounts had a different focus from the research questions addressed here. It is also acknowledged that there are probably many querns that were unavailable during the recording stages; it is likely that these were stored elsewhere or had not been retained. However, the sample size is large enough to produce significant results. Of the 59 lava querns from Vindolanda, 31 are upper stones or probably upper stones, 24 are lower or probably lower stones and 4 are unidentified.

Figure 8.32. Plan of the excavated area of Vindolanda showing the locations of the different period forts. The period 1 fort is outlined in red, period 2 and 3 in green, and the probable locations of period 4 and 5 are shown in blue. The third century stone fort and associated buildings is outlined in black (after Birley, 2009, Plate 6).

There were some issues regarding the quality of the data available for some of the recorded querns; some were stored outside without context information and there were several in the stores that were no longer labelled. This highlights more general issues that are faced by researchers in terms of how larger finds are retained and archived as these are not uncommon problems relating to querns or other bulk finds, such as building material (Shaffrey, 2006,

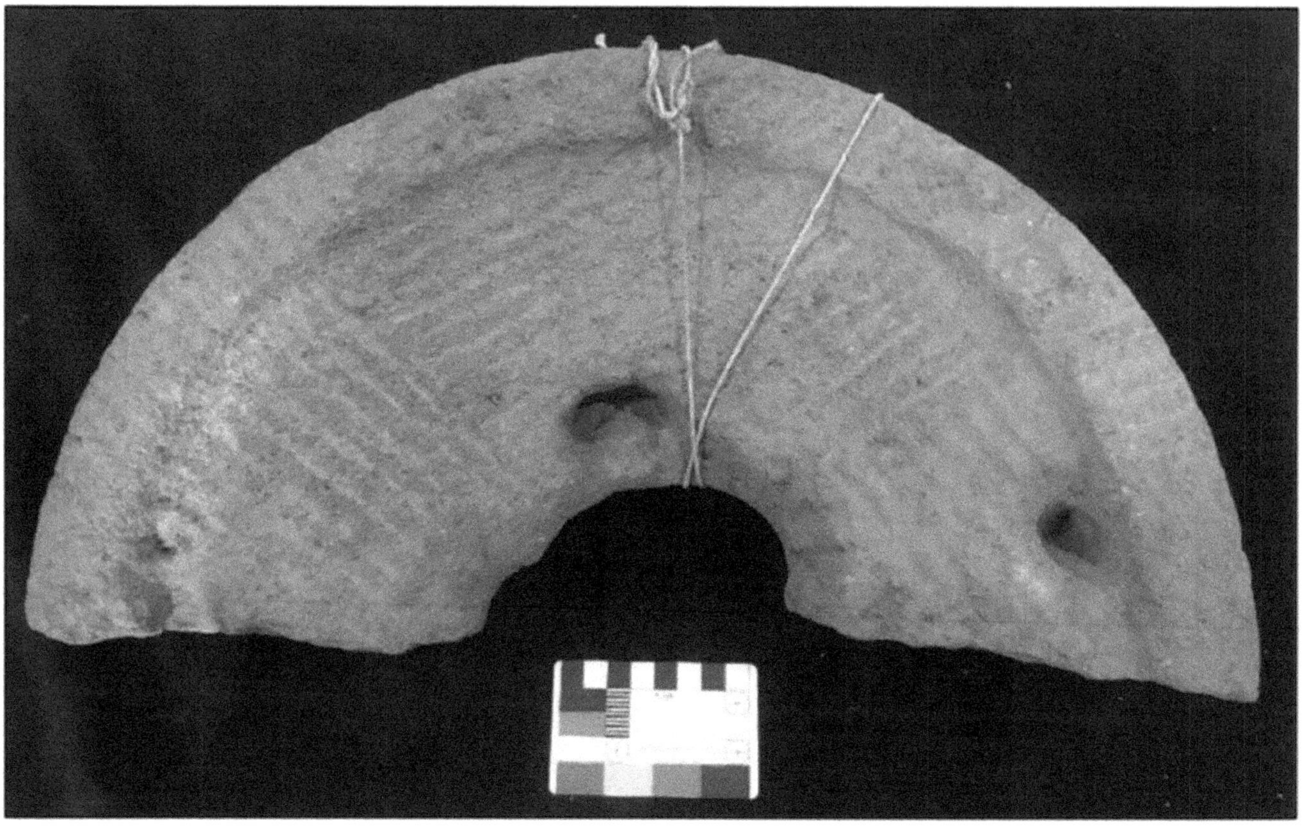

Figure 8.33. Upper lava quern stone showing annular raised kerb on the top surface (CAT 0014) (photo by author, ©Vindolanda Trust).

p. 3–4). However, these particular examples were not recovered from any other site and have been included on this basis. Although there are details that are impossible to extract from these objects due to the lack of findspot or data, they can be used to demonstrate presence or absence of particular quern types. The missing context information does also mean that there may be non-Roman querns within the dataset. As the extent of occupation at the site after the Roman period cannot be precisely determined, there is a possibility of later objects being present within the finds assemblages. Although many finds recovered from post-Roman contexts are likely to be residual, the absence of context information for some querns makes it difficult to assess each individual example to decide if this could be the case. Although this issue is not expected to be extensive at a predominantly Roman site such as this, it must be considered when interpreting the results.

As part of this study, the non-lava data has not been examined in great detail and its inclusion is to summarily provide a wider context for a more nunaced examination of lava querns at this site. As the main focus, there have been some aspects of data analysis that only involve lava; though if relevant and practicable, other querns have been considered and added to the dataset. As querns of all stone types are equally important when considering food production, stone supply and demand, socio-economic change and cultural practice, it is possible in a small-scale study such as this to widen the investigation slightly.

8.22. Vindolanda: Manufacture

As with the other case studies, manufacture has been explored through the examination of the function elements of the lava milling tools and the production of quantitative means of comparison. The results of the analysis are detailed below.

8.22.1. Kerbs

Of the 31 upper stone or probable upper stone examples, 24 had kerbs present in the form of an annular raised band around the top surface circumference (figure 8.33). Two upper stones without kerbs were recorded using information from 'The Roman Inscriptions of Britain.' Detailed descriptions of these objects were not provided, and so the presence of this feature cannot be ruled out. A further five were heavily weathered and degraded, making more subtle features difficult to identify with any certainty. Therefore, it is entirely possible that all querns in the dataset may once have had kerbs, though this cannot be proven. The observations indicate that a high level of standardisation existed across the lava quern assemblage at Vindolanda.

8.22.2. Diameter of Stone: Quern vs Millstone

To investigate whether lava querns could be distinguished from lava millstones at Vindolanda by examining diameters

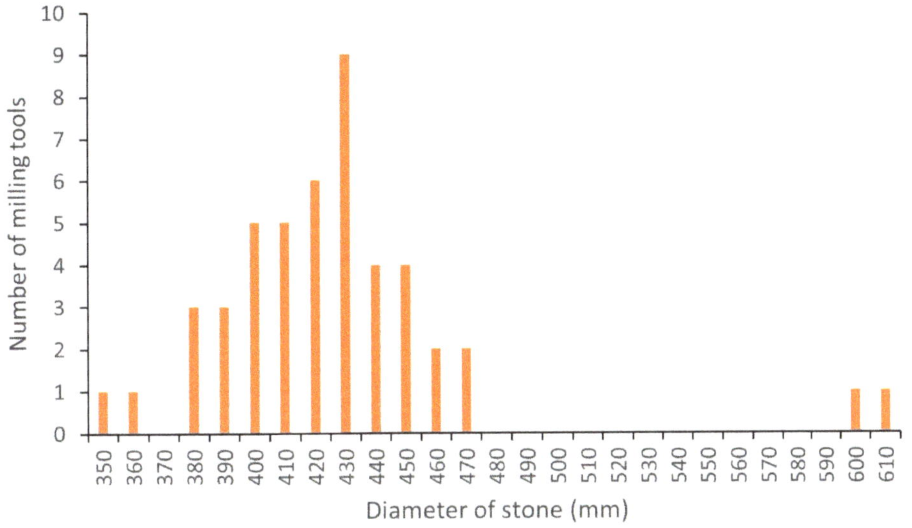

Figure 8.34. Distribution of all upper and lower stone diameters for Vindolanda lava milling tools, T=20.

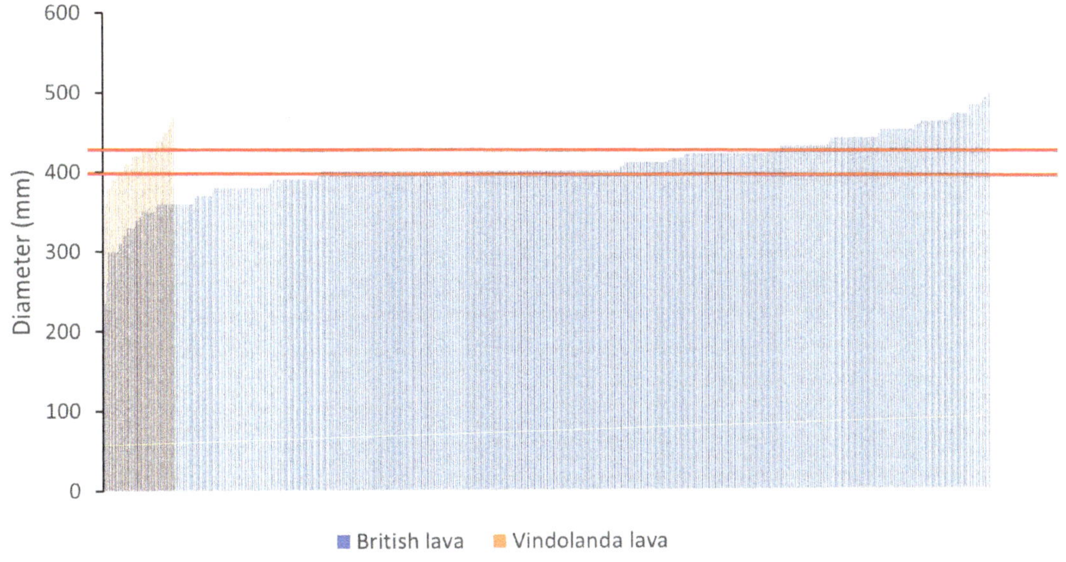

Figure 8.35. Chart showing the diameters of querns from Vindolanda (T=44) compared with those from the wider British dataset. The 25th and 75th percentiles have been indicated with red lines.

of the stones, the diameters for both upper and lower stones were plotted on a graph (figure 8.34). This included 20 stones from the assemblage for which this measurement was possible. The results show a clear differentiation between querns and millstones, with querns measuring between 380–470mm, and two millstones measuring 590–600mm. A separate group of slightly smaller querns exists, measuring 350–360mm, though both these diameters were estimated from fragments of querns, which may have produced inaccuracies. The most common diameter for a lava quern at Vindolanda is 430mm, and all other querns (excluding the two small anomalies) were within 50mm of this measurement. A high level of standardisation existed at the fort. The millstones are very clearly separate from the querns in terms of their size, which presents a different view to that seen at Housesteads where this division is more subtle. Perhaps the milling technology was different at Vindolanda, especially considering that it had a more suitable local water supply that could be exploited to power a water mill. This would have allowed for larger stones to be turned and less need for the smaller millstones that may have been driven using animal power. Chronology must also be considered, as Vindolanda was occupied before Housesteads and may have been initially using querns manufactured in earlier periods that were more highly standardised and smaller than their successors.

8.22.3. Quern Diameter

As has been identified from the distribution of quern diameters at Vindolanda, an elevated level of standardisation appears to have existed within the fort. However, it is worth exploring how well this compares with the standardised lava querns from elsewhere in the province. The diameters for the querns from Vindolanda have been plotted on a chart with those from the wider British dataset (figure 8.35), with the inclusion of lines to illustrate where the upper and lower quartiles for the lava

Table 8.8. Upper and lower quartiles for quern diameter for stones from Vindolanda compared with the wider British dataset

	Vindolanda	British dataset
25%	400	400
75%	430	420

querns from the whole British dataset exist. The upper and lower quartiles for the Vindolanda and British querns have also been presented in table 8.8. The results show that, in comparison, the Vindolanda querns are representative of those from elsewhere in Roman Britain. They do, however, have a wider range of sizes that are slightly larger than those from elsewhere in the province. The larger size has been found to occur also at Housesteads, though this difference is less exaggerated at Vindolanda.

8.22.4. Rynd

To identify whether there were any features of the Vindolanda querns that do not conform to the standardised types, the shapes of rynd chases were also examined to determine whether the rynd had been fitted to the upper stones in a consistent way. These had to be recorded carefully as wear of the stones can change how the rynd chase appears (figure 8.36). The most common form of rynd chase is that of a rectangular slot bedded into the upper surface of the upper quern stone that bridges the central eye perforation, which would have housed a flat rectangular bar to provide the pivot for rotation (see Chapter 2). From the 31 upper stones in the assemblage, all 11 that had evidence for a rynd chase were of the same form; that of a rectangular slot recessed into the top surface of the stone (CAT 0001, 0002, 0014, 0017, 0018, 0027, 0029, 0043, 0045, 0048 and 2317). It is possible that modification of this may have occurred on other examples to prolong the use-life of the stone, but this has been discussed in the section on reuse/modification.

8.22.5. Handle Fittings

Although the wider British dataset has shown that the most common handle fitting was the elbow type, other types of handle fitting may have been in use that are less archaeologically visible, such as the iron band type that has been observed on a quern from Housesteads (CAT 2284) and is likely to have been present on an example from Chesters (CAT 2175), and a further one from Corbridge (CAT 2179). The well-preserved assemblage from Vindolanda provides an opportunity to understand how prevalent these distinct types may have been across a large sample of lava querns from a military site. Overall, there were 15 upper querns with identifiable handle fittings. Of these, 13 were elbow types (CAT 0001, 0014, 0020, 0025, 0029, 0030, 0042, 0043, 0045, 0046, 0048, 2317 and 2319), with one clear iron band fitting (CAT 0032) and one probable iron band (CAT 0002). CAT 0032 presents a highly fragmented quern that has the remains of an iron band adhered to the outer circumference (figure 8.37). Fortunately, all parts of the upper stone had been preserved together, allowing for these to be reconstructed. However, it does show how less ideal preservation conditions would make this handle fitting type difficult to identify on fragmented lava. The other example, CAT 0002 is a complete upper stone that shows signs of wear but has no evidence for the presence of a handle (figure 8.38). Though an iron band cannot be proven, it is highly probable. This evidence suggests that elbow fittings were the most prevalent type, but that iron band fittings may have been more common than first thought and might be missed due to the difficulties involved in identifying them when querns are in fragmentary form.

The other observation relating to handle fittings involves the elbow type, which as described, were present on 13 upper stones. Of these, multiple elbow fittings were present on 6 stones (CAT 0001, 0029, 0030, 0032, 0043, 0048), with a further two where a second handle fitting had been started but not completed (CAT 0014, 0042). As these are later additions to the stones and do not relate to the manufacture process, they have been dealt with in the section on reuse/modification.

Figure 8.36. Three lava quern upper stones from Vindolanda showing differing levels of preservation and wear. Intact iron rynd *in-situ* within rynd chase (left, CAT 0002), intact rynd chase with iron rynd removed (centre, CAT 0045) and fully worn rynd chase with absent rynd (right, CAT 0001) (photo by author, ©Vindolanda Trust).

Figure 8.37. Fragmented upper stone with iron band handle fitting (CAT 0032) (photo by author, ©Vindolanda Trust).

Figure 8.38. Upper quern with signs of wear but no evidence for a handle fitting (CAT 0002) (photo by author, ©Vindolanda Trust).

8.23. Vindolanda: Distribution

Distribution analysis for the lava milling tools from Vindolanda have been explored spatially and chronologically. The details of the analysis are presented below.

8.23.1. Spatial Distribution

Unfortunately, due to the type of context information provided for the findspot of the querns, it is difficult to attribute specific locations to many of the objects. However, for 73 of the lava and non-lava querns, it has been possible to determine whether they were recovered from within the fort or from the *vicus*. Of these, 65 were recovered from the fort (89%), with the remaining 8 from the *vicus* (11%). It must be understood, however, that most excavation work has taken place within the area of the fort, with large areas of the *vicus* still unexplored. This does, unfortunately, make it difficult to draw significant conclusions from a comparison of distribution between

Table 8.9. Number of querns recovered from the fort compared to the vicus, with a further breakdown of the lava and non-lava examples

	Lava	Non-lava	Total
Fort	18 (28%)	47 (72%)	65 (89%)
Vicus	2 (25%)	6 (75%)	8 (11%)

the fort and the *vicus*. Of the 65 querns from the fort area, 18 were of lava (28%) and 47 (72%) were of other stone types. Within the vicus, two querns were of lava (25%) and the rest of other stone types (75%) (table 8.9). There is, therefore, little to separate the two areas from each other in terms of the distribution of lava for querns.

8.23.2. Chronological Distribution

In terms of chronological distribution, there were 85 querns with context data that allowed for them to be allocated a site period. The number of querns from each site period was plotted (figure 8.39) to obtain an overview of the distribution of all querns over time. In total, there were very few in the earlier periods, increasing slightly in period five and peaking in period seven, to fall away again in period nine. Although it might seem reasonable to assume that this relates directly to the level of food production and grain processing activities taking place at the site, it must be remembered that period six saw the construction of the first stone fort, with a second built-in period seven. The querns recovered from Vindolanda often occur within construction contexts; incorporated into foundations, flag floors or road surfaces (see analysis for re-use). Therefore, the querns assigned to these periods were not necessarily in use during that date range; they may simply have a higher chance of recovery as these periods and locations have been heavily researched and excavated and because the querns were deposited in areas and at periods when stone was more extensively used in construction. It seems unlikely that grain was not processed in the earlier periods. The high level of period six or seven querns could also be due to the recycling of stone for building projects, incorporating querns disposed of in previous periods.

When examining the difference between lava and non-lava types by chronological site period (figure 8.40), there are differences visible. Although the quantities are very small, there is a presence of lava querns at Vindolanda from the earliest period onwards. This gradually increases in period five, peaking in periods six and seven to fall away again in the later phases. There are no non-lava querns present until period four and five, increasing in period six to peak in period seven and eight before falling away. The overall number of lava querns is much smaller than that of other stone types, though the trend in numbers indicates a much earlier commencement and peak in use when compared to the non-lava querns. Interestingly, the non-lava querns peak as the lava begins to drop away, indicating that indigenous stone types may have begun to replace the imported querns. Unfortunately, due to the nature of the context information for the site and the difficulties involved in defining social context, there is little data that can be used to explore more detailed social distributions.

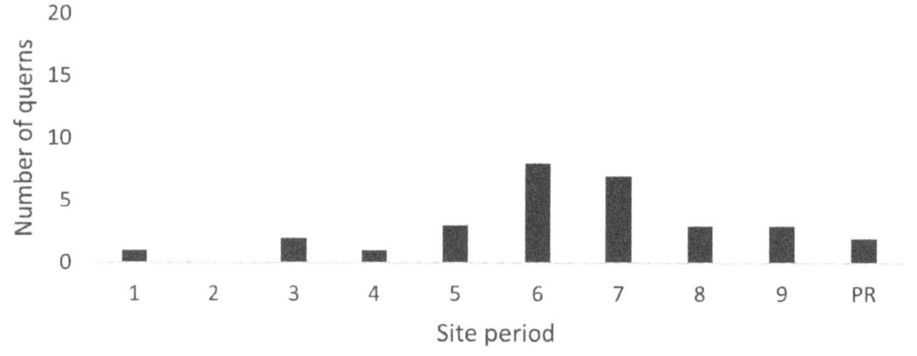

Figure 8.39. Chronological distribution of lava querns, T=30, by site period at Vindolanda.

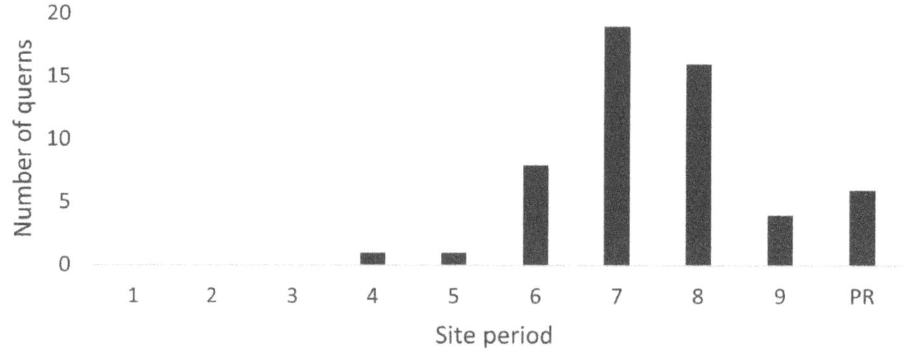

Figure 8.40. Chronological distribution of non-lava querns, T=55, by site period.

8.24. Vindolanda: Primary use

Primary use has been explored for the Vindolanda assemblage by analysing the lava milling tools for traces of wear, including measurements of the thickness of upper and lower stones. Further to this, decorative detail in the form of inscription has also been examined. The result of the analysis is presented below.

8.24.1. Wear Traces

There are many ways that use-wear analysis can be applied to the assemblage, but these rely on access to complete upper and lower stones, preferably as working pairs. Although there is one possible pair (CAT 0047 & 0048, figure 8.41), there are no others to compare them to. Furthermore, the extent of mending that has been completed on this quern, in the form of new handles being added has resulted in changing movements in the use of the quern. Consequently, earlier wear traces have become obscured by those that followed each phase of mending. The lower stone does have differential wear, and this can be interpreted as being a quern that was used in a fixed position. The heaviest area of wear on the upper stone is between the original and second handle-hole, though heavy wear continues after the original handle. This shows that it was mainly rotated in the anticlockwise direction, probably using the right hand (figure 8.42), and the range of movement appears to have been the same when either handle was in operation. Similar observations can be made of two other upper stones (CAT 0045 & 0001), both of which are also complete examples.

Figure 8.41. Possible pair of lava quern stones (CAT 0047 & 0048). A higher degree of wear can be observed on the top stone (photo by author, ©Vindolanda Trust).

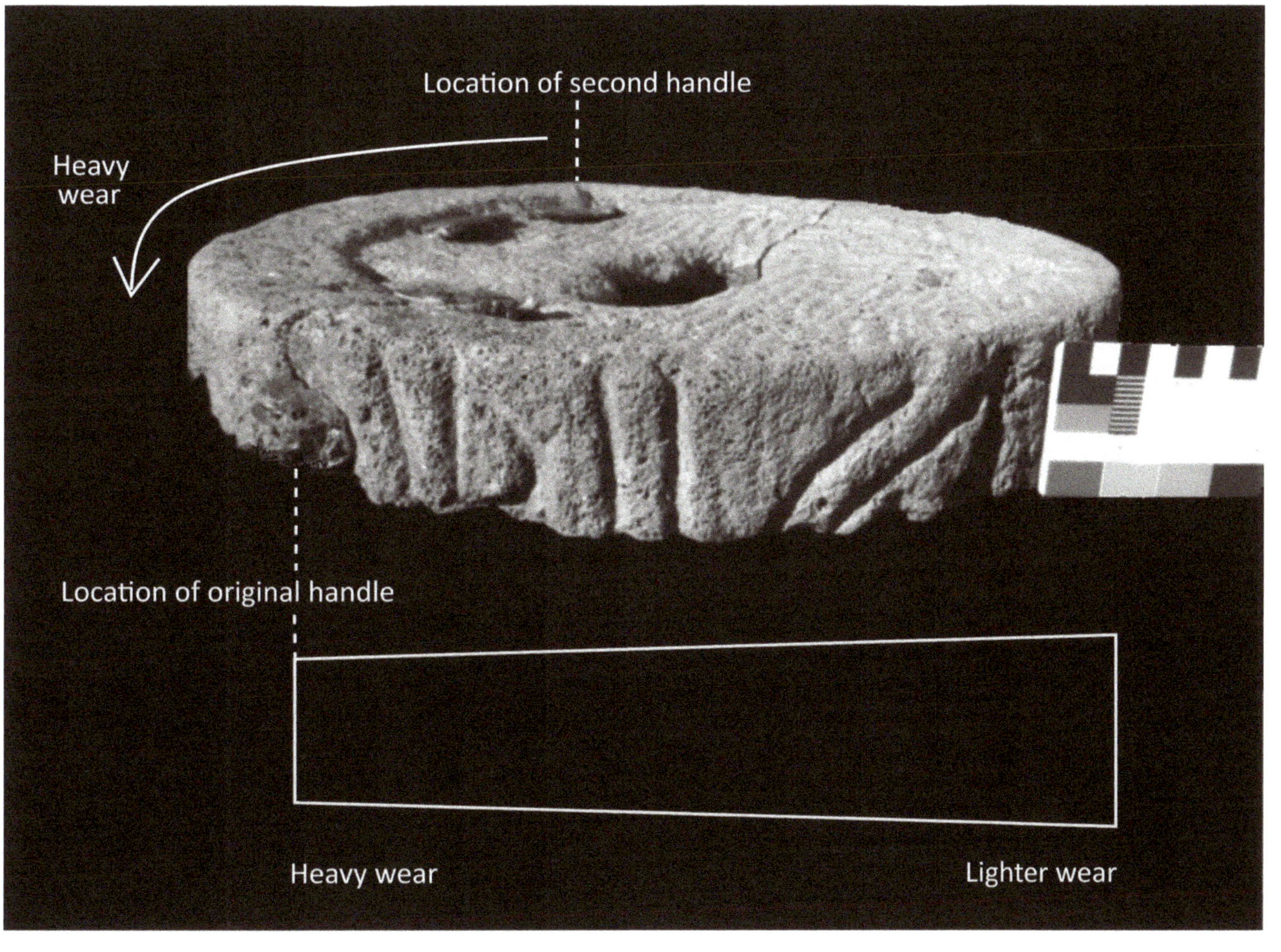

Figure 8.42. Upper quern stone (CAT 0048) showing heaviest wear after the original handle, showing it was probably rotated in a clockwise direction (photo by author, ©Vindolanda Trust).

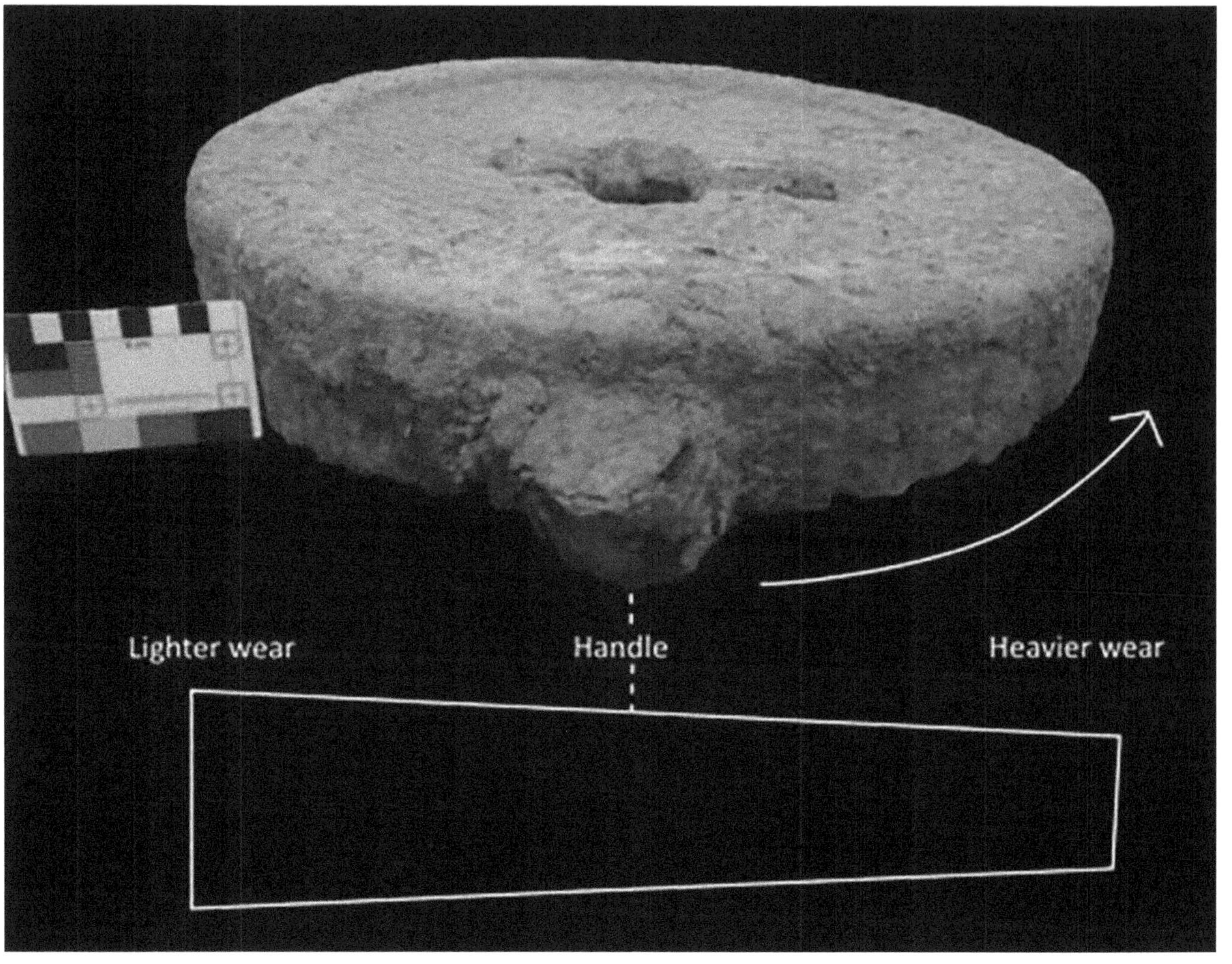

Figure 8.43. Quern upper stone with wear traces showing the direction of movement when in use (CAT 0045) (photo by author, ©Vindolanda Trust).

There was one complete upper stone that showed even wear and a similar thickness of stone for the whole circumference (CAT 0045, figure 8.43). The maximum thickness of this stone was 61mm, and only deviated from this by 4mm the whole way around despite a moderate level of degradation. This quern did not have a handle slot, though it had clearly been used and had worn down from an original thickness that would have been approximately 100mm. This stone is the example that is likely to have been fitted with an iron band handle. How this addition may have affected the operation of the quern is not fully known, but the even wear suggests that this occurred differently to querns with an elbow handle.

With regard to lower stones there were more complete examples within the assemblage compared to upper; seven showed signs of uneven wear, while two were evenly worn. This indicates that querns were often used in fixed positions at Vindolanda, and less often moved. There is one example of a lower non-lava quern recovered from the site that was fixed into a bench to maintain its position while in use (Bidwell, 1983, fig 29). Roman period querns from other sites have been found set into the floor, and this appears to have been a common occurrence (Cool, 2006, p. 74). The prevalence of complete examples in the dataset could be related to this method of fixing the stone into the floor and a higher chance of their recovery as a result. These may have been more likely to have remained in place once they ceased to be used, as was seen in the previously mentioned example that was set into a bench. Lower stones that were not fixed are more likely to have been disposed of in areas away from the main buildings of the fort complex, and are less likely, therefore, to have been found in excavations targeted at the fort itself.

8.24.2. Quern Thickness: Upper Stones

A total of 13 upper stones from the assemblage could be measured for thickness, and these are shown in figure 8.44, which also present the thickness for upper stones from all lava querns recorded in Roman Britain, excluding those from the site at 1 Poultry. The upper and lower quartiles for the British data are also indicated using red lines as a further means for comparison. The results show that many of the Vindolanda querns fall within the same range for thickness as 50% of those from the rest of Roman Britain; only one falls below, and eight are in the upper quartile for thickness. This shows that the quern

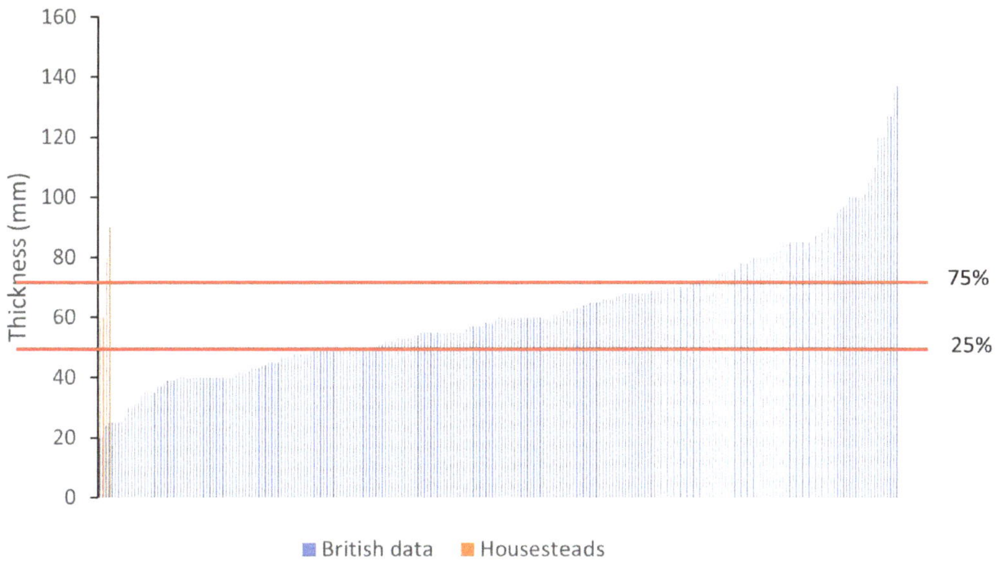

Figure 8.44. Chart showing the thickness of upper quern stones from Vindolanda compared with those from all Roman Britain, excluding the querns from the site at 1 Poultry (T=13).

Table 8.10. Table showing how the upper and lower quartiles for Vindolanda and Britain compare

	Britain	Vindolanda
25%	48	61
75%	73	83

stones from Vindolanda are typically thicker than those from elsewhere in Roman Britain, emphasised further in the table 8.10, which shows the upper and lower quartiles for the upper quern stone from Vindolanda compared with that of the British dataset. This follows the pattern seen at other military sites and relates to the better access to replacement querns that the military had, alongside the premature dumping of material during fort abandonment phases, such as the one that occurred at Vindolanda around AD 105.

8.24.3. Quern Thickness: Lower Stones

There were 14 lower stones in the Vindolanda dataset that could be measured for thickness, and these are presented in figure 8.45, which also shows the thickness for lower querns from the whole British dataset. Upper and lower quartiles for the British data are also shown for comparison. A similar outcome is seen in the data that was observed for the upper stones. One lower stone from Vindolanda occurs in the lower quartile for stone thickness, 5 in the upper quartile, while 7 occur within the same range as 50% of the querns from the wider dataset. The lower stones are, therefore, slightly thicker than those from elsewhere in Roman Britain. However, this difference is less pronounced than that for upper stones, as seen in table 8.11, which shows the upper and lower quartiles for Vindolanda lower stone thickness compared with that for

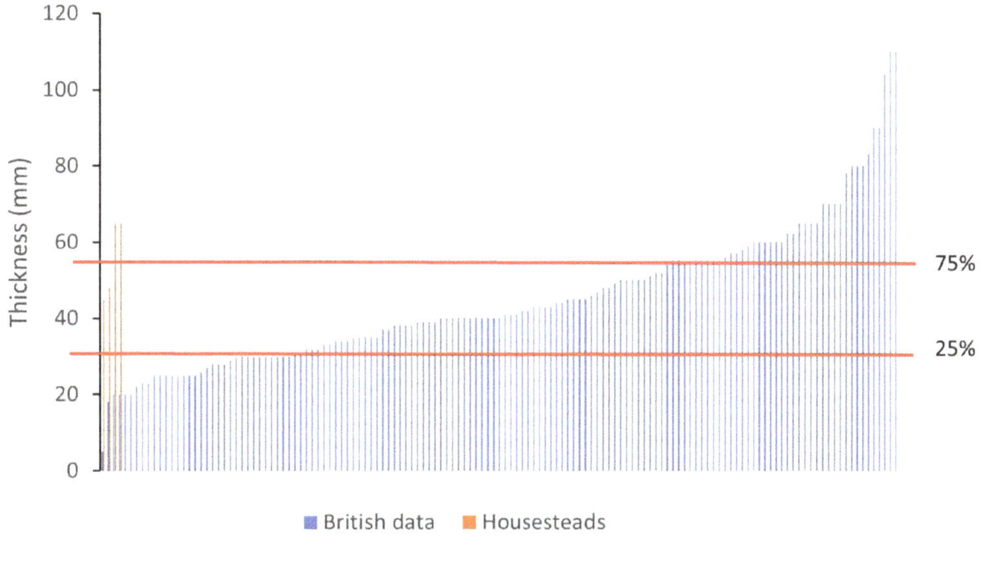

Figure 8.45. Thickness for lower stones from Vindolanda (T=14) compared with those from the wider British dataset, excluding those from the site at 1 Poultry. The upper and lower quartiles are indicated with red lines.

Table 8.11. Comparison between upper and lower quartiles for Britain and Vindolanda

	Britain	Vindolanda
25%	31	39
75%	55	56

the larger dataset. This is an interesting discrepancy, as it suggests that lower stones were worn to a similar extent to that seen at most other sites in Roman Britain, but the upper stones were deposited much sooner. How and why this trend exists is worth considering. It could relate to the use of fixed position lower stones, as suggested by the wear patterns on complete examples, or pairing old lower stones with new uppers once the more vulnerable original upper stones ceased to function.

8.24.4. Inscriptions

Inscriptions on lava querns are rare in Britain but occurred on three examples from Vindolanda that were recorded in person, with a further two added from published sources. It should be noted that this may not be representative of the extent to which inscriptions occur on lava querns, as there were many fragmented and degraded pieces in the assemblage that may also have been enhanced in such a way. However, as this was the only military site case study that had such inscriptions, Vindolanda should be considered as an exceptional site for this type of evidence; as a northern frontier site with an unusually large number of inscriptions, this might be expected. The two published examples were taken from Collingwood & Wright (1992, pp. 94–98, RIB. 2449.2), the first of which is a small fragment, with the following incised on the top surface of the upper stone,

ↄ Λ D

This has been interpreted as either meaning (centuria) Ad(…), to translate as 'century of Ad…', with Auditor being the most likely name (CAT 1190). The second example (CAT 1191), consists of another fragment which has, instead, been inscribed on the outside edge of the upper stone, around the circumference (figure 8.46). The letters inscribed are:

[…]ICTOR[..]

This could be [(centuria) V]ictor[is], to be translated as, [century of] Victor; or […V]ictor[is], to be interpreted as [property of] Victor (RIB 2449.14).

The three querns recorded in-person were taken from two complete and one fragment of upper stone. Although a specialist interpretation of these inscriptions is not achievable currently, it is still possible to describe them and attempt to understand potential meaning and context. The first example (CAT 0021) is the fragmented piece, and this does not present any clear letters or symbols. Although this piece has been inscribed, the level of degradation of the object has made it difficult to ascertain precisely what the figures could be (figure 8.47). What is identifiable is a possible 'V' and 'I', though whether this is part of a number or word cannot be established.

The next example (CAT 0045) can most certainly be defined as the inscription of a number, probably XXV. Though there is nothing to indicate what this may be in reference to, it would be safe to assume that this would relate to some form of military unit. The inscription is clearly visible, and wear has not been so heavy as to remove too much of the text at the bottom, as is often seen on other querns. The final inscribed quern (CAT 0048), although also a complete upper stone, was broken, mended in several places, and heavily worn; it had clearly come to the end of its use-life. The inscription was located around the outer circumference of the stone and had been damaged at the bottom as the quern was used and grinding surface worn down. The exact letters are hard to identify and are partly obscured by the broken handle socket that has been repaired with lead (figure 8.48). Behind this is a curved character, possibly an 'S' or 'C' followed by what could be an 'A'. There is clearly 'DII', though the inscribed figure after this is unclear in terms of lettering. This consists of two parallel angled lines and one connected vertical line (//I). This could, once again, be partly numerical in its content.

Overall, all the lava quern inscriptions appear to communicate the names and numbers of particular units

Figure 8.46. Inscription on side of lava quern from Vindolanda (CAT1191) (Collingwood & Wright, 1992, pp. 98, RIB. 2449.14)

Figure 8.47. Fragment of lava upper stone with inscription on outer circumference. A possible 'VI' is visible, though any other figures are indistinguishable (CAT 0021) (photo by author, ©Vindolanda Trust).

Figure 8.48. Inscribed upper lava quern stone from Vindolanda (CAT 0048) (photo by author, ©Vindolanda Trust).

and individuals. Those that are currently unpublished require further research. It should be noted that non-lava querns at Vindolanda were treated in a remarkably similar way to those of lava in terms of inscriptions and were also found to have been inscribed to identify ownership by an individual or a group (see examples in Collingwood & Wright (1992)). That all inscriptions were placed on upper stones is significant; this would have been the most visible location and was also the moving part of the quern, meaning that these aspects may have been important. That people went to the time and effort to add labels in this way reveals that these were personal objects, valued and to be taken care of. They were clearly not communal objects to be shared with everybody, but for the restricted use of those identified in the inscription; the quern was a physical manifestation of comradery and shared group identity, and it was important to be able to identify one group's property from another. That the lettering and labels are not standardised or always present shows that the addition of an inscription may also have been a matter of individual or group choice.

8.25. Vindolanda: Re-use/Modification

Reuse aside from that related to deposition context was difficult to identify, and there were no surfaces on the stones that indicate that they may have been used for

Figure 8.49. Upper quern showing wear at the grinding surface, which is worn down to the elbow handle fitting, making it unusable (CAT 0014) (photo by author, ©Vindolanda Trust).

Figure 8.50. Upper quern stone showing two elbow handle fittings, almost opposite each other. On the left, the grinding surface has been almost worn to the handle fitting and the one on the right may have been added to allow for continued use (CAT 0043) (photo by author, ©Vindolanda Trust).

other grinding, sharpening, or processing activities. All wear appeared to correlate with use as a grinding tool only. There were a considerable number of lava querns that were very well preserved, and so the absence of any evidence relating to this form of reuse likely reflects that this was not regular practice at the fort. There were, however, multiple examples of modification from the standardised form of a lava quern, which was mainly visible on upper stones; these being the part of the quern that houses most of the mechanical features of the device. Many of these modifications can be classed as repairs and were completed to prolong the use-life of the stone, the most common of these being the addition of new handle fittings.

This occurred on six stones (CAT 0001, 0029, 0030, 0032, 0043, 0048), with a further two where a second handle fitting had been started but not completed (CAT 0014, 0042). New handle fittings were of the elbow form and were added once the grinding surface had worn down to the extent that it interfered with the original handle (figure 8.49). As wear tends to be most extensive around the handle area, new handles appear to have been added on the opposite side of the stone or at right angles to it, where the stone was thicker due to this differential wear (figure 8.50). Multiple handles are not common for lava querns and have only been recorded on two other occasions: one from Carlisle (CAT 0381) and another from Usk (CAT 2347). However, complete upper querns are rare, and the handle slot is the most vulnerable part of the stone where breakages are likely to occur. This means that this phenomenon is unlikely to be identifiable on upper querns that are fragmented, which describes 99% of the total lava quern assemblage (2696 out of 2723 querns). The excellent preservation of the material from Vindolanda highlights the extent to which mending may have occurred and is unlikely to be a site-specific practice, especially when considering the extreme levels of wear seen at some rural and southern sites.

Modification by adding extra handle fittings appears to have taken place by people with mixed stone working skill levels. This is especially apparent on stones where modification work had been abandoned due to a failed attempt, when compared to those where such fixes were successful. One of the best examples of a less skilled quern mend is that of CAT 0048, which has two handles and a failed attempt at a handle (figure 8.51 and 8.52). The first handle (1) has been worn through at the bottom, prompting a new one to be fitted at right angles to the original (2). For some reason, this was abandoned, which may have been

Figure 8.51. Upper stone CAT 0048 that has 2 handle fittings and one abandoned attempt at adding a handle (photo by author, ©Vindolanda Trust).

Figure 8.52. Side view of handle fittings,1 and 3, and attempted fitting 2 on CAT 0048 (photo by author, ©Vindolanda Trust).

due to it being placed too high on the side of the stone. The third handle (3) was then added close by and still retains its iron fitting. Similarly, an abandoned attempt at a handle fitting has occurred on CAT 0042, but this has occurred at an angle that had little chance of success and the hole at the side never joined the one on the top surface.

Other evidence for modification has also been observed within the assemblage which, again, appears to have been completed to compensate for wear and extend the use-life of the object. Referring again to CAT 0048, it would appear that a fix was attempted to reinforce a weak point in the stone where the original handle was positioned. Shallow holes on either side of the vulnerable handle have been added, perhaps with the aim of bridging the area with an iron fitting or equivalent material to bond the stone back together. Elsewhere on the stone, deliberate damage to the kerb is similar to that seen for the upper quern from Housesteads with the iron bar fitting (CAT 2284). As the rynd chase has been worn away, the idea may have been to add an iron bar to house the spindle as a replacement for the original rynd. However, these interepretations are speculative and based on modified querns that still retain their metal fittings.

Other possible identifications for the presence of an iron bar fitting exist for CAT 0025, 0030, 0032 and 0041. Evidence for each varies, as these are fragmentary upper stones that do not present the complete set-up as seen at Housesteads. CAT 0025 has a recessed rectangular slot in the upper surface that is position adjacent to the kerb (figure 8.53). This is much further out than an ordinary rynd chase, and may have instead accomodated an iron bar fitting. Similarly, CAT 0030 has two shallow adjacent holes in the upper surface that may have held the two ends of an iron loop for this type of fitting. One of these holes still contains iron and lead (figure 8.54). CAT 0032 has the remains of an iron bar adhered to the upper surface (figure 8.55). This stone had lead 'plugs' that would have fixed part of an iron mechanism to the stone. As this stone had an iron band handle fitting, the lead may have been used to fix the iron bar. Finally, CAT 0041 has a plug of lead with iron projecting from it, in a way that resembles the fitting of the iron loop for an iron bar fitting (figure 8.56). It seems probable that this type of modification may have been more prevalent than first thought. Evidence for this has not been observed at other sites thus far, and could be limited to military sites. However, now that a near-complete parallel exists for comparison, further examples might be identified in the future.

Figure 8.53. CAT 0025 that has a recessed rectangular slot that may have accommodated an iron bar fitting like that at Housesteads (2284) (photo by author, ©Vindolanda Trust).

Figure 8.54. CAT 0030 showing shallow holes in the upper surface of the stone that may have housed the ends of an iron loop as part of an iron bar fitting (photo by author, ©Vindolanda Trust).

8.26. Vindolanda: Deposition

There are some noticeably clear instances of quern reuse and the scale of this can be presented by examining the types of feature from which querns are recovered (figure 8.57). The prevalence of querns incorporated into floors or roads and within structures, including ovens and corn dryers, demonstrates the extent to which these objects were recycled. In terms of the distinction between re-use of lava compared with non-lava, there does not appear to be much of a difference in the relative numbers of querns reused in particular ways. This indicates a lack of preference for stone type, but also shows that the military had access to a variety of types. Whether or not there was a socio-cultural dimension to reuse in this way is difficult to determine with any certainty and it would be simple to describe this as being a purely functional activity relating to the continued use of superior quality stone.

In terms of when these querns became recycled building material, of the thirteen lava querns reused in this way,

Figure 8.55. CAT 0032 showing possible iron bar adhered to the upper surface of the stone. A lead plug is visible under the iron on the stone as a white/grey shape to the left of the iron corrosion product. A further two lead plugs may have fitted the iron loop ends (above) (photo by author, ©Vindolanda Trust).

Figure 8.56. CAT 0041 showing end of possible iron loop for an iron bar fitting (photo by author, ©Vindolanda Trust).

all were small fragments consisting of a quarter or less of a whole quern stone; except one, which was a half. Although this may partly be due to post-depositional processes, such as weathering, the pieces of quern selected had come to the end of their primary use-lives by the time they were incorporated into structures and floors. Most of the pieces were no more than a few centimetres thick and the majority were upper stones that were more susceptible to damage and wear. Of the 39 identifiable quern stones in the dataset, 22 incorporated into floors or structures were upper stones. Lower stones seem to have had a longer use life and are less likely to be recovered (figure 8.58); these may have been re-paired with new upper stones to allow for continued use. If stones were specially selected for this kind of disposal, it would be expected that there would be a greater disparity, but the ratio of upper to lower stones within this context is not dissimilar to the ratio within the whole dataset. It seems logical to deduce that those recycled and disposed of in this way must have been from those available from the normal 'pool' of broken querns.

8.27. Vindolanda: Conclusion

To summarise, this chapter has examined four military site case studies: Corbridge and Chesters that were excavated as part of older projects, yielding fewer querns

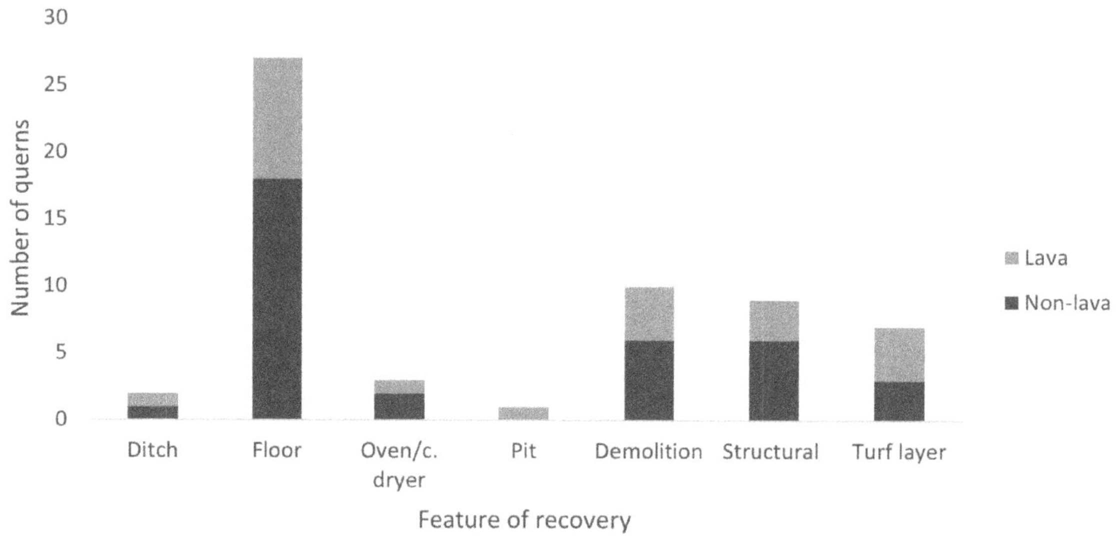

Figure 8.57. Number of querns recovered from different types of feature and split into lava or non-lava types, T=59.

Figure 8.58. Percentage of upper to lower stones found in structural or flooring contexts, (T=13), compared to that of the whole assemblage (T=55).

and millstones due to probable limited retention of stone artefacts; Housesteads and Vindolanda that were subject to long durations of archaeological investigation and have produced data and finds to modern standards for recent excavations. Vindolanda has produced the largest dataset and best-preserved finds from across the four sites. This elevated level of preservation has allowed for detailed analysis to be carried out on typology, use-wear, and quern stone modification. Furthermore, the observations made in relation to the assemblage from Vindolanda enables better identification of similar features on fragmentary stones from other sites.

As grain was a vital part of the military (and civilian) diet (Davies, 1971, p. 122–132), the ability to process grain into edible food products would have formed an essential part of day-to-day activities. The capacity of grain stores at both legionary and auxiliary forts, both in Britain and elsewhere in the Empire illustrate this importance (Haverfield, 1920; Rickman & Rickman, 1971; Richardson, 2004). In some periods, a certain level of respect was associated with the frugality of the military diet. This is suggested by Herodian (IV, vii) in his account of the Emperor Caracalla, who is described as having led the life of an ordinary soldier, both processing his grain ration and baking his own bread. The solidarity Caracalla shared with his troops with such behaviour may not be a true reflection of reality but does highlight the culture of military food consumption and the way that troops were expected to conduct themselves.

This sense of solidarity can be seen in the assemblage of querns recovered from Vindolanda in the inscriptions found on some of the upper quern stones, which have also been noted on those of non-lava. These tend to relate to unit names or numbers (a *centuria, turma* or *contubernium*), indicating group ownership and use. These were not communal objects but were designated for the use of specific groups of people and labelled accordingly. It was clearly of some importance to ensure that the group was able to identify their own quern, and personalisation of the object would have allowed each group to claim joint ownership and use of the tool. If they did grind their own grain by hand, this would have been an object that a group of soldiers would have had regular contact with, and it may have held significant meanings associated with camaraderie, group identity and shared experience. The coming together of people to prepare, cook and consume food is a way that strong social bonds are formed, reinforcing a sense of cohesion and shared aims (Fischler, 1988; Carroll, 2005); something that must have been a desirable outcome for soldiers who were required to live, fight, and potentially die side-by-side. This concept is amplified by the evidence of quern mending that appears to have been a common occurrence at Vindolanda and is likely to have been the case at other sites, though evidence is less apparent. These were clearly valued objects and worth repairing, sometimes being fixed multiple times before being disposed of. The need to repair could suggest that the supply of querns to the fort was not stable, but this seems unlikely when considering the evidence that we have for the large-scale movement of goods in and out of Vindolanda (Bowman & Thomas, 1983), and military supply in general. It may be possible, instead, to consider whether lava querns were preferred over other stone types, or that the shared ownership of a significant and meaningful object may have been an experience that was worth preserving.

Small-scale food processing on a group level may not have been the only means by which this activity was undertaken. The presence of two lava millstones at Vindolanda, the evidence for a possible watermill at Chesters, coupled with the small lava millstone/large quern from the site of the fort, and the unusually large lava querns at Housesteads suggest otherwise. Moritz (1958, pp. 124–130) states that larger sized milling stones, of between 65–80cm in diameter, were operated by several people or by animals as part of grain processing on a large scale, as evidenced by inscriptions on the millstones that refer to *centuria* or *turma*. By the same measure, Jacobi (1914) has suggested that querns of approximately 40cm in diameter were for the use of each *contubernium*, also based on quern inscriptions evidence. That differently sized milling tools existed is not in question, but care must be taken when determining that these were used in such a specific way. The inscriptions found on lava milling tools at Vindolanda need to be examined in more detail, and meanings of these correlated with the diameter of the object. This research provides the possibility for such a study to be completed, which would need to be widened further to include more examples to understand how food processing may have been organised between and within units of soldiers.

Furthermore, the distribution analysis of the querns at Housesteads has provided a glimpse into how such activities may have been spatially organised within a fort setting. Although querns are unlikely to be recovered from their context of use, the clustering of lava milling tools in the area around the ramparts raises some important questions, especially when correlated with the structural evidence for ovens in these areas of the fort. Other fort locations have similarly provided finds evidence for food preparation activities being carried out in fort periphery areas, such as amphorae and mortaria. This emphasises the need to carry out more in-depth analysis of finds evidence, including environmental samples, within such fort areas in the future.

Analysis of quern stone thickness has also produced some interesting outcomes. Overall, the results correlate with the conclusions relating to quern thickness for all Romano-British military sites, in that lava querns appear to have sometimes been deposited well before full use had been made of them, and that they are generally thicker at the point of deposition than at rural and urban sites (see Chapter 5). The possible reasons for this are varied and may relate to the types of sites where lava querns were entering circulation, and where supplies of lava querns were abundant enough for them to be replaced easily. As the Roman military had good access to imported goods, earlier disposal of lava querns does seem more likely in these contexts. However, the periodical abandonment of fortifications along the frontier, and the redeployment of troops may also provide a reason for patterns seen. Such mass movement of troops often coincided with the dumping of military equipment, and this may have resulted in thicker or less used objects being deposited prior to full use being made of them. If attacks on forts and subsequent destruction of occupation areas also occurred, a similar pattern of deposition might be expected.

From the four case study sites, both Chesters and Corbridge yielded unusually thick stones measuring over 100mm: two upper and one lower stone from Corbridge, and one upper stone from Chesters. There were only 9 sites across Britain that yielded 106 upper stones measuring 100mm or more, and 4 sites with 14 lower stones of this thickness or greater. Of these, 90 of the upper stones and 11 of the lower stones were recovered from the site at 1 Poultry. It is likely that these represent stones that were either unused, or little used prior to deposition. Therefore, due to the high density of such stones, 1 Poultry, or its immediate vicinity, has been interpreted as a location where the centralised distribution of lava querns may have occurred. Breakage on arrival into London may have resulted in a high volume of unused querns being deposited, which is supported by the fact that the bulk of those deposited were upper stones that are more vulnerable to breakage. This pattern is consistent with the ratio of upper to lower stone of equal thickness recovered at other sites across Roman Britain, with there being 16 upper stones and 3 lower.

Consequently, serious consideration should be given as to whether Corbridge was a centralised distribution location for lava milling tools at the northern frontier. Although a small number of thick quern stones were recovered from the site, the rarity of them elsewhere in Roman Britain suggests that these are exceptional. Other sites receiving lava imports may have broken down and

Table 8.12. Comparison of upper and lower quartiles for quern diameter across the four sites compared with that of the wider British dataset

	25%	75%	No querns
Corbridge	385	408	8
Chesters	375	400	6
Housesteads	423	460	10
Vindolanda	400	438	44
British dataset	400	420	553

reused such querns, which would remove them from the archaeological record. This must not be discounted as it would have implications for how findspots for thicker querns are interpreted. However, the central position of Corbridge within the Stanegate system, its clear role as a distribution centre for military supply and the fact that it was accessible via river and road all support the theory that it may have held a vital role in lava quern distribution across the military sites at or near Hadrian's Wall.

The differences seen in quern diameter between sites is another result that is worth considering. Looking at the upper and lower quartiles for quern diameter across the four sites compared with the figures for the wider British dataset (table 8.12), the querns from Vindolanda most closely resemble those from elsewhere in the province. Corbridge and Chesters have yielded querns with slightly smaller diameters to those from the wider dataset, while the querns from Housesteads are significantly larger. For Housesteads, Chesters and Corbridge, the datasets are small, and this means that anomalies in the data are more likely to skew the upper and lower percentiles. However, the quern diameters from Housesteads were consistent in being large and were, therefore, different to those seen at the other three sites. Reasons for this might relate to the size of the garrison stationed there and the need for larger milling tools to accommodate the needs of a large population. It might also be a result of different chronologies of manufacture and lava quern supply if lava quern size were a feature that changed over time, or different manufacture quarries/ supply networks.

Interpreting different supply networks for lava between the sites is a complex undertaking, and one that is hampered by small dataset sizes and the differential discard and retention policies for finds that existed during the various excavation projects across the four sites. It is interesting to note that all of the initial occupiers of these military sites had strong connections to Belgic Gaul or Germania prior to being stationed in Britannia, either due to the origins of the unit or from prior postings in the region: At Corbridge, the *ala Petriana* were probably at *Mogontiacum* (Mainz) previously (Spaul, 1994, pp. 180–182; Clauss, et al., 2022); the *ala Augusta ob virtutem* at Chesters are likely to have been in Germania in the first century, before being moved to *Calunium* (Lancaster) and then to Chesters (Spaul, 1994, p. 55–57); Vindolanda was initially occupied by the first cohort of Tungrians from Belgic Gaul, who had been at Mons Graupius before being garrisoned at the fort (Birley, 2009, p. 45–62); while Housesteads probably received the first cohort of Tungrians at some stage after they departed Vindolanda (Crow, 1995, pp. 57–59; Birley, et al., 2013, pp. 291–293). These troops and their quartermasters would, therefore, have been very familiar with the use of Mayen lava in milling tools prior to being posted to Britannia, as these were already distributed at military sites in Germania and Gallia Belgica (Gluhak & Hofmeister, 2011, p. 1616). It is probable that the initial movement of troops into Britain during AD 43 saw the introduction of lava carried into the province by the troops themselves. Although the military case study sites are later in date, continued use of lava may have been part of military culture- a familiar tool and material in an unfamiliar landscape. Furthermore, it negated the immediate need to discover and exploit suitable indigenous stone supplies for vital equipment, which could be tackled after the province was fully secure and integrated into the Roman empire.

9

Conclusion

This research presents the first holistic study of imported lava milling tools in Roman Britain. This has been achieved by compiling a database of a substantial corpus of Romano-British material, including 2,707 lava milling tools from 564 sites in Roman Britain and data from a further 601 sites where lava was absent but milling tools of other lithologies were present. In-person data that involved examining the querns and millstones first-hand was conducted for the bulk of the case studies, including 18 sites: Chichester and 11 sites in its hinterland, Verulamium, Vindolanda, Corbridge, Housesteads, Binchester and Chesters. The querns recorded for these sites comprise 301 lava and 127 non-lava examples. The rest of the dataset was compiled using information taken from Ruth Shaffrey's database, and from a systematic sweep of 'grey literature' reports from developer led excavations and published excavation monographs, reports, and county journals. Effective use has been made of the Rural Settlement of Roman Britain Project database as a means of identifying reports containing quern data. Repeated Covid lockdowns prevented access to more museums and archives for in-person data collection and have limited the choice of case study sites. This is most notable for London and Colchester, both of which were significant sites in terms of lava milling tool introduction, distribution, and consumption. Nonetheless, issues have been mitigated as much as possible with the volume of data collated and that the analyses incorporate a representative sample of lava milling tools data. The survey has established key patterns of distribution, consumption, and chronology, and provides a basis for further milling tool data to be added in the future.

This work contributes data relating to previously unknown aspects of lava milling tool use such as distribution and deposition in the province, whilst also examining the physical characteristics and functional elements of this significant group of material culture. Context has been provided for this study in the form of a critical discussion of the stone trade, and North Sea trade, while the extensive work completed on Mayen quarries in Germany has been summarised to provide a view of the manufacture stages of lava milling tools (Chapter 2).

As a summary, analysis of the data collated has yielded the following new information:

- A high degree of standardisation exists within Romano-British assemblages and that almost all identifiable lava querns can be classified as a Röder type 4 (Chapter 2).
- Some variation exists in British assemblages, but these occurrences are uncommon (Chapter 5).
- As an imported product, lava milling tools have shown some similarities with that of other imported goods and commodities, such as writing equipment. They appear in greatest numbers at military and urban sites.
- Lava also fulfilled a need for milling tools in regions where suitable natural stone sources were less available, specifically the east and parts of the south-east of England.
- In the east, all types of rural sites consumed lava, but elsewhere people of villas and roadside settlements were more likely to have access to this stone type than those inhabiting lower status rural sites.
- Data analysis of stone diameter has shown that the division between hand operated querns and mechanically driven mills is most clear at 500mm.
- Lava millstones show similar restricted distributions to that of lava querns, being most prevalent at urban and military sites. This may relate to centralised processing occurring in areas with higher populations/more elite consumers.
- Chronological change in lava milling tool use has shown that these objects were introduced in the first century AD and that a peak in deposition occurred from the second to third centuries (Chapter 5).
- Imports may have reduced or stopped by the mid-third century.
- The change in distribution over time correlates with what is known of military movements in the regions where forts or camps were present, emphasising the connection between the military and lava milling tool consumption.
- Comparison of stone thickness measurements between military, rural and urban sites has shown that querns were worn significantly thinner at rural sites than at urban and military sites, showing possible unequal access to imported goods at different site types.
- Lava may have been incorporated into ritual pit deposits at Silchester and Verulamium.
- Some deposition patterns are observable in the form of querns reused as building material and road surfacing on military sites, and deposition in ditches and enclosures at rural sites.
- At Housesteads, lava querns were deposited in the rampart area of the fort, where ovens and bakehouses were situated, and other evidence of food preparation is prevalent (Chapter 6).

Overall, this study has shown that systematic use of measurements to explore specific life stages, such as quern stone thickness for primary use, has the potential to produce interesting patterns that have real world meaning.

Results have shown the archaeological potential that lava querns and millstones have if time is invested in their recording and analysis. Unfortunately, a current lack of consistency exists in finds reports for lava milling tools, and this extends to those of indigenous stone types also. One of the most important recommendations for future recording would be, as a minimum, to distinguish lava from other quern stone types when cataloguing milling tools, as there are many clear differences in terms of economic and social considerations that would apply to imported vs indigenous material culture groupings. It is concerning how many times during data collection this basic distinction had not been made and a total number of milling tools is provided with no attempt at identification on this most basic scale. It is understandable that geological determinations are not always possible, but lava is easily distinguishable from sandstone and conglomerate and should be recorded independently. Quantification has been another issue and is especially complicated at sites where high levels of fragmentation are apparent. It is recommended that, where possible, attempts should be made to refit lava quern fragments to gain an estimate of the number of individual objects and to identify and count how many separate objects there are using typological features, size, and material as a way of differentiating between them. Where this is not possible, all fragments recovered from a single archaeological feature should be counted as one object, with the provision of a total number of fragments/fragment groups and the total weight. Though this is not a perfect system, it does at least create consistency in recording methodology and the ability to fully scrutinise the data on an individual site level.

Initial identifications should include whether the stone is an upper or lower, which can be determined by the curvature of the grinding surface or the presence of features that are typical of either stone. This should be followed by a brief description of any features, such as handle slots, kerbs, central holes, iron fittings or dressing patterns. Correct terminology for identification is provided in Chapter 2. Where an object differs from the standardised type, more attention should be placed on describing the stone and understanding how and why it differs. Where possible, measurements should be taken of the measured or estimated diameter of the milling tool, which can be completed using the same method and tools that are utilised for estimating diameters of pot rim sherds. Whether the measurements were measured or estimated should always be indicated. Diameters are vitally important for identifying millstones from hand operated querns. This distinction is important to make as it allows us to identify locations where intensive centralised milling practices were occurring, a fundamental identifier of economic activity, population size and organisation of labour. Measurement of stone thickness should also be recorded where a clear edge exists, and this should be taken at the edge of the stone regardless of whether it is an upper or lower stone. If there is no edge, if a diameter cannot be measured, or if features are not identifiable, this should be indicated in the report.

Building on the work of this study, a number of suggestions for potential future work can be made. One of these would be a systematic campaign of thin section and geochemical analysis on lava querns from Romano-British contexts. Though this research originally aimed to complete these investigations, restrictions imposed due to Covid over the past few years has made such work impossible. The work would form a substantial future project and should utilise the previous work completed by Gluhak and Hofmeister (2009; 2010) as a guide. The results of such a study would provide more detail pertaining to the specific exchange networks that were in place between the Rhineland and Britain during the Roman era, how these changed over time and how different parts of the province may have been supplied from different quarries and lava sources. Other future work should also aim to tackle some of the larger missing datasets, including London and Colchester, though published data from these areas have been included in the dataset. These locations could be highly significant in understanding the initial introduction of lava milling tools into the province. Their absence is regrettable but was caused by the lack of access to these collections due to Covid restrictions at the time of data collection.

Overall, this study has shown that the distribution and consumption of lava milling tools was predominantly an early Roman period phenomenon, and that the main consumers of imported lava products resided in the urban and rural areas of the south-east and the east of Britain, and the military sites of the north. The dataset produced as part of this project can be further interrogated to answer different research questions and, it is hoped, added to. As with any research, the outcomes here provide a base of knowledge from which new theories and interpretations can be developed to challenge or support those presented here. I personally welcome others to continue to explore this fascinating body of material culture, as there is much more to be learnt.

Bibliography

Alberti, M., 2018. The Construction, Use, and Discard of Female Identities: Interpreting Spindle Whorls at Vindolanda and Corbridge. *Theoretical Roman Archaeology Journal,* 1(1), pp. 1–16.

Alcock, J. P., 2001. *Fod in Roman Britain.* 1st ed. Stroud: Tempus.

Allen, J. R., 2012. A Rhenish Pompeiian-style Mill from Early Roman Silchester (Calleva Atrebatum). *Britannia,* 43(1), pp. 260–268.

Allen, J. R., 2018. The Whetstones from Roman Silchester at Reading Museum: The Vitorian-Edwardian Excavations. *Proceedings of the Hampshire Field Club Archaeological Society,* 73(1), pp. 156–158.

Allen, J. R. L., 2014. *Whetstones from Roman Silchester (Calleva Atrebatum), North Hampshire: Character, manufacture, provenace and use.* 1st ed. Oxford: BAR Publishing.

Allen, J. R. L., forthcoming, 2022. Whetstones in Roman Britain: Character, Distribution, Provenance and Industries. *Britannia.*

Allen, J. R. L., Rose, E. J. & Fulford, M., 2003. Reuse of Roman stone in the Reedham area of East Norfolk: intimations of a possible 'lost' Roman fort. *Britannia,* 34(1), pp. 129–141.

Allen, M. et al., 2018. *The Rural Settlement of Roman Britain: an online resource [data-set].* [Online] Available at: https://doi.org/10.5284/1030449 [Accessed 27th February 2022].

Allen, M. & Lodwick, L., 2017. Agricultural Strategies in Roman Britain. In: M. Fulford & N. Holbrook, eds. *The Rural Economy of Roman Britain.* London: Society for the Promotion of Roman Studies, pp. 142–177.

Allison, P., 2006. Mapping for gender: Interpreting artefact distribution inside 1st and 2nd century forts in Roman Germany. *Archaeological Dialogues,* 13(1), pp. 1–20.

Amouric, H., 1997. L'anille et les meules. In: D. Garcia & D. Meeks, eds. *Techniques et économie antiques et médiévales. Le temps de l'innovation: Colloque d'Aix-en-Provence, 21–23 mai 1996.* Paris: Travaux du Centre Camille-Jullian, 21, Editions Errance, pp. 39–47.

Ancient World mapping Centre, 2021. *Resources.* [Online] Available at: http://awmc.unc.edu/wordpress/map-files/ [Accessed 18th August 2020].

Anderson, T. J., Grenne, T. & Soler, J. M. F., 2011. Volcanic Quern and Millstone Quarries in Cabo de Gata (Almería) and Campo de Calatrava (Ciudad Real), Spain. In: D. Williams & D. Peacock, eds. *Bread for the People. The Archaeology of Mills and Milling: Proceedings of a Colloquium Held in the British School or Rome, 4th–7th November 2009.* Oxford: BAR Publishing International Series 2274, pp. 151–167.

Antonelli, F., Nappi, G. & Lazzarini, L., 2001. Roman millstones from Orvieto (Italy): petrographic and geochemical data for a new archaeometric contribution. *Archaeometry,* 43(2), pp. 89–167.

Appadurai, A., 1986. *The social life of things: commodities in cultural perspective.* 1st ed. Cambridge: Cambridge University Press.

Baatz, D., 1995. Die Wassermühle bei Vitruv X 5,2. Ein archäologischer Kommentar. *Saalburg Jahrbuch,* 48(1), pp. 5–18.

Bang, P. F., 2008. *The Roman Bazaar: A comparative study of trade and markets in a tributary empire.* 1st ed. Cambridge: Cambridge University Press.

Barry, C., Burnham, J. & Wacher, S., 1990. *The Small Towns of Roman Britain.* 1st ed. Berkley: University of California Press.

Bastow, M. E., 1999. The Botanical Material. In: P. Abramson, D. S. Berg & M. R. Fossick, eds. *Roman Castleford. Excavations 1974–85, Volume II. The Structural and Environmental Evidence.* Wakefield: Yorkshire Archaeology Service, pp. 163–122.

Beger, B., 1938. *Bundesarchiv, Bild 135-BB-152-11.* [Online] Available at: https://commons.wikimedia.org/wiki/File:Bundesarchiv_Bild_135-BB-152-11,_Tibetexpedition,_Tibeter_mit_Handm%C3%BChle.jpg [Accessed 4th September 2023].

Belmont, A. & Mangartz, F., 2006. *Millstone Quarries. Research, Protection and Valorization of a European Industrial Heritage (Antiquity-21st Century), Colloque international Grenoble, 22–25 septembre 2005, Maison des Sciences de l'Homme-Alpes.* 1st ed. Mainz: RGZM.

Bémont, C., Jeanlin, M. & Lahanier, C., 1993. *Les figurines en terre cuite gallo-romaines.* 1st ed. Paris: Maison des Sciences de l'Homme.

Bidwell, P., 1983. *The Roman Fort of Vindolanda at Chesterholm, Northumberland.* 1st ed. London: English Heritage.

Bidwell, P., 1999. *Hadrian's Wall 1989–99.* 1st ed. Carlisle: Cumberland & Westmorland Antiquarian & Archaeological Society.

Bidwell, P. T., 1985. *The Roman Fort of Vindolanda.* 1st ed. London: Historic Buildings and Monuments Commission for England.

Bijker, W., 1997. *Of Bicycles, Bakelites, and Bulbs: Toward a Theory of Sociotechnical Change.* 1st ed. London: MIT Press.

Birley, A., Birley, A. R. & de Bernardo Stempel, P., 2013. A Dedication by the "Cohors I Tungrorum" at Vindolanda to a Hitherto Unknown Goddess. *Zeitschrift für Papyrologie und Epigraphik,* 186(1), pp. 287–300.

Birley, A. R., 2016. The complexity of intramural and extramural relationships on the northern frontier of Roman Britain- a Vindolanda case study. In: S. Hoss & A. Whitmaore, eds. *Small finds and ancient social practice in the north-west provinces of the Roman Empire.* Oxford: Oxbow, pp. 223–262.

Birley, E. B., 1961. *Research on Hadrian's Wall.* 1st ed. Kendal: Titus Wilson & Son.

Birley, E., Birley, R. & Birley, A., 1993. *Vindolanda Research Reports, New Series. Volume II: The early wooden forts. Reports on the auxilliaries, the writing tablets, inscriptions, brands and graffitti.* 1st ed. Hexham: Roman Army Museum Publications.

Birley, R., 2009. *Vindolanda: A Roman frontier fort on Hadrian's Wall.* 1st ed. Stroud: Amberley Publishing.

Bishop, M. C. & Dore, J. N., 1988. *Corbridge: Excavations of the Roman fort and town 1947–80.* 1st ed. London: Historic Buildings and Monuments Commission for England.

Bjoertvedt, E., 2016. *Aquincum millstone.* [Online] Available at: https://commons.wikimedia.org/wiki/File:Aquincum_millstone_IMG_0967.jpg [Accessed 23rd September 2023].

Blagg, T. F. C., 1976. Tools and Techniques of the Roman Stonemason in Britain. *Britannia,* 7(1), pp. 152–172.

Blagg, T. F. C., 1980. Roman Civil and Military Architecture in the Province of Britain: Aspects of Patronage, Influence and Craft Organization. *World Archaeology,* 12(1), pp. 27–42.

Blumell, L. H., 2007. Beware of Bandits! Banditry and Land Travel in the Roamn Empire. *Journeys,* 8(1–2), pp. 1–20.

Bogaers, J. E., 1967. Reviewed Work(s): The Civitas Capitals of Roman Britain by J. S. Wacher. *The Journal of Roman Studies,* 57(1), pp. 230–234.

Boon, G. C., 1969. Belgic and Roman Silchester. *Archaeologia,* 102(1), pp. 1–81.

Bosanquet, R. C., 1904. The Roman camp at Housesteads. *Archaeologia Aeliana,* 2(1), pp. 193–300.

Bowman, A. K., 1974. Roman Military Records from Vindolanda. *Britannia,* 5(1), pp. 360–373.

Bowman, A. K. & Thomas, J. D., 1975. The Vindolanda Writing Tablets and Their Significance: An Interim Report. *Historia: Zeitschrift für Alte Geschichte,* 24(3), pp. 463–478.

Bowman, A. K. & Thomas, J. D., 1983. *Vindolanda: The Latin Writing Tablets.* 1st ed. London: Society for the Promotion of Roman Studies.

Breeze, D., 2018. The Value of Studying Roman Frontiers. *Theoretical Roman Archaeology Journal,* 1(1), pp. 1–17.

Breeze, D. J. & Dobson, B., 1976. *Hadrian's Wall.* 1st ed. London: Penguin Books.

Breeze, D. J. & Dobson, B., 1987. *Hadrian's Wall.* 1st ed. London: Penguin.

Brien, D. L., 2019. Object biography: Writing the lives of objects, artefacts and things. In: J. Parnell, ed. *New and Experimental Approaches to Writing Lives.* London: Bloomsbury Publishing, pp. 34–46.

Brindle, T., 2017. Miling and Grinding. In: M. Fulford & N. Holbrook, eds. *The Rural Economy of Roman Britain.* London: Society for the Promotion of Roman Studies, pp. 71–72.

Brindle, T. & Smith, A., 2017. Artefact Production: Stone, Jet and Glass. In: M. Fulford & N. Holbrook, eds. *The Rural Economy of Roman Britain.* London: Society for the Promotion of Roman Studies, pp. 198–200.

Brück, J., 2006. Fragmentation, Personhood and the Social Construction of Technology in Middle and Late Bronze Age Britain. *Cambridge Archaeological Journal,* 16(3), pp. 297–315.

Buckland, P., 1981. The early dispersal of insect pests of stored products as indicated by archaeological records. *Journal of stored products research,* 17(1), pp. 1–12.

Buckley, D. G. & Major, H., 1983. Quernstones. In: *Colchester Archaeological Report 2: The Roman Small Finds from Excavations in Colchester 1971–79.* Colchester: Colchester Archaeological Trust, pp. 73–76.

Buckley, D. & Major, H., 2016. Quern Stones. In: S. Lucy & C. Evans, eds. *Romano-British Settlement and Cemeteries: Mucking Excavations by Margaret and Tom Jones, 1965–1978.* Oxford: Oxbow Books, pp. 131–138.

Buffone, L., Lorenzoni, S., Pallara, M. & Zanettin, E., 2003. The millstones of ancient Pompei: A petro-archaeometric study. *European Journal of Minerology,* 15(1), pp. 15–207.

Burnham, B., Hunter, F. & Fitzpatrick, A., 2004. Roman Britain in 2003. *Britannia,* 35(1), pp. 253–316.

Burridge, K., 1969. *Tangu Traditions.* 1st ed. Oxford: Clarendon Press.

Burström, N. M., 2014. Things in the Eye of the Beholder: A Humanistic Perspective on Archaeological Object Biographies. *Norwegian Archaeological Review,* 47(1), pp. 65–82.

Butler, C. & Lyne, M., 2001. *The Roman Pottery Production Site at Wickham Barn, Chiltington, East Sussex.* 1st ed. Oxford: BAR Publishing British Series 323.

Campbell, D. B., 2010. Women in Roman forts." Residents visitors or barred from entry. *Ancient Warfare,* 4(6), pp. 48–53.

Carroll, M., 2005. The preparation and consumption of food as a contributing factor towards communal identity in the Roman army. In: Z. Visy, ed. *Limes XIX. Acts of the XIXth International Congress of Roman Frontier Studies.* Pécs: University of Pécs Press, pp. 363–372.

Chapman, H., Unpublished. Querns and Mills. In: P. D. Brown & H. Chapman, eds. *Roman Technology.* London: Museum of London.

Clauss, M., Kolb, A., Slaby, W. A. & Woitas, B., 2022. *Epigraphik-Datenbank Clauss/Slaby. CIL XIII 6820.* [Online] Available at: https://db.edcs.eu/epigr/epi_einzel.php?s_sprache=en&p_belegstelle=CIL+13%2C+06820&r_sortierung=Belegstelle [Accessed 24th February 2022].

Cleere, H., 1978. Roman harbours in Britain south of Hadrian's Wall. In: J. du Plat Taylor & H. Cleere, eds. *Roman Shipping and Trade: Britain and the Rhine provinces.* London: Council for British Archaeology, pp. 36–40.

Collingwood, R. G. & Wright, R. P., 1992. Volume II, Fascicule 4: Instrumentum Domesticum. In: S. S. Frere & R. S. O. Tomlin, eds. *The Roman Inscriptions of Britain.* Stroud: Alan Sutton Publishing, pp. 94–100.

Cool, H. E. M., 2003. Introduction. In: P. Wilson & J. Price, eds. *Aspects of Industry in Roman Yorkshire and the North.* Oxford: Oxbow, pp. 1–11.

Cool, H. E. M., 2006. *Eating and Drinking in Roman Britain.* 1st ed. Cambridge: Cambridge University Press.

Corder, P., 1943. A Decorated Lava Quern from Verulamium. *The Antiquaries Journal,* 23(3–4), p. 158.

Craven, J. A., 2012. *MUGA Pitch, Hartismere School, Eye. Post-Excavation Assessment Report 2012/145.,* Bury St Edmunds: Suffolk County Council Archaeological Service.

Crawford, O. G. & Röder, J., 1955. The Quern Quarries of Mayen in the Eifel. *Antiquity,* 29(1), pp. 68–77.

Creighton, J., 2006. *Britannia: The Creation of a Roman Province.* 1st ed. London: Routledge.

Creighton, J. & Fry, R., 2016. *Silchester : changing visions of a Roman town : integrating geophysics and archaeology : the results of the Silchester mapping project, 2005–10.* 1st ed. London: Society for the Promotion of Roman Studies.

Crow, J., 1995. *Housesteads.* 1st ed. London: BT Batsford.

Crummy, N. & Eckardt, H., 2003. Regional Identities and Technologies of the Self: Nail-Cleaners in Roman Britain. *The Archaeological Journal,* 160(1), pp. 44–69.

Crummy, P., 1977. Colchester: The Roman Fortress and the Development of the Colonia. *Britannia,* 8(1), pp. 65–105.

Crummy, P., 1997. *City of Victory: The story of Colchester- Britain's first Roman town.* 1st ed. Colchester: Colchester Archaeological Trust.

Curle, J., 1911. *A Roman Frontier Post and its People: The Fort of Newstead in the Parish of Melrose.* 1st ed. Glasgow: Univeristy of Glasgow.

Curwen, C. E., 1937. Querns. *Antiquity,* 11(1), pp. 133–151.

Dannehl, K., 2009. Object biographies from production to consumption. In: K. Harvey, ed. *History and Material Culture.* London: Routledge, pp. 123–138.

Davies, R. W., 1971. The Roman Military Diet. *Britannia,* 2(1), pp. 122–142.

Dawson, M., 1994. *A Late Roman Cemetery at Bletsoe. Monograph 1,* Bedford: Bedfordshire County Council.

Dawson, M.-J., 2019. *Rethinking small town market status in Roman Britain: a review of the data for five case studies in the Thames Valley region.* [Online] Available at: https://kar.kent.ac.uk/76363/1/9Volume%201.pdf [Accessed 23rd February 2022].

de Beaune, S. A., 2004. The Invention of Technology. Prehistory and Conginition. *Current Anthropology,* 45(1), pp. 139–162.

Deiss, J. J., 1966. *Herculaneum: Italy's buried treasure.* 1st ed. London: Thomas Y. Crowell Company.

Derochette, J. M., 2010. *Banne d'ordanche, Monts d'Ore.* [Online] Available at: http://jm-derochette.be/volcanic_rocks/lame_43.htm [Accessed 25th April 2022].

Deutsche Forschungsgemeinschaft, 2019. *Roman Port Andernach.* [Online] Available at: http://www.spp-haefen.de/en/projects/der-rhein-als-europaeische-verkehrsachse/die-teilprojekte/roman-port-andernach/ [Accessed 2nd April 2019].

Dimbleby, G. W., 1978. A Pollen Analysis from Verulamium. *Hertfordshire Archaeology,* 6(1), pp. 112–115.

Dobson, B. & Breeze, D. J., 2000. *Hadrian's Wall.* 4th ed. London: Penguin.

Down, A., 1989a. *Chichester Excavations 7.* 1st ed. Chichester: Phillimore.

Dreesen, R. et al., 2014. Unraveling geological and geographical provenances of lithic materials during Roman times in Belgium: a fruitful collaboration between geologists and archaeologists. *European Geologist,* 38(1), pp. 10–14.

Duncan-Jones, R., 1977. *The Economy of the Roman Empire: Quantitative Studies.* Revised edition ed. Cambridge: Cambridge University Press.

Eaton, T., 2000. *Plundering the Past: Roman Stonework in Medieval Britain.* 1st ed. Cheltenham: Tempus.

Eckardt, H., 2005. The social distribution of Roman artefacts: the case of nail-cleaners and brooches in

Britain. *Journal of Roman Archaeology,* 18(1), pp. 139–160.

Eckardt, H., 2006. The Character, Chronology, and Use of the Late Roman Pits: The Silchester finds assemblage. In: M. Fulford, A. Clarke & H. Eckardt, eds. *Life and Labour in Late Roman Silchester: Excavations in Insula IX since 1997.* London: Society for the Promotion of Roman Studies, pp. 221–248.

Eckardt, H., 2015. *Objects and Identities: Roman Britain in the north-western provinces.* 1st ed. Oxford: Oxford University Press.

Eckardt, H. & Williams, H., 2003. Objects without a past? The use of Roman objects in early Anglo-Saxon graves. In: 1st, ed. *Archaeologies of Remembrance.* London: Plenum, pp. 141–170.

Edwards, B. J. & Breeze, D. J., 2000. *The Twelfth pilgrimage of Hadrian's Wall, 1999 : a report.* 1st ed. Kendal: Cumberland and Westmorland Antiquarian and Archaeological Society.

English Heritage, 2016. *History of Corbridge Town.* [Online] Available at: https://www.english-heritage.org.uk/visit/places/corbridge-roman-town-hadrians-wall/history-and-stories/history/ [Accessed 27th January 2022].

Evans, E., 1994. Military Architects and Building Design in Roman Britain. *Britannia,* 25(1), pp. 143–164.

Evans, J., 2002. Pottery from Thornbrough Farm. In: P. R. Wilson, ed. *Cataractonium. Roman Catterick and its Hinterland: Excavations and Research, 1958–1997. Part I.* York: Council for British Archaeological Research Report, pp. 479–476.

Evans, J., 2007. The Roman pottery from Wroxeter's hinterland. In: V. L. Gaffney, R. H. White & H. Goodchild, eds. *Wroxeter, the Cornovii, and the Urban Process: Final Report on the Wroxeter Hinterland Project 1994–1997. Vol. 1, Researching the Hinterland.* Portsmouth: Journal of Roman Archaeology Supplementary Series 68, pp. 146–167.

Evans, R., 2017. *Mass and Elite in the Greek and Roman Worlds: From Sparta to Late Antiquity.* 1st ed. Abingdon: Taylor and Francis.

Evers, K. G., 2011. *The Vindolanda tablets and the ancient economy.* 1st ed. Oxford: BAR Publishing British Series 544.

Fenwick, V., 1978. *The Graveney boat: a 10th Century find from Kent.* 1st ed. Oxford: BAR Publishing.

Finley, M., 1999. *The Ancient Economy.* 3rd, renewed by trustees of Darwin College, University of Cambridge ed. London: University of California Press.

Firth, R., 2012. *Primitive Economics of the New Zealand Maori.* 2nd ed. London: Routledge.

Fischer, M. L., 1998. *Marble Studies: Roman Palestine and the Marble Trade.* 1st ed. Konstanz: Xenia.

Fischler, C., 1988. Food, Self and Identity. *Anthropology of Food,* 27(2), pp. 275–292.

Fitzpatrick, A. P., 2017. 'The famous volcanic rock of the Eifel': Was it imported to Britain in prehistory?. In: R. Shaffrey, ed. *Written in Stone; Papers on the function, form and provenancing of prehistoric stone objects in memory of Fiona Roe.* Oxford: Oxbow Books, pp. 195–214.

Frere, S., 1983. *Verulamium Excavations: Volume 2.* 1st ed. London: The Society of Antiquaries of London .

Frere, S. S. & St Joseph, J. K., 1974. The Roman Fortress at Longthorpe. *Britannia,* 5(1), pp. 1–129.

Fryer, J., 1978. *Ports in Roman Britain,* Univeristy of Keele: Unpublished PhD thesis.

Fry, J., 2009. A Late Bronze Age, Anglo-Saxon and medieval Settlement Site at Manston Road, Ramsgate: Stone. In: P. Andrews, et al. eds. *Kentish Sites and Sites of Kent. A Miscellany of four archaeological excavations.* Salisbury: Wessex Archaeology, pp. 228–229.

Fulford, M., 1984. *Silchester Defences 1974–80.* 1st ed. London: Society for the Promotion of Roman Studies.

Fulford, M., 1989. *The Silchester Amphitheatre: Excavations of 1979–85.* 1st ed. London: Society for the Promotion of Roman Studies.

Fulford, M., 2001. Links With the Past: Pervasive 'ritual' behaviour in Roman Britain. *Britannia,* 32(1), pp. 199–218.

Fulford, M., 2008. Nero and Britain: The Palace of the Client King at Calleva and Imperial Policy towards the Province after Boudicca. *Britannia,* 39(1), pp. 1–13.

Fulford, M., 2020a. Introduction. In: M. Fulford, A. Clarke, E. Durham & N. Pankhurst, eds. *Silchester Insula IX: The Claudio-Neronian Occupation of the Iron Age Oppidum.* London: Society for the Promotion of Roman Studies, pp. 3–10.

Fulford, M., 2021. *Silchester Revealed: The Iron Age and Roman Town of Calleva.* 1st ed. Oxford: Windgather Press.

Fulford, M. & Bird, J., 1975. Imported Pottery from Germany in the Late Roman Period. *Britannia,* 6(1), pp. 171–181.

Fulford, M. & Brindle, T., 2016. Introduction. In: M. Fulford & N. Holbrook, eds. *The Rural Settlement of Roman Britain.* London: Society for the Promotion of Roman Studies, pp. 1–16.

Fulford, M. & Clarke, A., 2009. *Silchester Insula IX: The 'Town Life' Project 2006–2009,* Reading: University of Reading.

Fulford, M. & Clarke, A., 2011. *Silchester, City in Transition. The Mid-Roman Occupation of Insula IX c. A.D. 125–250/300: A report on excavations undertaken since 1997. Britannia Monograph 25.* 1st ed. London: Society for the Promotion of Roman Studies.

Fulford, M. et al., 2018b. *Silchester Roman Town: The baths 2018.* 1st ed. Reading: University of Reading.

Fulford, M., Clarke, A., Durham, E. & Pankhurst, N., 2018a. *Late Iron Age Calleva : the pre-conquest occupation at Silchester Insula IX.* 1st ed. London: Society for the Promotion of Roman Studies.

Fulford, M., Clarke, A., Durham, E. & Pankhurst, N., 2020. *Silchester Insula IX : the Claudio-Neronian occupation of the Iron Age oppidum : the early Roman occupation at Silchester Insula IX.* 1st ed. London: Society for the Promotion of Roman Studies.

Fulford, M. et al., 2019. *Silchester Roman Town: The baths 2019.* 1st ed. Reading: University of Reading.

Fulford, M. et al., 2019. *Silchester Roman Town: The Baths 2019,* Reading: University of Reading.

Fulford, M., Clarke, A. & Eckardt, H., 2006. *Life and Labour in Late Roman Silchester: Excavations in Insula IX since 1997.* 1st ed. London: Society for the Promotion of Roman Studies.

Fulford, M., Clarke, A., Eckardt, H. & Shaffrey, R., 2002. Victorian Excavation Methodology: The Society of Antiquaries at Silchester in 1893. *The Antiquaries Journal,* 82(1), pp. 285–306.

Fulford, M. et al., 2017. *Silchester Insula XXX: Temple excavation 2017.* 1st ed. Reading: University of Reading.

Fulford, M., Clarke, A. & Lucas, S., 2011. *Silchester Insula IX: The 'Town Life' Project 2011,* Reading: Univeristy of Reading.

Fulford, M. & Machin, S., 2021. Building Britannia: Pre-Flavian Private and Public Construction across Southern Britain. *Britannia,* Volume doi:10.1017/S0068113X2100009X, pp. 1–19.

Fulford, M., Pankhurst, N., Wheeler, D. & Machin, S., 2017. *The Silchester Environs Project: The Roman tilery and pottery industry at Little London, Pamber 2017,* Reading: University of Reading.

Fulford, M. & Timby, J., 2000. *Late Iron age and Roman Silchester: Excavations on the site of the Forum-Basilica, 1977, 1980–86.* 1st ed. London: Society for the Promotion of Roman Studies.

Gallala, W., Younes, A., Ouazaa, N. L. & Hadjzobir, S., 2018. Roman Millstones of Carthage (Tunisia): a Geoarchaeological Study Using Petrological and Geochemical Methods. *Geoheritage,* 10(1), pp. 673–686.

Gallardo, F., Castro, V. & Miranda, P., 1999. Riders on the storm: rock art in the Atacama Desert. *World Archaeology,* 31(2), pp. 225–242.

Gallery V Exhibits, 1998. *Appalachia Through the Eyes of Doris Ulmann: A Photograph Exhibit.* [Online] Available at: http://community.berea.edu/galleryv/GrindCn.HTML [Accessed 5th March 2019].

Garrow, D., 2012. Odd Deposits and Average Practice: A critical history of the concept of structured deposition. *Archaeological Dialogues,* 19(2), pp. 85–115.

Giljohann, R., Hunold, A. & Wenzel, S., 2017. Rural Life and Industry between the Eifel and the Rhine. In: M. Reddé, ed. *Gallia Rustica 1: les campagnes du nord-est de la Gaule, de la fin de l'âge du fer à l'antiquité tardive.* Bordeaux: Université Bordeaux Montaigne, pp. 125–152.

Gillam, J. & Greene, K., 1981. Roman Pottery and the Economy. In: Anderson & Anderson, eds. *Roman Pottery Research in Britain and North-West Europe.* Oxford: British Archaeological Research, pp. 1–25.

Glauben, A. M., 2012. Der vicus von Mayen (Lkr. Mayen-Koblenz).. In: M. Grünewald & S. Wenzel, eds. *Römische Landnutzung in der Eifel- neue Ausgrabungen und Forschungen, Tagung in Mayen, vom 3. bis zum 6. November 2011.* Mainz: RGZM, pp. 87–98.

Gluhak, T., 2010. Mühlen für das römische Militär. Geochemische Herkunftsanalysen von Mühlsteinen aus den augusteischen Lagern Haltern, Oberaden, Anreppen und Barkhausen. *Archäologisches Korrespondenzblatt,* 40(1), pp. 273–284.

Gluhak, T. M. & Hofmeister, W., 2009. Roman lava quarries in the Eifel region (Germany): geochemical data for millstone provenance studies. *Journal of Archaeological Science,* 36(1), pp. 1774–1782.

Gluhak, T. M. & Hofmeister, W., 2011. Geochemical Provenance Analyses of Roman Lava Millstones North of the Alps: a Study of their Distribution and Implications for the Beginning of Roman Lava Quarrying in the Eifel Region (Germany). *Journal of Archaeological Science,* 38(1), pp. 1603–1620.

Goldsworthy, A. K., 1998. *The Roman Army at War: 100 BC–AD 200.* 1st ed. Oxford: Clarendon Press.

Goodburn, R. & Grew, F., 1984. Objects of Stone. In: S. Frere, ed. *Verulamium Excavations III.* Oxford: Oxford University Committee for Archaeology, pp. 77–82.

Gosden, C. & Marshall, Y., 1999. The Cultural Biography of Objects. *World Archaeology,* 31(2), pp. 169–178.

Green , C., 2017. Querns and Millstones in Late Iron Age and Roman London and South-East England. In: D. Bird, ed. *Agriculture and Industry in South-Eastern Roman Britain.* Oxford: Oxbow Books, pp. 156–179.

Greene, E., 2013. Female networks in the military communities of the Roman west: a view from the Vindolanda tablets. In: E. Hemelrijk & G. Woolf, eds. *Women and the Roman City in the Latin West.* Leiden: Brill, pp. 369–390.

Greene, K., 1986. *The Archaeology of the Roman Economy.* 1st ed. Berkeley: California Univeristy Press.

Groenman-van Waateringe, W., 1997. Classical Authors and the Diet of Roman Soldiers: true or false?. In: W. Groenman-van Waateringe, B. L. van Beek, W. J. H. Willems & S. L. Wynia, eds. *Roman Frontier Studies 1995.* Oxford: Oxbow Books, pp. 261–265.

Hackworth Petersen, L., 2003. The Baker, His Tomb, His Wife, and Her Breadbasket: The Monument of Eurysaces in Rome. *The Art Bulletin,* 85(2), pp. 230–257.

Hajnalová, M. & Rajtár, J., 2009. Supply and subsistence: the Roman fort at Iža. In: W. S. Hanson, ed. *The army and frontiers of Rome." Papers offered to David Breeze on the occasion of his sixty-fifth birthday and his retirement from Historic Scotland.* Portsmouth: Journal of Roman Archaeology, pp. 194–207.

Hamilakis, Y., 1999. Stories from exile: fragments from the cultural biography of the Parthenon (or 'Elgin') Marbles. *World Archaeology,* 31(2), pp. 303–320.

Hamilton, S., 2002. Between Ritual and Routine: interpreting British Prehistoric pottery production and distribution. In: A. Woodward & J. D. Hill, eds. *Prehistoric Britain: the ceramic basis.* Oxford: Oxbow Books, pp. 38–53.

Hansen, S. S. & Larsen, A.-C., 2000. Miniature querns and millstones from Shetland's Scandinavian Past. *Acta Archaeologica,* 71(1), pp. 105–121.

Hanson, W. S., Daniels, C. M., Dore, J. N. & Gillam, J. P., 1979. The Agricolan Supply Base at Red House, Corbridge. *Archaeologia Aeliana (5th Series),* 7(1), pp. 1–98.

Harding, D. W., 2009. *The Iron Age Round-House: Later Prehistoric Building in Britain and Beyond.* 1st ed. Oxford: OUP Oxford.

Harman, G., 2002. *Tool-being: Heidegger and the metaphysics of objects.* 1st ed. Chicago: Open Court.

Hatherley, C., 2020. *Doune Roman fort, Stirlingshire: excavations in 1999, 2008 and 2010. Scottish Archaeological Internet Reports 92.* [Online] Available at: https://doi.org/10.9750/issn.2056-7421.2020.92 [Accessed 25th February 2022].

Haverfield, F., 1920. *The Provisioning of Roman Forts.* 1st ed. Highgate: Titus Wilson & Son.

Hayward, K. M., 2009. *Roman Quarrying and Supply on the Periphery: A geological study of first-century funerary monuments and monumental architecture.* 1st ed. Oxford: BAR Publishing British Series 500.

Herva, V.-P., 2005. The life of buildings: Minoan building deposits in an ecological perspective. *Oxford Journal of Archaeology,* 24(1), pp. 215–227.

Herz, N. & Wenner, D., 1981. Tracing the Origin of Marble. *Archaeology,* 34(1), pp. 14–21.

Heslop, D. H., 2008. *Patterns of Quern Production, Acquisition and Deposition: A Corpus of Beehive Querns from Northern Yorkshire and Southern Durham.* 1st ed. Leeds: Yorkshire Archaeological Society.

Hill, J. D., 1995. *Ritual and Rubbish in the Iron Age of Wessex. BAR British Series 242.* 1st ed. Oxford: BAR Publishing.

Hill, J. & Rowsome, P., 2011. *Roman London and the Walbrook Stream Crossing: Excavations at 1 Poultry and Vicinity, City of London. MOLA Monograph 37.* 1st ed. London: Museum of London Archaeology.

Hingley, R., 1989. *Rural settlement in Roman Britain.* 1st ed. London: Seaby.

Hirt, A. M., 2010. *Imperial Mines and Quarries in the Roman World: Organizational Aspects 27 BC–AD 235.* 1st ed. Oxford: Oxford University Press.

Historic England, 1999. *Reculver Saxon Shore fort, Anglo-Saxon monastery and associated remains.* [Online] Available at: https://historicengland.org.uk/listing/the-list/list-entry/1018784 [Accessed 9th September 2021].

Hodder, I., 2012. *Entangled: An archaeology of the relationships between humans and things.* 1st ed. Chichester: Wiley-Blackwell.

Hodgson, N., 1999. *History of Chesters Roman Fort.* [Online] Available at: https://www.english-heritage.org.uk/visit/places/chesters-roman-fort-and-museum-hadrians-wall/history/#footnote-4 [Accessed 3rd February 2022].

Hodgson, N., 2008. The Development of the Roman Site at Corbridge from the First to Third Centuries AD. *Archaeologia Aeliana,* 37(1), pp. 47–92.

Hodgson, N., 2009. *Hadrian's Wall 1999–2009.* 1st ed. Kendal: Cumberland & Westmorland Antiquarian & Archaeological Society.

Hodgson, N., 2014. The accomodation of soldiers' wives in Roman fort barracks. on Hadrian's Wall and beyond. In: R. Collins & F. McIntosh, eds. *Life in the Limes: Studies of people and objects of the Roman frontiers.* Oxford: Oxbow Books, pp. 18–28.

Horden, P. & Purcell, N., 2000. *The Corrupting Sea: A study of Mediterranean history.* 1st ed. Oxford: Blackwell.

Hörter, F. & Röder, M. J., 1951. Die Geschichte der Basaltlava-Industrie von Mayen und Niedermendig Teil I: Vor. und Frühgeschichte. *Sonderdruck Jahrbuch: für Geschichte und Kultur des Mittelrheins und seiner Nachbargebiete,* 1(1), pp. 1–35.

Hunn, J. R., 1992. The Verulamium Oppidum and its Landscape in the Late Iron Age. *The Aracheological Journal,* 149(1), pp. 39–68.

Ingrem, C., 2006. The Animal Bone. In: M. Fulford, A. Clarke & H. Eckardt, eds. *Life and Labour in Late Roman Silchester: Excavations in Insula IX since 1997.* London: Society for the Promotion of Roman Studies, pp. 167–188.

Jaccottey, L. & Longepierre, S., 2011. Pompeian Millstones in France. In: D. Williams & D. Peacock, eds. *Bread for the People. The Archaeology of Mills and Milling: Proceedings of a Colloquium Held in the British School or Rome, 4th–7th November 2009.* Oxford: BAR Publishing International Series 2274, pp. 97–116.

Jacobi, H., 1914. Römische Getreidemühlen. *Saalburg Jahrbuch,* 3(1), pp. 75–96.

Jodry, F., 2011. First Century Querns of the Roman Army in the Light of Modern Texts. In: D. Williams & D. Peacock, eds. *Bread for the People. The Archaeology of Mills and Milling: Proceedings of a Colloquium Held in the British School or Rome, 4th–7th November 2009.* Oxford: BAR Publishing International Series 2274, pp. 85–92.

Jones, M., 2000. The Plant Remains. In: 1st, ed. *Late Iron Age and Roman Silchester: Excavations of the site of the Forum Basilica 1977, 1980–86.* London: Society for the Promotion of Roman studies, pp. 505–512.

Jones, M. J., 2004. Cities and Urban Life. In: M. Todd, ed. *A Companion to Roman Britain.* Oxford: Blackwell, pp. 162–192.

Jongman, W., 2007. The Early Roman Empire: Consumption. In: W. Scheidel, I. Morris & P. R. Saller, eds. *The Cambridge Economic history of the Greco-Roman World.* Cambridge: Cambridge University Press, pp. 592–618.

Joyce, R. A. & Gillespie, S. D., 2015. *Things in motion: Object histories, biographies and itineraries.* 1st ed. Sante Fe: SAR Press.

Joy, J., 2009. Reinvigorating object biography: reproducing the drama of object lives. *World Archaeology,* 41(4), pp. 540–556.

Kling, J., 2008. The underground Millstone Quarries at Niedermendig and Mayen. History, Research, Threats and Valuation. In: J. Silvertant & J. Ni Mháirtin, eds. *What Lies Beneath: Proceedings of the 3rd International Symposium on Archaeological Mining History.* Maastricht: Institute Europe Subterranea, pp. 138–169.

Kopytoff, I., 1986. The cultural biography of things: commoditization as process. In: A. Appadurai, ed. *The Social Life of things: commodities in cultural perspective.* Cambridge: Cambridge University Press, pp. 64–91.

Köstner, E., 2012. Stad, Land, Fluss: Rechtliche aspekte der landnutzung in der Eifel nach dem Gallischen krieg. In: M. Grünewald & S. Wenzel, eds. *Römische Landnutzung in der Eifel. neue Ausgrabungen und Forschungen, Tagung in Mayen, vom 3. bis zum 6. November 2011.* Mainz: Römisch-Germanischen Zentralmuseums, pp. 73–85.

Künow, J., 1980. *Negotiator et Vectura: Händler und Transport im freien Germanien.* 1st ed. Marburg: Universität Marburg.

Lakin, D., 1997. *Summerton Way, Thamesmead, London, SE8: A post-excavation assessment,* London: Museum of London Archaeological Service.

Laurence, R. & Trifilò, F., 2015. The global and the local in the Roman empire: connectivity and mobility from an urban perspective. In: M. Pitts & M. J. Versluys, eds. *Globalisation and the Roman world: world history, connectivity and material culture.* Cambridge: Cambridge University Press, pp. 99–122.

Lepareux-Couturier, S., 2014. Complex dressing patterns on grinding surfaces of rotary querns and millstones from Antiquity in the Paris Basin, France: state of research and perspectives. *AmS-Skrifter,* 24(1), pp. 149–158.

Lundock, J., 2015. *A Study of the Deposition and Distribution of Copper Alloy Vessels in Roman Britain.* 1st ed. Oxford: Archaeopress.

MacGregor, G., 1999. Making sense of the past in the present: a sensory analysis of carved stone balls. *World Archaeology,* 31(2), pp. 258–271.

MacKie, E. W., 2007. Rotary Quernstones. In: W. S. Hanson, ed. *Elginhaugh: a Flavian fort and its annexe.* London: Society for the Promotion of Roman Studies, pp. 492–505.

Malinowski, B., 1926. *Crime and Custom in Savage Society.* 1st ed. London: Routledge.

Malmendier, U., 2002. *Societas publicanorum: staatliche Wirtschaftsaktivitäten in den Händen privater Unternehmer.* 1st ed. Köln: Böhlau Verlag.

Mangartz, F., 2007. Een 'logboek' voor De Meern 1: beschrijving en determinatie van de herkomst van het natuursteen. In: J. M. Morel & E. Jansma, eds. *Een Romeinse Rijnaak, gevonden in Utrecht-De Meern: Resultaten van het onderzoek naar de platbodem 'De Meern 1'.* Amersfoort: Rapportage Archeologische Monumentenzorg, pp. 246–255, Access Policy: http://creativecommons.org/licenses/by-sa/4.0.

Mangartz, F., 2008. *Römischer Basaltlava-Abbau zwischen Eifel und Rhein: Monographien des Römisch-Germanisches Zentralmuseum 7; Vulkanpark Forschungen 7.* 1st ed. Mainz: RGZM.

Mangartz, F., 2012. Römerzeitliche Mühlsteinproduktion in den Grubenfeldern des Bellerberg-Vulkans bei Mayen (LKR. Mayen-Koblenz). In: M. Grünewald & S. Wenzel, eds. *Römische Landnutzung in der Eifel: Neue Ausgrabungen und Forschungen, Tagung in Mayen, vom 3. bis zum 6. November 2011.* Mainz: RGZM, pp. 1–24.

Mattingly, D., 2006. *An Imperial Posession: Britain in the Roman Empire.* 1st ed. London: Penguin Books.

Maxwell, G. S., 1976. The Roman Road East of Castledykes. *Glasgow Archaeological Journal,* 4(1), pp. 93–98.

McAlister, D., 2013. Childhood in Viking and HibernoScandinavian Dublin 800–1100. In: D. Hadley & L. T. Harkel, eds. *Everyday Life in Viking Towns: Social approaches to towns in England and Ireland 800–1100.* Oxford: Oxbow, pp. 86–102.

McIntosh, F. C., 2019. *The Clayton Collection: An archaeological appraisal of a 19th century collection of Roman artefacts from Hadrian's Wall.* 1st ed. Oxford: BAR Publishing.

McLaren, D. & Hunter, F., 2008. New aspects of rotary querns in Scotlan. *Proceedings of the Society of Antiquaries Scotland,* 138(1), pp. 105–128.

Meskell, L., 2005. Introduction: Object Orientations. In: L. Meskell, ed. *Archaeologies of Materiality.* Oxford: Blackwell Publishing, pp. 1–17.

Millett, M., 1990. *The Romanization of Britain.* 1st ed. Cambridge: Cambridge University Press.

Millett, M. & Wilmott, T., 2003. Rethinking Roman Richborough. In: P. Wilson, ed. *The Archaeology of Roman Towns.* Oxford: Oxbow Books, pp. 184–194.

Moore, T., 2007. Perceiving communities: Exchange, landcapes and social networks in the later Iron Age of western Britain. *Oxford Journal of Archaeology,* 26(1), pp. 79–102.

Moore, T., 2017. Beyond Iron Age 'towns': Examining oppida as examples of low-density urbanism. *Oxford Journal of Archaeology,* 36(3), pp. 287–305.

Moritz, L. A., 1956. Vitruvius' water-mill. *Classical Review,* 6(3–4), pp. 193–196.

Moritz, L. A., 1958. *Grain Mills and Flour in Classical Antiquity.* 1st ed. Oxford: Clarendon Press.

Morris, F. M., 2010. *North Sea and Channel Connectivity during the Late Iron Age and Roman Period (175/150 BC–AD 409).* 1st ed. Oxford: BAR .

Muldowney, M., 2008. *Land Adjacent to Beech House, Exning,* Bury St Edmunds: Suffolk County Council Archaeological Service.

Neal, D., Hunn, J. & Wardle, A., 1990. *Excavation of the Iron Age, Roman and Medieval settlement at Gorhambury, St Albans.* 1st ed. Swindon: English Heritage.

Niblett, R., 1999. *The Excavation of a Ceremonial Site at Folly Lane St Albans.* 1st ed. London: The Society for the Promotion of Roman Studies.

Niblett, R., 2001. *Verulamium: The Roman City of St Albans.* 1st ed. Stroud: Tempus.

Niblett, R., Manning, W. & Saunders, C., 2006. Verulamium: Excavations within the Roman town 1986–88. *Britannia,* 37(1), pp. 53–188.

Oesterwind, B. C., 2012. Spätlatenezeit und frühe Römische Kaiserzeit in und um Mayen (Lkr. Mayen-Koblenz). In: M. Grünewald & S. Wenzel, eds. *Römische Landnutzung in der Eifel- neue Ausgrabungen und Forschungen, Tagung in Mayen, vom 3. bis zum 6. November 2011.* Mainz: RGZM, pp. 49–72.

Olsen, B., 2010. *In defense of things: Archaeology and the ontology of objects.* 1st ed. Plymouth: Rowman & Littlefield.

Parkhouse, J., 1997. The Distribution and Exchange of Mayen Lava Quernstones in Early Medieval Northwestern Europe. In: G. De Boe & F. Verhaeghe, eds. *Exchange and Trade: Papers of the Medieval Europe Brugge 1997 conference Vol 3.* Zellik: Instituut voor het Archeologisch Patrimonium, pp. 97–106.

Peacock, D., 1987. Iron Age and Roman quern production at Lodsworth, West Sussex. *Antiquaries Journal,* 67(1), pp. 61–85.

Peacock, D., 1989. The Mills of Pompeii. *Antiquity,* 63(1), pp. 205–214.

Peacock, D., 2013. *The Stone of Life: Querns, Mills and Flour Production in Europe up to c. AD 500.* 1st ed. Southampton: The Highfield Press.

Peacock, D. P., 1980. The Roman millstone trade: a petrological sketch. *World Archaeology,* 12(1), pp. 43–53.

Peacock, D. P., 1998. *The Archaeology of Stone: A report for English Heritage.* 1st ed. Swindon: English Heritage.

Peacock, D. P. S. & Williams, D., 1986. *Amphorae and the Roman Economy: An introductory guide.* 1st ed. London: Longman.

Peacock, D., Williams, S. & Jones, S., 2007. Basalt as ships' ballast and the Roman incense trade. In: D. P. S. Peacock & D. Williams, eds. *Food for the gods : new light on the ancient incense trade.* Oxford: Oxbow, pp. 28–70.

Pearce, J., 2002. Food as a substance and symbol in the Roman army: a case study from Vindolanda. In: P. Freeman, J. Bennett, Z. T. Fiema & B. Hoffmann, eds. *Limes XVIII. Proceedings of the XVIIIth International Congress of Roman Frontier Studies held in Amman, Jordan (September 2000), Volume II.* Oxford: British Archaeological Research Reports International Series 1084, pp. 931–944.

Pearson, A., 2006. *The Work of Giants: Stone and quarrying in Roman Britain.* 1st ed. Stroud: Tempus.

Pearson, A. F., 1999. Building Anderita: Late Roman Coastal Defences and the Construction of the Saxon Shore Fort at Pevensey. *Oxford Journal of Archaeology,* 18(1), pp. 95–117.

Peers, L., 1999. Many tender ties: the shifting contexts and meanings of the S Black bag. *World Archaeology,* 31(2), pp. 288–302.

Pitts, M., 2014. Reconsidering Britain's first urban communities. *Journal of Roman Archaeology,* 27(1), pp. 133–173.

Polanyi, K., 1968. *Primitive, Archaic and Modern Economies: Essays of Karl Polanyi, Edited by George Dalton.* 1st ed. New York: Garden City.

Reniere, S., 2018. *Sourcing the Stone in Roman Northern Gaul: on the provenance, use and socio-economics of stone tools in a stoneless landscape: the case of the northern civitas Menapiorum.* 1st ed. Gent: Universiteit Gent.

Reniere, S. et al., 2016. Querns and mills during Roman times at the northern frontier of the Roman Empire (Belgium, northern France, southern Netherlands, western Germany): Unravelling geological and geographical provenances, a multidisciplinary research project. *Journal of Lithic Studies,* pp. 403–428.

Renzulli, A. et al., 2002. Provenance and trade of volcanic rock millstones from Etruscan-Celtic and Roman archaeological sites in Central Italy. *European Journal of Mineralogy,* 14(1), pp. 175–783.

Richardson, A., 2004. Granaries and garrisons in Roman forts. *Oxford Journal of Archaeology,* 23(4), pp. 429–442.

Rickman, G. & Rickman, G. R., 1971. *Roman Granaries and Store Buildings.* 1st ed. Cambridge: Cambridge University Press.

Rippon, S., 2008. Coastal Trade in Roman Britain: the Investigation of Crandon Bridge, Somerset, a Romano-British Trans-shipment Port beside the Severn Estuary. *Britannia,* 39(1), pp. 85–144.

Rippon, S., Smart, C. & Pears, B., 2015. *The Fields of Britannia: Continuity and Change in the Late Roman and Early Medieval Landscape.* 1st ed. Oxford: Oxford University Press.

Robinson, M., 2011. The Macroscopic Plant and Invertebrate Remains. In: 1st, ed. *Silchester: City in Transition: The Mid-Roman occupation of Insula IX c. AD 125–250/300. A report on excavations undertaken since 1997.* London: Society for the Promotion of Roman Studies, pp. 281–293.

Röder, J., 1953. Zur Lavaindustrie von Mayen und Volvic. *Germania,* 31(1), pp. 24–7.

Röder, J., 1955. Die Lavasteinbrüche von Volvic. *Jahrbuch für Geschichte und Kultur des Mittelrheins und seiner Nachbargebiete 1954/55,* 6(7), pp. 33–41.

Röder, J., 1961. Neue Ausgrabungen in Andernach. *Germania,* 39(1), pp. 208–213.

Rogers, A., 2013. *Water and Roman Urbanism: Towns, Waterscapes, Land Transformation and Experience in Roman Britain.* 1st ed. Boston: Brill.

Roll-stone, 2009. *Mayener Basaltlava: https://creativecommons.org/liceses/by-sa/3.0/de/deed.en.* [Online] Available at: https://commons.wikimedia.org/wiki/File:Mayener_Basaltlava.jpg [Accessed 6th September 2023].

Rushworth, A., 2009. *Housesteads Roman Fort- The Grandest Station.* 1st ed. Swindon: English Heritage.

Russell, B., 2013. *The Economics of the Roman Stone Trade.* 1st ed. Oxford: Oxford University Press.

Salisbury, R. B. & Rebay-Salisbury, K., 2017. Processes of theory: from production sequences and process to chaînes opératoires and object biographies. In: *Material chains in late prehistoric Europe and the Mediterranean.* Bordeaux: Ausonius Editions, pp. 15–29.

Sands, R. & Marlière, E., 2020. Produce, Repair, Reuse, Adapt and Recycle: The multiple biographies of a Roman barrel. *European Journal of Archaeology,* 23(3), pp. 356–380.

Saunders, N. J., 1999. Biographies of brilliance: pearls, transformations of matter and being c. AD 1492. *World Archaeology,* 31(2), pp. 243–257.

Schäfer, K., 2000. Drehscheibe des antiken Steinhandels. In: *Steinbruch und Bergwerk. Denkmäler römischer Technikgeschichte zwischen Eifel und Rhein.* Mainz: Verlag des Römisch-Germanischen Zentralmuseums, pp. 83–109.

Sefryn, P., 2013. Archaeologies of the Postindustrial Body. In: P. Graves-Brown, R. Harrison & A. Piccini, eds. *The Oxford Handbook of the Archaeology of the Contemporary World.* Oxford: Oxford University Press, pp. 685–699.

Seip, L. P., 1999. Transformations of meaning: the life history of the Naxalk mask. *World Archaeology,* 31(2), pp. 272–287.

Shaffrey, R., 2003. The Rotary Querns from the Society of Antiquaries' Excavations at Silchester, 1890–1909. *Britannia,* 34(1), pp. 143–174.

Shaffrey, R., 2006. *Grinding and Milling: A Study of Romano-British Rotary Querns and Millstones Made from Old Red Sandstone.* 1st ed. Oxford: BAR Publishing British Series 409.

Shaffrey, R., 2015. Intensive Milling Practices in the Romano-British Landscape of Southern England: Using newly established criteria for distinguishing millstones from rotary querns. *Britannia,* 46(1), pp. 55–92.

Shaffrey, R., 2019. The Movement of Ideas in the Late Iron Age and Early Roman Britain: an imported rotary quern design in south-western England. *Britannia,* 50(1), pp. 393–402.

Shaffrey, R., 2021. *Feeding Roman Silchester: Querns and Millstones in and around Roman towns.* [Online] Available at: https://www.cambridge.org/core/journals/britannia/article/feeding-roman-silchester-querns-and-millstones-in-and-around-roman-towns/FAE81E498D5A01F7667237E412F3CDFF [Accessed 9th September 2021].

Shaffrey, R. & Gluhak, T., 2020. Provenenancing lava querns: petrographic and geochemical analysis of three early Roman querns from Kent. *Council For British Arcaheology Newsletter: South East,* 53(1), pp. 1–3.

Shaffrey, R. & Roe, F., 2011. The Widening Use of Lodsworth Stone: Neolithic to Romano-British quern distribution. In: D. Williams & D. Peacock, eds. *Bread for the People: The archaeology of mills and milling.* Oxford: BAR Publishing International Series 2274, pp. 309–324.

Sijpesteijn, P. J., 1987. *Customs duties in Graeco-Roman Egypt.* 1st ed. Amsterdam: Terra Publishing.

Simpson, F. G., 1976. *Watermills and Military Works on Hadrian's Wall.* 1st ed. Kendal: Titus Wilson & Son.

Smith , A., 2016a. The East. In: M. Fulford & N. Holbrook, eds. *The Rural Settlement of Roman Britain.* London: Society for the Promotion of Roman Studies, pp. 208–241.

Smith, A., 2016b. The Central Belt. In: M. Fulford & N. Holbrook, eds. *The Rural Settlement of Roman Britain.* London: Society for the Promotion of Roman Studies, pp. 141–207.

Smithsonian Institution, 2021. *East Eifel Volcanic Field.* [Online] Available at: https://volcano.si.edu/volcano.cfm?vn=210812#:~:text=Activity%20at%20the%20East%20Eifel,from%200.45%20to%200.35%20Ma.

Spaul, J. E. H., 1994. *The Auxiliary Cavalry Units of the Pre-Diocletianic Imperial Roman Army.* 1st ed. Andover: Nectoreca Press.

St Albans and Hertfordshire Architectural and Archaeological Society, 2019. *Plan of Verulamium.* [Online] Available at: https://www.stalbanshistory.org/archaeology/the-roman-city-of-verulamium/map-of-verulamium [Accessed 27th September 2021].

Starley, D., 2003. Analysis of Ferrous Metalworking Evidence. In: F. Hammer, ed. *Industry in North-West Roman Southwark: Excavations, 1986–8.* London: Museum of London Archaeology, pp. 131–140.

Stead, I. M., 1976. *Excavations at Winterton Roman villa and other Roman sites in north Lincolnshire,* s.l.: Inspectorate of Ancient Monuments, unpublished works.

Stead, I. M. & Rigby, V., 1989. *Verulamium: The King Harry Lane Site.* 1st ed. London: English Heritage.

Steensgaard, N., 1981. Violence and the Rise of Capitalism: Frederic C. Lane's Theory of Protection and Tribute. *Review (Fernand Braudel Center),* 2(1), pp. 247–273.

Strabo, 1923. *Geography, Volume II: Books 3–5. Translated by Horace Leonard Jones.* 1st ed. Havard: Harvard University Press.

Swift, E., 2012. Object biography, reuse and recycling in the Late to Post-Roman transition period and beyond. *Britannia,* 43(1), pp. 167–215.

Swift, E., 2014. Design, function and use-wear in spoons: reconstructing everyday Roman social practice. *Journal of Roman Archaeology ,* 27(1), pp. 203–237.

Swift, E., 2017. *Roman Artefacts and Society: Design, Behaviour and Experience.* 1st ed. Oxford: Oxford Univeristy Press.

Taylor, J., 2007. *An atlas of Roman Rural Settlement in England.* 1st ed. York: Council for British Archaeology.

Thomas, R. I., 2014. Roman Naukratis and its Alexandrian context. *British Museum Studies in Ancient Egypt and Sudan,* 21(1), pp. 193–218.

Thurmond, D. L., 2006. *A Handbook of Food Processing in Classical Rome: For Her Bounty No Winter.* 1st ed. Leiden: Brill.

Todd, G., 2016. *Bronze Age Milling Stone & Saddle Quern.* [Online] Available at: https://www.flickr.com/photos/101561334@N08/29554958323 [Accessed 25th September 2023].

Topping, P., 2008. Landscape narratives: the South-East Cheviots Project. *Proceedings of the Prehistoric Society,* 74(1), pp. 323–364.

Tucker, G. D., 1982. Millstones north and south of the Scottish border. *Industrial Archaeology Review,* 6(3), pp. 186–193.

Tyers, P., 1996a. *Potsherd: Atlas of Roman Pottery- BB2.* [Online] Available at: http://potsherd.net/atlas/Ware/BB2 [Accessed 27th March 2020].

Tyers, P., 1996b. *Argonne Ware.* [Online] Available at: http://potsherd.net/atlas/Ware/ARGO [Accessed 9th September 2020].

Tyers, P., 1996c. *Potsherd: Atlas of Roman Pottery-Mayenware.* [Online] Available at: https://potsherd.net/atlas/Ware/MAYN [Accessed 14th September 2021].

van Driel-Murray, C., 1995. Gender in Question. *Theoretical Roman Archaeology: Second Conference Proceedings,* 14(1), pp. 3–21.

van Driel-Murray, C., 2001. Vindolanda and the Dating of Roman Footwear. *Britannia,* 32(1), pp. 185–197.

Vanderbilt, S., 2020. *Roman Inscriptions of Britain: RIB 586.* [Online] Available at: https://romaninscriptionsofbritain.org/inscriptions/586 [Accessed 25th February 2022].

Wacher, J., 1995. *The Towns of Roman Britain.* 2nd ed. Abingdon: Routledge.

Ward-Perkins, J. B., 1980a. The Marble Trade and its Organisation: Evidence from Nicomedia. In: J. H. D'Arms & E. C. Kopff, eds. *The Seabourne Commerce of Ancient Rome: Studies in archaeology and history.* Rome: MAAR, pp. 325–336.

Ward-Perkins, J. B., 1992. Materials, Quarries and Transportation. In: H. Dodge & B. Ward-Perkins, eds. *Marble in Antiquity: Collected papers of J.B Ward-Perkins.* London: The British School of Rome, pp. 23–30.

Watts, M., 2002. *The Archaeology of Mills and Milling.* 1st ed. Stroud: Tempus.

Watts, M., Watts & Sue, 2016. *The Application of Animal and Water Power: from the Mediterranean to Britain.* [Online] Available at: https://millsarchive.org/explore/features-and-articles/entry/171161/from-quern-to-computer-the-history-of-flour-milling/11664 [Accessed 29th January 2018].

Watts, S., 2014. *The Life and Death of Querns: The Deposition and Use-contexts of Querns in South-Western England from the Neolithic to the Iron Age.* 1st ed. Southampton: Highfield Press.

Watts, S., 2019. Rotary Querns, Rubbing Stones and a Stone Bowl. In: A. Poulter, ed. *The Transition to Late Antiquity on the lower Danube: Excavations and survey at Dichin, a Late Roman to early Byzantine Fort and a Roman aqueduct.* Oxford: Oxbow Books, pp. 391–408.

Webster, G., 2004. *Boudica: The British revolt against Rome AD 60.* 1st ed. London: Routledge.

Weinandt, 2010. *Panorama Ettringer Lay: https://creativecommons.org/licenses/by-sa/3.0/de/deed.en.* [Online] Available at: https://commons.wikimedia.org/wiki/File:Panorama_Ettringer_Lay.jpg [Accessed 7th September 2023].

Welfare, A. & Campbell, E., 1995. The milling stones. The petrology of the querns. In: W. H. Manning, J. Price & J. Webster, eds. *Report on the Excavations at Usk, 1965–1976. The Roman Small Finds.* Cardiff: University of Wales Press, pp. 214–237.

Welsby, D. A., 2009. The Stonework, Brick and Tile. In: A. Rushworth, ed. *Housesteads Roman Fort- The Grandest Station.* Swindon: English Heritage, pp. 339–361.

Wenzel, S., 2011. Villa und Burgus von Obermendig 'Im Winkel' (LKR. Mayen-Koblenz) im kontext der Römischen Besiedlung des Segbachtals. In: M. Grünewald & S. Wenzel, eds. *Römische Landnutzung in der Eifel: Neue Ausgrabungen und Forschungen.* Mainz: Römisch-Germanisches Zentralmuseum, pp. 131–158.

Wenzel, S., 2019. Mühlsteine aus den Graubungen des Jahres 2014 in Wederath- Belginum. In: D. Sauer, N. Hasslinger & R. Cordie, eds. *Was aßen Kelten und Römer? Umwelt, Landwirtschaft und Ernährung westlich des Rheins.* Morbach: Archäologiepark Belginum, pp. 161–168.

Wenzel, S., 2020. Lava rotary querns of 'Iron Age type' in Roman times. *Journal of lithic studies: Proceedings of the 2nd Meeting of the Association for Ground Stone,* 7(3), p. doi: https://doi.org/10.2218/jls.3096.

Wenzel, S., 2021. Villa und Burgus von Mendig, 'Im Winkel'. In: S. Wenzel, M. Grünewald & R. Giljohann, eds. *Römische Landnutzung im antiken Industrierevier der Osteifel.* 1st ed. Heidelberg: Propylaeum, pp. 3–151.

Wheeler, R. E. M., 1932. A Prehistoric Metropolis: The first Verulamium. *Antiquity,* 6(22), pp. 133–147.

Wheeler, R. E. M., 1936. *Verulamium: A Belgic and two Roman cities.* 1st ed. London: Society of Antiquaries of London.

Wheeler, R. E. M., 1938. Verulamium Again. *The Antiquaries Journal,* 18(2), pp. 137–141.

White, D., 1963. A Survey of Millstones from Morgantina. *American Journal of Archaeology,* 67(2), pp. 199–206.

Williams, D. & Peacock, D., 2011. Pompeian Style Mills in Britain. In: D. Williams & D. Peacock, eds. *Bread for the People. The Archaeology of Mills and Milling: Proceedings of a Colloquium Held in the British School or Rome, 4th–7th November 2009.* Oxford: BAR Publishing International Series 2274, pp. 117–122.

Williams, R. J. & Zeepvat, R. J., 1994. *Bancroft. A late Bronze Age/Iron Age settlement, Roman villa and temple-Mausoleum.* s.l.:Buckinghamshire Archaeological Society Monograph no. 7, Buckinghamshire Archaeological Society.

Williams-Thorpe, O., 1988a. Provenancing and archaeology of Roman millstones from the Mediterranean Area. *Journal of Archaeological Science,* 15(1), pp. 253–305.

Williams-Thorpe, O., 1988b. The provenance of donkey mills from Roman Britain. *Archaeometry,* 30(2), pp. 89–275.

Williams-Thorpe, O. & Thorpe, R. S., 1989. Provenancing and Archaeology of Roman Millstones from Sardinia (Italy). *Oxford Journal of Archaeology,* 8(1), pp. 89–117.

Williams-Thorpe, O. & Thorpe, R. S., 1993. Geochemistry and trade of Eastern Mediterranean millstones from the Neolithic to Roman periods. *Journal of Archaeological Science,* 20(1), pp. 263–320.

Williams-Thorpe, O., Thorpe, R. S., Elliott, C. & Xenophontos, C., 1991. Archaeology, geochemistry, and trade of igneous rock millstones in Cyprus during the Late Bronze Age to Roman periods. *Geoarchaeology,* 6(1), pp. 27–60.

Wilson, A., 2010. The Roman Watermills. In: P. Bennett, I. Riddler & C. Sparey-Green, eds. *The Roman Watermills and Settlement at Ickham, Kent.* Canterbury: Cabterbury Archaeological Trust, pp. 42–67.

Wilson, A. & Schorle, K., 2009. A Bakers Funerary Relief from Rome. *Papers of the British School at Rome,* 77(1), pp. 101–123.

Woolf, A., 1999. Adventus, Patrocinium, and the Urban Landscape in Late Roman Britain. In: A. Leslie, ed. *Theoretical Roman Archaeology and Architecture: The Third Conference Proceedings.* Glasgow: Cruithne Press, pp. 33–47.

Woolf, G., 2007. Rethinking the Oppida. *Oxford Journal of Archaeology,* 12(2), pp. 222–234.

Xeidakis, G. S. & Varagouli, E. G., 1997. Design and Construction of Roman Roads: The Case of Via Egnatia in the Aegean Thrace, Northern Greece. *Environmental and Engineering Geoscience,* 3(1), pp. 123–132.

Young, R. & Simmonds, T., 1995. Marginality and the nature of later prehistoric upland settlement in the north of England. *Landscape History,* 17(1), pp. 5–16.

Zaharia, A., 2013. Technical Norms and Legal Rules on the Construction and Administration of Roman Roads. *Contemporary Readings in Law and Social Justice,* 5(2), pp. 382–387.

www.ingramcontent.com/pod-product-compliance
Ingram Content Group UK Ltd.
Pitfield, Milton Keynes, MK11 3LW, UK
UKHW061848041026
15591UKWH00015B/968

9 781407 360812